A Science of Impurity

Water Analysis in Nineteenth Century Britain

A Science of Impurity

Water Analysis in Nineteenth Century Britain

Christopher Hamlin

UNIVERSITY OF CALIFORNIA PRESS

Berkeley and Los Angeles

University of California Press
Berkeley and Los Angeles, California

Library of Congress Cataloging-in-Publication Data

Hamlin, Christopher, 1951–
     A science of impurity : water analysis in nineteenth century
     Britain / Christopher Hamlin.
       p.   cm.
     Includes bibliographical references (p.
     ISBN 0-520-07088-7 (alk. paper)
     1. Drinking water–Great Britain–Analysis–History–19th century.
     I. Title.
TD257.H35   1990 363.6'1'094109034–dc20   90-10827

Printed in Great Britain

'You praise no waters but have stood your Test
And under chemic Tortures Truth confess'd'

from Diederick Wessel Linden
A Treatise on ... Chalybeat Waters (1755)

Contents

Acknowledgments

I thank June Fullmer, Victor Hilts, Aaron Ihde, Ernan McMullin, Anne Hardy, Alan Rocke, Robert Siegfried, Joel Tarr, Mick Worboys for conversations, support, and critical readings of chapters. I think often of Bill Coleman in this regard and wish he were still here to thank too. I thank Bill Luckin, John Pickstone, and Roy Porter for their support and criticism when it was most needed. I thank Fern Hamlin for all of these things and for an immense number of other things, in the preparation of this book and otherwise. I am grateful to librarians and archivists at the American Philosophical Society, the Greater London Records Office, the Public Records Office, and the Institution of Civil Engineers for putting into my hands materials I would doubtless never have seen, and in many cases never have known about. The *Lancet*, the Royal Society of Chemistry, the *Philosophical Magazine*, and the Public Records Office have graciously allowed reproduction of illustrations.

A Note on Notes

In most cases complete citations are given in the Bibliographic Essay. An asterisk in end of chapter notes indicates cases where full citations are given earlier in end of chapter notes.

Commonly Used Acronyms and Abbreviations

Official Inquiries

RPPC = Rivers Pollution Prevention Commission (1865 and 1868)
RCMWS = Royal Commission Metropolitan Water Supply (1828, 1892–3)
RCWS = Royal Commission on Water Supply, 1867–9

Journals

BMJ = British Medical Journal
CN = Chemical News
FWA = Food, Water, and Air
JPH&SR = Journal of Public Health and Sanitary Review
JRSA = Journal of the Royal Society of Arts
MPICE = Minutes of Proceedings, Institution of Civil Engineers
SR = The Sanitary Record
TSIGB = Transactions, Sanitary Institute of Great Britain

Organisations and Miscellaneous

GBH = General Board of Health
LCC = London County Council
LGB = Local Government Board
MO = Medical Officer
MOPC = Medical Officer of the Privy Council
PRO = Public Record Office
PSC = Previous Sewage Contamination
SPA = Society of Public Analysts

Introduction

This book is about the application of science to one area of public decision making, the determination of water quality. It is a universal concern, though one which has until recently seemed unproblematic to most of us in the industrialized west. The quality of our water was one of those things we had forgotten how to worry about; we could rely, we felt, on the authority of our authorities: on the bacteriologists and engineers with their cultures and chlorine, on the managers and lawyers who would ensure that enough good water continued to reach us. Many things now threaten that trust: a new generation of toxic wastes, growing demand for water in arid regions, the prospect of changing climate and, in the United Kingdom, the sale of water supplies to private investors. All these matters have become public issues and no longer do we assume that good and safe water will always come from the tap. Some buy bottled water, an ironic throwback to the days of the water carrier, others equip their taps with the newest in domestic filters, an appliance common in the nineteenth century, others drink anything but water, a practice that spurred nineteenth century temperance reformers to insist that public authorities provide good, pure water for all.

We are thus at the end of a period of sanguine ignorance, and water has become another of the many aspects of our world in which we fear crisis. To resolve such questions we look to technical solutions, but when we look to those who usually supply such authority we find some who reassure us that all will be well and others who seem bent on deepening our anxiety. Thrust into the midst of technical controversies on matters about which we know little, we are left to worry both about the supposed dangers themselves and about the rationality of our fear.

The trust which we once had for our water and for many other aspects of our environment was an achievement of the great public health campaign of the last two centuries. Beginning in the late 1830s what would become the modern system of public health administration began to take shape in Britain. Among numerous other aspects of the natural and social environment, provision of safe water came to be seen as a responsibility of government. While at the beginning of the century water came from rivers, springs, or shallow wells (or, in the case of the wealthy, from the mains of private water companies), by its end it came through mains, often owned by municipal governments, from the reservoirs constructed by them at great expense.[1]

The trust we had come to hold was due not only to the water itself, but to the scientific authority that sanctioned that water, to those who certified that it would be pure and plentiful. The nineteenth century was also a time of massive growth of science, of its clear emergence as a profession and, most importantly here, of its utilization in public decision-making. Scientists defined long-term possibilities, rationalized the running of the ship of state, grappled with technical complexities in a way which no government of amateurs possibly could.

In most accounts, the growth of science, the provision of public services like water supplies, and the recognition of responsibility for the health of the public have been closely integrated within a network of mutual cause and effect, together constituting social progress. The public health movement was touted as the scientific answer to the grave urban problems of the day, and the subsequent progress of public health administration, from Edwin Chadwick to John Simon, and on to Simon's late-century successors, is seen as a transformation guided by science (at least in the cases where the experts were not stymied by the cheeseparing bureaucrats of the Treasury).[2] Thus the number of scientists grew because more were needed and they became professionalized as they became social authorities, on whose word matters of individual liberty, public policy, and the distribution of vast capital were decided.

In fact, however, while we have good accounts of some aspects of sanitary science[3] and good accounts of public health administration[4], we know relatively little about how science guided public health;[5] we have mainly the claims of administrators that theirs were scientific administrations.

Water analysis is an ideal area for exploring the relations between science and public health administration for water matters are so central in the story of sanitary achievement. The drinking of what was little better than dilute sewage at the beginning of the century led to repeated invasions of cholera and typhoid, and to the famous mid-century investigations of John Snow and William Budd, who demonstrated the link between bad water and outbreaks of these diseases. Science, in the form of bacteriology, is held to have finally resolved the problem, as the great Robert Koch and his disciples quickly detected the microorganisms responsible for these diseases in the early 1880s.

Informed public policy then became possible. The bacteriological enlightenment is thus seen as the great watershed in environmental medicine, separating a pre-scientific period in which medicine could offer little more than a false cultural authority from the contemporary period of scientific precision where the authority is real. It has been, both for writers and readers of histories of public health, the occasion for a sigh of relief: safety at last.[6]

For reasons I develop below, I believe this is an unsatisfactory depiction, as much in its inconsistency with the historical record as in its perspective toward the relations of science, social concerns, and public policy. The story of the relations between science and public health was more complicated and contingent, a matter more of the opportune intersection of these two contexts than of their co-evolution. First science: for most of the nineteenth century Britain was not an easy place to live for non well-to-do people who wished to occupy themselves with basic scientific research. Even if one were lucky enough to secure a professorship of some sort, such a position was likely to be more important as an index of prestige and a basis for further contacts than for the income that came with it. Some scientists did live fairly well, but they did so by stringing together a number of remunerative posts: as consultants, witnesses, authors, entrepreneurs, as well as teachers. Most of the chemists, who are the main characters of this book, had such careers. As well as actual products (e.g. electroplating works), or services (e.g. fertilizer analysis), they hoped to sell authority: they would become members of Coleridge's clerisy, the profession on which society depended for the cultural authority over certain problems, and they claimed an epistemic warrant for that status.[7] In connection with the determination of the medicinal qualities of mineral waters, chemists had been fighting for such status in matters of water quality long before the quality of public water supplies became an issue in the 1820s.

Thus part of the story is one of aggressive and successful discipline-promotion, the struggle of a group of experts to acquire authority, regardless of the state of their art at the time. We might be tempted to see them as charlatans, for prior to the 1890s they were claiming to be analysing waters without (as we know now) any correct (or even very definite) idea of what components or contaminants of waters had active effects. Yet just as historians have come to recognize that the quack doctor played an important social role, and sold his patients what was to them a real service, so too we need to recognize that the authority sold by these chemists was a real and a valued commodity.[8]

The second context, of public health, is more complicated. Many historians would probably agree that the history of the efforts of governments to safeguard health belongs at least as much to the history of ideas, of politics, and of social policy, as it does to the history of applied science. Yet too frequently we assume that public health improvement was a coherent enterprise, its scope well-defined, its goals clear, with minor disagreements occasionally existing only as to means. Water policy has been seen in this context. Knowing what we do of the relations between pure water and disease and (until recently) confident in the universality and obviousness of the arrangements of our society for supplying water, it is hard to see the securing of better water as anything other than an obvious and essential way of lowering mortality. Yet what social actions were necessary, and equally what standards would apply to water, were continually matters of conflict. Thus the achievement in public health was a genuinely political achievement, forged from a peculiar assortment of ideology, institutions, political circumstance, and perceptions of nature. Science, because it was expected to yield a single correct answer to any question, was an ideal to which to appeal for legitimacy, but people with opposing proposals could summon it, and in most cases it supplied them with predictions and assessments suitable for advancing their proposals. It was an idiom for argument, and a way of discovering arguments, much less than a way of resolving them.

In water analysis, for example, even after the coming of bacteriology, the patterns of activity and even of innovations reflect the history of the politics of water supply, not the history of epidemiological recognition of water-borne diseases. It is true that the question that analysts were to answer was one of what would be the effects on health of consuming certain waters. Yet their answers meant as much with regard to control of public water supplies as they did with regard to informing consumers or doctors whether their water was safe to drink. To persuade the rate-payers of a town, or a parliamentary select-committee, that the water was satisfactory, was equally a way of acknowledging that existing conditions were acceptable; likewise, the argument that a water was bad was usually part of a plea to transfer ownership of the waterworks (usually from private to public control), or to undertake major capital expenditure, or to exact legal penalties from those responsible for its condition.[9]

I suggest then that development of the kinds of water standards we now have (or of any standard of environmental quality) was not the result of scientific discovery, but that scientific arguments were wielded on all sides in an effort to obtain whatever set of standards various parties regarded as desirable. This remained the case after the coming of bacteriology. Even when techniques were available for detecting the microbes responsible for typhoid and cholera, the answer to the ultimate question of 'is the water safe to drink' depended on how much trust one was willing to assign to analytical techniques, and this in turn continued to be considered in terms of a host of other questions: were present supplies good enough, not just in terms of quality and with regard to health, but in terms of quantity and with regard to industry? How were multiple uses and claims on water to be reconciled? Compromises in distributing benefits and risks were impossible to avoid, and bacteriologists engaged in debates about the certainty and significance of their results that parallel those of chemists, and even those of the mineral water analysts a century earlier.

The story outlined thus far—a profession on the make, social and political questions in scientific disguise—may seem a familiar one to a generation of historians and sociologists of science who have emphasized the frailty of scientific knowledge as enthusiastically as their predecessors emphasized its robustness.[10] Yet with these questions we jump back to the present, for they make clear that the problem of making rational policy in an environment of scientific uncertainty is much the same now as then. Our need for an authority in which to ground our decisions is as acute as the Victorians' was, and we too look to science, as representative of natural truths, as the source of that authority. To attend only to the undeniable realities of aggressive discipline-promotion or the struggles for water rights or even to the cultural construction of concepts of purity will not be enough, for if we are not careful such inquiries will trivialize the efforts of the past and provide no useful guidance for the present.[11] What we need to do, using history both as sounding board and guide, is to explore general issues of the relations between science and policy in a way that is anthropological and philosophical, as well as historical.

Two decades ago Alvin Weinberg coined the label 'trans-scientific,' for problems that could be stated in the terms of science, but were not scientifically soluble, an apt characterization for the problems that faced nineteenth century water analysts.[12] Weinberg, himself a successful scientist and administrator of science, looked to various political and legal mechanisms to resolve these kinds of problems. These would utilize science, but in what way science would supplement, complement or displace other forms of making decisions Weinberg did not say.

To this problem of what science does, did, can do, or must do, very many answers have been offered. Three seem especially helpful here, though none resolves the problem that arises in Weinberg's article. The first comes from the anthropologist Mary Douglas, whose analyses of the social construction of pollution and risks have influenced the current generation of historians and sociologists of science. All societies manufactured for themselves boundaries, represented in terms of God, money, time, and nature, which defined for them the circumstances in which social action was necessary or environmental circumstances intolerable. While the boundaries themselves were ultimately arbitrary (at least to outsiders), their maintenance was vital to social solidarity. The tenacity with which peoples throughout the world clung to irrational pollution taboos could thus be understood as a real and admirable effort to maintain one's cosmology, and hence one's identity. Applied to our own society, Douglas' perspective was taken to indicate that the limits, possibilities, and necessities that had been sold to the public as uniquely privileged results of scientific rationality could be shown to be as time- and culture-specific as those of any other society.[13] But because this recognition was to help fuel a liberation from arbitrary authority, these critics tended to be much less sympathetic than Douglas to the need to maintain the boundaries that provided identity.[14]

Where Douglas' perspective offered little help was with the questions of how authorities came to be, and of what to do without one. So strongly did she insist on the necessity of pollution taboos, for example, that the prospect of a society rent by conflict over what its environmental standards ought to be represented a chaos too appalling to be contemplated. Yet this was the case in nineteenth century Britain; it was a time of change in which both permissible uses of public resources and mechanisms for governing that use changed significantly. Working on what has been called the 'revolution in government' question, British historians have gone far in working out the details of this transformation. The switch from government by deference and custom to government by a professional and scientific civil service has been seen as a mixture of the drawing into government of followers of Jeremy Bentham's notions of rationalized public administration and of the response (sometimes by opportunistic officials) to the development (and discovery) of unprecedented social and technical problems. Much of the historical writing on public health belongs to this context, where, it is argued, the discovery of conditions of public danger mandated concerted action.[15]

For these historians the bringing of science, or more broadly, expertise, into government was the interesting problem. Yet frequently their trust in the ideal of science as a neutral means of resolving conflicts and determining policy led them to take an uncritical attitude toward the actual activities of scientists; so long as it was scientists who were in the positions of policy-making, the policy they made could be assumed, in some vague sense, to bear some manner of higher epistemic warrant.[16] The implications of this history are clearly antithetical to those of Douglas' anthropology: in this view, the authority that government came to possess by the end of the century was far from arbitrary, it was no less than the manifestation of social progress.

Yet nineteenth century scientists were quite adept at exploiting the ideal of science toward their own ends, and this leads to the third perspective, one less well-developed, but which seems especially appropriate for understanding the involvement of nineteenth century British scientists in water matters. In his study of the first half-century of the Royal Institution, Morris Berman showed how the ideal of scientific objectivity became in the early nineteenth century a hallmark of responsible decision-making and the participation of scientists accordingly indispensable.[17] Toward this achievement Berman was cynical: the ideal that there was one truth, and that science would therefore uncover the answer to any question, was the means with which industrial society could 'smooth over structural contradictions.' Conflicts over power and struggles for justice could in this way be neutralized by being redefined as technical questions. In water matters scientists were not univocal, and the ideal of science cannot be seen mainly as a means of oppression by a dominating class. Yet the recognition of the enormous symbolic importance of

having an ideal of neutrality to legitimate policy is of central importance. This ideal was as powerful in maintaining social order (and
far more flexible) than the codes of impermissible behaviours about
which Douglas wrote. And while it may have come in with the rise of
utilitarianism, its application to water matters, and to many other
issues of technical policy, took place in a very different social setting
than that which interested the revolution in government historians.

Water policy belonged to the context of Parliament, particularly
the select-committee system, and to the courts, rather than to the
context of the civil service. These were the structures of decision-
making worked out for road and canal projects in the eighteenth century, adapted to railways and municipal improvements in the nineteenth. In all these enterprises Parliament set terms for the purchase
of rights of property by entrepreneurs and others who claimed to be
acting for the public good. It was a transitional means of decision-
making, an organized way of eliminating traditional common law
rights over use of the environment that were seen to be interfering
with the public good. Traditionally, the keeper of an ancient mill
had the right to flood the riverbanks upstream and to abstract a certain portion of water for the mill race, even if the river was thereby
made unsuitable for other purposes, from floating boats to draining
a town. But under pressure from a public health authority wishing
to drain lands or to acquire gathering grounds for water supplies,
Parliament might eliminate those rights, compensating those who
were made to yield them.

While this context has been characterized by historians of canal
and railway projects, its significance as one of the principal means
of bringing science into British government has not been recognized.
Above all it was a context of conflict. Both proponents of a project
and those parties resisting it (there were often many, and their resistance was often primarily an attempt to gain higher compensation)
were represented by counsel before parliamentary select-committees,
and the witnesses the committees heard, including the scientists, engineers, and medical men, were those chosen by each party to present
its case. Science was a rich and expressive idiom of that conflict, one
characterized by the ideal that there was a best answer, a natural
truth, for any question, and yet possessing vast flexibility, being capable indeed of giving expression to very nearly any argument one
wished to advance.[18]

It is to such a context that water analysis belongs. With a few exceptions analyses were done not by the disinterested public experts

charged with managing the people's health, but by those engaged in policy conflicts in which representatives of both sides typically claimed to be representing true science and defending the public health. Hence far from representing an elite, unified in their pursuit of science and insulated from the worlds of politics and speculation, chemists felt the tensions over water quality at least as acutely as they were felt in the world at large. While they might agree that water assessment was a matter for chemistry, they disagreed intensely on what constituted an adequate analysis, on which processes were reliable, on what skills an analyst had to possess, on how results were to be interpreted, and what public responses they indicated.

However flexible, the idiom that science provided was by no means arbitrary. Its rules and boundaries were provided respectively by rules of inference and by contemporary medical and chemical theories. Yet these concepts and ways of arguing—what are usually seen as the stuff of science itself—did not lead to resolutions, for science was in fact only to provide the arguments in such a context. But chemists—at least the best of them—were doing more than dressing up what were usually the blatantly self-interested proposals of speculators in an arcane and impressive language. Their testimony and analyses were effective precisely because they were able to show that contemporary understanding of nature made possible, plausible, or necessary certain consequences which those who hired them wished to demonstrate, say that water running in a river would invariably become pure, for example.

We can gain a sense of the possibilities of this idiom by considering the problem of water analysis itself—as it was understood in the nineteenth century, and, indeed, as it is understood today. The central question of water analysis seems a simple one: is there anything bad in the water? There are really two questions here, one of determining composition and one of assessing harmfulness, with the second the more important. It may seem that the first question, of composition, must be answered first, yet some of the most prominent analysts, like Edward Frankland (chapters 6 and 7), frequently worked in the opposite order, assessing water on other factors. Analysis might confirm their diagnoses, but its main function was to symbolize to the public the validity of the assessment.

The most important conflicts that arose in answering these questions took place over the assumptions one had to make. Take the case of simply finding out what is in the water, for example. Here three issues arise.

(1) How does one know that one has distinguished all the entities that exist in the water that ought to be detectable using the analytical scheme one is using? For example, mineral water chemists were concerned with the various inorganic salts a water contained. But they could never be confident that what they isolated as a particular chemical species, say sodium chloride, was not in fact a mixture of various species which had not yet been distinguished. Hence a chemist's claim to have made a complete analysis of a water was equally a statement that all chemical species had been discovered and could be distinguished. Bacteriology presented a similar issue. During the '80s and early '90s most bacteriological water analysts were willing to admit that their medium of choice, gelatine–peptone, was not suited to the growth of all microbe species. In practice, however, they tended to treat the colonies that grew as corresponding to the actual microbe population of the water.

(2) How does one know that analytical operations do not change the material being analysed in some way, and if one assumes such changes do happen, how does one determine what changes they are? In 1815 the Scottish chemist John Murray proposed that rearrangement of acids and bases went on in mineral water samples during analysis; what a patient drank might be a quite different mixture of salts from what the chemist discovered on analysis. But Murray saw no way to confirm his idea; as he pointed out, any intervention to establish the composition at a particular stage was equally open to the charge that it altered the sample. Similar criticisms were raised with regard both to processes for determining the organic matter in potable waters and to bacteriological techniques.

(3) How does one know that one has chosen the appropriate analytical scheme, that one is analysing water on the right level? During the century analysts were interested in telling four or five distinct stories of what actually was in the water: at first it was inorganic salts; then various parameters relating to organic matter, living and dead; then numbers of bacteria, and finally species of bacteria. When the question of what was in the water was raised, questioner and analyst were usually thinking in terms of one of these schemes. But it was not always clear which one was appropriate to the questions at hand.

As for assessment, key questions had to do with whether one knew what the active medicinal or pathogenic entities in waters were, and, if they were known, whether they could be reliably detected. For the most part, the mineral water chemists active in the early part of the century claimed they did know the identities of the active medicinal ingredients and could readily detect them. The potable water analysts who succeeded them usually admitted that they were not sure what caused water-borne diseases and had grave doubts that the entities could be reliably detected. For most of the century confusion about how to understand disease causation (and hence how to demonstrate that one had discovered the cause or even a cause of a particular disease) made it unclear how to interpret the information provided by analysis. Had such questions been raised solely with regard to epidemic diseases, epidemiology might have provided means of resolving these sorts of disagreements, but both mineral water physicians and sanitarians were at least as interested in chronic conditions, where it was practically impossible to single out the effects of a single cause from a host of others.

Most potable water analysts did not even claim to be directly measuring the harmful entities water might contain, but based their assessments on various sorts of 'indicator' arguments. In modern chemistry an indicator is some substance that in some 'visible way shows the condition of ... some system.'[19] Something they measured, the water analysts claimed, bore a definite relation to the whatever-it-was that caused water-borne disease. Even after discovery of the cholera and typhoid microbes, indicator arguments remained important, since the tests for detecting these very infrequent contaminants were tricky and subject to too many false negatives (cases in which a negative result is obtained when the pathogen is actually present in the water from which the sample has been taken).

Several types of indicator arguments were used. Some chemists conceived the organic matter they measured as containing (or even being) the harmful substance though the extent of its harmfulness might vary from time to time, being often below the threshold. Others viewed the entities they measured as an innocuous matrix for the harmful entities, even though the harmful entities might only rarely be present and hence the tests would give many false positives (cases in which the indicator would be present, yet the dangerous entity absent). A few others, particularly microscopists, held out hope of discovering some entity that had nearly a one-to-one correlation with the dangerous matter.

A consequence of the use of indicators was a great deal of controversy as to how much significance should be assigned to a particular finding. Were signs of sewage contamination alone sufficient to condemn a water? How weak might indicators be and yet still warrant being taken seriously? If one based one's advice on the finding of indicators known to give frequent false positives, how was one to keep the public from becoming complacent? And if one ignored such indicators, what was the point of analysis? And if a water-borne epidemic struck in such a case, was not the analyst responsible? How chemists responded to such dilemmas depended on their own values, the strategies they took in dealing with the public, the contexts in which they were working, and the vested interests they were working for.

It can be seen from this outline that water analysts regularly faced central problems (and paradoxes) of the philosophy of science, problems of causation or correlation, of realism or operationalism, of distinguishing fact from theory, of whether observation involves intervention. They were also regularly confronting central problems of political philosophy (what was to be the role of the scientist in government?) and ethics (what responsibility did the water analyst hold to the water drinking public?). However much the resolutions they found to these problems reflected the immediate circumstances of the case at hand and the interests of the client who was sponsoring the science, the questions were real questions that arise and will continue to arise whenever societies grapple with great issues of public policy.

This book takes the following course. The first two chapters are on mineral water analysis, mainly in Britain and mainly between 1780 and 1850. The first concentrates on methodological and epistemic matters, the second on social and ideological contexts. Chapter 3 takes up the beginnings of concern for the quality of potable water, focusing on controversies over the quality of London's water in the years around 1828. Chapter 4 considers the conflict between chemical and microscopical methods of analysis that occurred in connection with the London water controversy of 1849–52. Chapter 5 deals with the impact of Justus von Liebig's conception of the zymotic process of disease on the theory and practice of water analysis during the late '50s and early '60s. It suggests how markedly different from previous conceptions of impurity were the zymotic poisons Liebig envisioned. Chapters 6 and 7 are concerned with the central role of Edward Frankland. The former chronicles his career as a water scientist and explains how he came in the late 1860s to the radical positions he took, while the latter is concerned with the reactions of other water scientists to Frankland, particularly in the '70s and early '80s. Chapter 8 deals with the emergence of a new context for water analysis and a new group of water analysts, the public analysts and local medical officers, who began in the 1870s to bring water assessment into their work in an important way. Chapter 9 deals with the transformation of the germ theory into the science of bacteriology and is concerned with debates during the mid 1880s on what meaning if any could rightly be assigned to the number of bacterial colonies that appeared on a plate in which a small sample of water had been cultured. Chapter 10 takes up the incorporation of ecological and determinative bacteriology into water quality evaluation. It shows how limited was the utility of bacteriological techniques to those most concerned with water quality. The conclusion returns to the issue of expertise and raises the question of what constituted progress in water analysis and the larger question of what constitutes satisfactory authority in technological controversies.

1   J A Hassan, 'The Growth and Impact of the British Water Industry,' pp 531–47.
2   Royston Lambert, *Sir John Simon, 1816–1904, and English Social Administration*; S E Finer, *The Life and Times of Sir Edwin Chadwick*; R A Lewis, *Edwin Chadwick and the Public Health Movement, 1832–1854*; R M MacLeod, 'The Frustration of State Medicine, 1880–1899,' *Medical History* 11 (1967): 15–40; Jeanne L Brand, *Doctors and the State: The British Medical Profession and Government Action in Public Health, 1870–1912* (Baltimore: Johns Hopkins University Press, 1965); and Anthony S Wohl, *Endangered Lives: Public Health in Victorian Britain*. This perspective is also evident in many older works.
3   M Pelling, *Cholera, Fever and English Medicine, 1825–1865* is the standard work on the pathological theories that guided the sanitarians. See also John M Eyler, *Victorian Social Medicine: The Ideas and Methods of William Farr* (Baltimore: Johns Hopkins University Press, 1979).
4   See reference 2 and also Jeanne L Brand, 'John Simon and the Local Government Board Bureaucrats, 1871–76,' *Bull. Hist. Med.* 37 (1963): 184–94; Roy M MacLeod, *Treasury Control and Social Administration: A study of Establishment Growth at the Local Government Board, 1871–1905*, Occasional Papers in Social Administration 23 (London: Bell, 1968).
5   But see Bill Luckin, *Pollution and Control: A Social History of the Thames in the Nineteenth Century* (Bristol: Adam Hilger, 1986).

6   This view is characteristic of an earlier generation of historians of public health (e.g. C-E A Winslow, *The Conquest of Epidemic Disease: A Chapter in the History of Ideas* [1943; rpt. Madison: University of Wisconsin Press, 1980]); even for the modern school of critical social historians of medicine, it can be argued that bacteriology still marks the major watershed between modern and pre-modern medicine (cf Roy Porter, *Disease, Medicine, and Society in England, 1550–1860* (London: MacMillan, 1987), pp 25–7.

7   Roberts and Bud, *Science versus Practice*; Russell, Coley, and Roberts, *Chemists by Profession*; J Morrell and A Thackray, *Gentlemen of Science: Early Years of the British Association for the Advancement of Science* (Oxford: Clarendon, 1981), pp 17–21; I Inkster, 'Introduction: Aspects of the History of Science and Science Culture in Britain, 1780–1850 and Beyond' in I Inkster and J Morrell, eds, *Metropolis and Province, Science and British Culture 1780–1850* (London: Hutchinson, 1983).

8   W Bynum and R Porter, eds, *Medical Fringe and Medical Orthodoxy, 1750–1850* (London: Croom Helm, 1986).

9   Arthur Silverthorne, *London and Provincial Water Supplies* (London: Crosby, Lockwood, 1884); J H Balfour Browne, *Water Supply* (London: MacMillan, 1880); Francis Sheppard, *London, 1808–1870, The Infernal Wen* (London: Secker and Warburg, 1971).

10   Barry Barnes, *About Science* (Oxford: Basil Blackwell, 1985); B Barnes and D Edge, *Science in Context: Readings in the Sociology of Science* (Milton Keynes: Open University Press, 1982).

11   Some of the more radical sociologists of science have begun to make this argument. See H Rose, 'Hyper-Reflexivity: A New Danger for the Counter-Movements,' in H Nowotny and H Rose, eds, *Counter-Movements in the Sciences: The Sociology of Alternatives to Big Science* (Dordrecht: Reidel, 1979), pp 277–89; and S Woolgar, 'Irony in the Social Study of Science' in K Knorr-Cetina and Michael Mulkay, eds, *Science Observed: Perspectives in the Social Study of Science* (London: Sage, 1983), pp 239–66.

12   A Weinberg, 'Science and Transcience,' *Minerva* 10 (1972): 209–22.

13   M Douglas, *Purity and Danger: An Analysis of the Concepts of Pollution and Taboo* (London: Routledge and Kegan Paul, 1966); M Douglas, 'Environments at Risk' in B Barnes and D Edge, *Science in Context,** pp 260–75; M Douglas and A Wildavsky, *Risk and Culture: An Essay on the Selection of Technical and Environmental Dangers* (Berkeley: University of California Press, 1982), esp. pp 85–95.

14   See Thomas Haskell, ed, *The Authority of Experts: Studies in History and Theory* (Bloomington: Indiana University Press, 1984), esp. the introduction and M S Larson, 'The Production of Expertise and the Constitution of Expert Power,' pp 28–80.

15 For the most recent review of the literature see Roy MacLeod, ed, *Government and Expertise: Specialists, Administrators, and Professionals, 1860–1919* (Cambridge: Cambridge University Press, 1988). See also Gillian Sutherland, ed, *Studies in the Growth of Nineteenth Century Government* (Totowa, NJ: Rowman and Littlefield, 1972) and Oliver MacDonagh's original paper, 'The Nineteenth Century Revolution in Government: A Reappraisal' *Historical J.* 1 (1958): 52–67.

16 Cf Roy M MacLeod, 'Government and Resource Conservation: The Salmon Acts Administration, 1860–1886,' *J. Br. Studies* 7 (1968): 115–50; *idem*, 'The Alkali Acts Administration, 1863–84: The Emergence of the Civil Scientist,' *Victorian Studies* 9 (1965): 85–112.

17 M Berman, *Social Change and Scientific Organization, the Royal Institution, 1794–1844* (Ithaca: Cornell University Press, 1978).

18 Contrast H Parris, *The Government and the Railways in the Nineteenth Century* (London: Routledge and Kegan Paul, 1965), which takes the MacDonagh view, with Richard S Lambert, *The Railway King, 1800–1871: A Study of George Hudson and the Business Morals of His Times* (London: G Allen and Unwin, 1934) and Anthony Burton, *The Canal Builders* (London: Eyre Methuen, 1972), which focus on technical controversy. See also for various aspects of this issue J Fullmer, 'Technology, Chemistry, and the Law in Early Nineteenth Century England,' *Technology and Culture* 21 (1980): 1–28; F Clifford, *A History of Private Bill Legislation*, 2 vols (1885; rpt. London: Cass, 1968); F T K Pentelow, *River Purification: A Legal and Scientific view of the Past 100 Years, being the Buckland Lectures for 1952* (London: Edward Arnold, 1953); Clement Higgins, *A Treatise on the Law Relating to the Pollution and Obstruction of Watercourses* (London: Stevens and Haynes, 1877).

19 Funk and Wagnall's *Standard College Dictionary* (1963), s.v. 'indicator'.

1 The Most Difficult Operation in Chemistry: The Analysis of Mineral Waters

I had rather trust to the ... remedies of old women and nurses, than to ... most of the writers ... upon mineral waters; for, such, from some motive or other, are rarely to be confined between the narrow limits of truth and right reason.[1]

Charles Lucas

Early in Robert Smith Surtees' novel *Handley Cross*, 'one Roger Swizzle,' an apothecary of modest means, learns of a mineral spring at Handley Cross 'capable of "curing everything" '. Swizzle analyses the spring, 'and finding the ingredients he expected' sets himself up with great success as an 'experimental (q.v. quack) practitioner,' recommending a regimen for wealthy dyspeptics which combines drinking the waters with eating whatever one wants. Swizzle's initiative quickly transforms Handley Cross from an insignificant hamlet into a major spa.[2]

What Surtees was satirizing was happening, or at least being attempted, in towns throughout Britain during the first half of the nineteenth century. During the eighteenth century the market in mineral waters had been dominated by a few towns, like Epsom, Bath, and Harrogate. But by the end of the century their dominance was being challenged by a host of little known places where some astute doctor, landowner, or local booster sought to cash in on the virtues of a spring which might well have been used to cure locals (people and animals) since time immemorial.[3] For the town that succeeded in promoting its waters, there was plenty of money to be made. We are familiar only with the successful—towns like Cheltenham and Leamington—but many others failed to attract enough wealthy patrons and lapsed into obscurity.[4]

We are most familiar with these springs through the literary and social history of the Georgian age, when they became places of resort for the aristocracy and for flocks of their imitators. Historians of the spa have rightly paid attention mainly to the social life, recognizing that this had more to do with the attractions of spas than did concentrations of dissolved salts.[5] Yet ultimately claims of medicinal properties and of a chemical composition that could account for those properties underlay the prosperity of any spa.

Had one joined a conversation of chemists or doctors discussing the qualities of water before 1850, they would almost certainly have been referring to the waters of these various therapeutic springs. Medicinal powers were claimed for thousands of springs throughout Europe, and their waters possessed a great range of compositions and properties. Some were hot, others unusually cold; some smelt strongly, often of hydrogen sulphide, others had the bitter taste of dissolved chlorides and sulphates, or the sprightliness of dissolved carbonic acid. There was a similar range of medical effects: many saline springs worked as gentle purgatives; sulphurous springs were recommended for skin conditions; chalybeate or iron-containing waters restored patients to former vigour.

A vital part of making a claim for the virtues of one's mineral water was a chemical analysis of that water. This book, which is mainly about the analysis of potable (drinking) water, begins with an extended consideration of mineral water analysis because in a great many ways the sort of scientific enterprise that mineral water analysis was—in terms of its practitioners, range of techniques, conventions of inference, and its social setting amidst political and medical controversy—carried over into the science of potable water analysis that began to develop around 1850.

In the Final Analysis?

In 1840 Dr Augustus Bozzi Granville of London, medical reformer and controversialist, set out to survey the 'spas of England,' intending to publish a companion to his recently published volumes on *The Spas of Germany*. Granville recognized the intense competition among resorts; wherever he went proprietors prevailed upon him to endorse the claims of the place. But he was skeptical. In some cases

he suspected outright fraud—at Cheltenham he feared that taps
to different waters actually came from a single source; at Radipole
he doubted claims about sulphur content.[6] But more common and
more troubling than outright frauds were conflicting claims of the
composition of mineral waters. When it came to evaluating these
Granville was puzzled. Multiple analyses of the same water, even
when done by analysts of the 'first respectability,' differed enormously both in the constituents reported and in the quantities of
those constituents.[7] It was not clear which, if any, analyses were to
be believed nor how one was to judge claims of medicinal effectiveness.

It is Granville, the gossipy chronicler of spa living and spa management, who best records the importance analytical chemistry had
come to have in warranting the claims of proprietors of mineral
waters. Granville commented, for example, on the custom of distributing cards printed with an analysis of the water to patrons of
a spring. Though he questioned many of these analyses and recognized that they frequently were used solely for publicity, Granville
still accepted analysis as the only way to authenticate claims made
for a mineral water. At Woodhill and Hockley Green he advised entrepreneurs trying to market unknown yet potent springs of the necessity of a thorough analysis, even though the spring's powers might
be well accepted by local medical men.[8] Elsewhere, as at Scarborough, Granville worried that even well accepted waters might lose
popularity if the public became aware how much chemists differed in
their assessments of composition. He wrote of a Scarborough public
meeting where it was pointed out that 'of five successive analyses
which had appeared in the course of perhaps two thirds of a century,
there was not one that did not differ from the rest in every essential particular .... Some too admitted ingredients as present in the
waters, which the others had not even mentioned.'[9] Conditions were
much the same at Ashby de la Zouch, where Granville considered
four divergent analyses and was 'sadly puzzled as to which to adopt,'
and at Stratford, Gloucester, and Bath.[10] Doubts about composition
led directly to doubts about medical efficacy, Granville recognized.
His solution in each case was another analysis, an unimpeachable
one, done 'by a professed chemist, a man of undoubted eminence,
whose name and well-known experience in the difficult art of properly analysing mineral waters shall stamp ever after the analysis ...
with an authority and authenticity never to be questioned.'[11]

The problem of course was who to trust with this final analysis:

what so vexed Granville was that the divergent analyses were all too frequently done by chemists he respected, men like Richard Phillips, W T Brande, C G B Daubeny, J F Daniell, Charles Scudamore, Frederick Accum, and others. Granville did have favourites—the Leeds Quaker William West, and two German chemists, A Walcker and Edward Schweitzer, who were connected with F A Struve's artificial mineral water establishment in Brighton—but he was not forthcoming when it came to why their analyses (or any other particular analysis) were to be favoured.[12]

A twofold problem faced those who like Granville hoped to offer the public an impartial assessment of the properties of mineral waters. First there was the problem of the social context of mineral water inquiries. Most of what was written about springs, and most of the analyses done of their waters, was frankly partisan, the product of those with financial interests in the springs. Often the writers or analysts were local medical men—the Roger Swizzles—hoping to promote use of a nearby spring and hence to gain the practice of those who could be attracted to take its waters. In some cases this partisanship probably involved outright fraud; more often it involved only a convenient liberality in interpretation.

Second was the problem that there was legitimate scientific uncertainty with regard to the analysis and medicinal use of mineral waters: before about 1860 there was no consensus about the proper procedure for analysing mineral waters, no *Standard Methods*. Instead there were many who claimed authority. There was also disagreement as to the medical significance of results. While most authorities agreed on a general classification—mineral waters were either sulphurous, saline, chalybeate, or carbonated—there was substantial disagreement as to which compounds in which quantities had which sorts of medicinal effects and even whether or not chemical investigations were capable of determining medicinal effects at all.

No matter how much it might have been exacerbated by the expansion of resorts during the first half of the century, the confusion that confronted Granville was not unique to Britain in the 1840s. Both problems were bound up with one another, and neither was new. Forty-five years before Granville, William Saunders had characterized the corpus of tracts on particular mineral waters:

> We shall find in many of them a great fund of chemical knowledge, and excellent medical observations; but we shall also, in the greater number of these works, meet with certain modes of treating the subject, which may fairly be brought under candid criticism. Some of the

writers (especially those who have shewn themselves skilful and zeal-
ous chemists) have, I think, sometimes refined [sic] too much on the
science, and have endeavoured to transfer the same accuracy of dis-
crimination which experimental chemistry affords, to the explanation
of minute effects produced on the living body by various substances
during their stay in its complicated organs. Others again, have endeavoured to throw a veil of mystery over the whole subject, and
professing to disregard all the information which chemistry affords,
they have studiously avoided any attempt to explain the effects produced by certain mineral waters by a review of their contents, and
have strongly favoured the ideal of their being *specifics* prepared by
the hand of nature, against some of the most formidable and obstinate diseases with which the human race is afflicted.[13]

Eighty-five years earlier, the physicians Charles Lucas and Diederick Linden had railed at the pretension and corruption of mineral
water physicians and chemists in similar treatises on mineral waters.
The 'most pompous' of the numerous tracts on mineral waters were
written, Lucas noted, by men 'living and practicing upon the spot,
not always competent judges of the subject, but always interested
in the fame of the particular water, which was their idol.'[14] While
Lucas was willing to accept in principle the claim that mineral waters had medicinal potency, he felt that their use was completely
devoid of legitimate medical rationale: physicians were viciously attacking one another all the while being ignorant of the properties
of waters.[15] At Bath (toward which he was particularly harsh) and
elsewhere wealthy invalids were fleeced by mercenary physicians, yet
they ignored the advice they paid for, insisting on taking the waters without regard to season or constitution.[16] In some cases any
beneficial effects the waters might have had were undermined by
supplementary doses of physic doctors prescribed.[17] Ultimately the
spas were nothing but gathering grounds for sycophants, Lucas concluded, and it was futile to wish otherwise. 'Forms, fashions, and
flattery rule the world,' he wrote, 'and a man may as well refuse to
eat modish stinking wild fowl or venison at a great man's feast, be
insensible to the beauty of his mistress, hound or horse, or disrelish
any other prevailing vice or folly, as decline drinking of his favourite
spring, or deny having received benefit of it.'[18]

Whatever the dominance of ignorant fashion, proprietors of unfre-
quented spas still looked to chemistry for vindication. Linden wrote
of making analyses at Islington, where 'the Proprietor ... with great
Frankness and Chearfulness, granted me the leave to make a few

Experiments on the *Fountain-Head*. He told me that he had long wished-for, and had often desired and invited gentlemen from the Faculty [i.e. the Royal College of Physicians] to make *Experiments*, that the Public, by their means, might be satisfied, that this water was a Natural, and not an Artificial Compound.'[19]

In addition to denouncing the corruption of mineral water chemistry, Lucas was also critical of common standards of knowledge and practice among chemical analysts, and of what might be called the epistemic integrity of the science: he questioned the meaning of the concept of acidity, noted the arbitrariness of classifications, the indiscriminate use of the term 'sulphur.'[20] Linden similarly worried about the persistence of speculative atomism in mineral water analysis, as 'too apt ... to amplify our conceits ... and reasonings upon them, beyond their due limits.' Such doctrines did not provide any illumination to the physician.[21]

In sum, their complaints suggest that the problems which plagued mineral water analysis in the mid eighteenth century were the same problems which had plagued the art and science of chemistry, along with the rest of the medical crafts and professions, for more than two centuries. Noel Coley has called eighteenth century mineral water analysis 'a subject in the uncharted hinterland between chemistry and medicine not quite respectable in either.'[22] Yet it should be noted that neither of those fields was wholly respectable itself. Learned physicians clung to ancient monopolies and still accused one another of killing patients with antimony or bloodletting. Quacks roamed Europe promising to restore sight or safely remove bladder stones. Alchemists still touted miracle medicines as well as gold making schemes. At the same time the intellectual foundations of chemistry were confused, insecure, and continually in flux. Aristotelian forms, alchemical essences, Paracelsian principles, subtle fluids, corpuscles and atoms, and acids, bases, and salts—all were put forward to explain chemical phenomena. Lucas himself fused seemingly inconsistent systems of explanation with the assertion that 'there is ... constant succession of creation or generation ... into principles and back out into substance.'[23] Thus, both intellectually, in terms of what they thought they were up to, and socially, in terms of where they fit into the fabric of society, mineral water chemists in the late eighteenth and early nineteenth centuries were perpetuating a colourful, if not entirely honourable heritage. The following sections take up the chemists' attempts to cast off much of that heritage by seeking a solid foundation for mineral water analysis, and by seeking to make the mineral water chemist an honoured and indispensable professional. The second issue, of the social context of mineral water analysis, I take up in the next chapter.

Mineral Water Analysis Before Bergman

In part, the chaos Granville, Lucas, and Linden described was a consequence of the need to answer important medico-scientific questions through use of a science that was simultaneously trying to settle on a set of basic units and arrive at a consensus about the nature of acids, bases, and salts and the properties of aqueous solutions. Exacerbating these problems were difficulties peculiar to mineral water analysis and the great range of incompatible techniques chemists had developed over the years for determining the properties of mineral waters.

A claim frequently made at the end of the eighteenth century was that mineral water analysis was 'the most difficult operation in chemistry.' As the 1797 edition of the *Encyclopedia Britannica* explained it,

> Almost all mineral waters contain several different substances, which being united with water may form with each other numberless compounds: Frequently some of the principles of mineral waters are in so small quantity, that they can scarcely be perceived; although they may have some influence on the virtues of the water, and also on the other principles contained in the water.—The chemical operations used in the analysis of mineral waters, may sometimes occasion essential changes in the substances that are to be discovered. And also, these waters are capable of suffering very considerable changes by motion, by rest, and by exposure to air.[24]

To understand why there was so much perplexity it may help briefly to contrast eighteenth century views of the dissolved state with modern views. With some exceptions, eighteenth century chemists looked upon solutions in much the same way we look upon mixtures. A salt in solution was essentially little different from a salt in its crystalline form; it was simply broken up more finely, perhaps into its constituent particles. Likewise its medical activity was usually conceived of as a direct activity; within the body, the salt acted in a qualitatively similar way to its action outside the body. Hence in the eighteenth century view, little in the way of complex chemical change went on either in the solution or within the imbiber's body.

With the rise of physical chemistry, and in particular the dissociation theory of Svante Arrhenius in the 1890s, these views changed radically. In the case of ionic compounds, the kinds of salts that most of the mineral water chemists were interested in, the process of solution itself came to be understood as a chemical change in which the salts dissociated into component ions: common salt, for example, into Na^+ and Cl^-, gypsum into the calcium ion Ca^{2+} and the sulphate radical SO_4^{2-}. Moreover, the dissolved state came to be recognized as a continually active system of ions coming together to form compounds and then coming apart, maintaining an equilibrium so long as physical conditions did not change. Changes in chemical or physical conditions, such as driving off dissolved carbon dioxide or changing temperature or driving off some of the solvent, would change this equilibrium, perhaps forcing some ions to remain in salts and fall out of solution. What all this meant was that the substances that were held to be in mineral waters were a great deal less stable than eighteenth century analysts usually held them to be. What was in the water was an artefact of the particular conditions under which the water happened to exist at the time of analysis; the way it acted in the body was due more to the interaction of water and body than to the properties of the stable salts that analysts recovered in analysis. In a sense then they were searching for an entity which, at least as they conceived it, did not exist. They were looking for stable substances where an inquiry into the responses of ions to changing conditions might have been more illuminating. As we shall see, even by the mid 1830s analysts had begun to recognize a problem; by that time it had become common to express results in terms of acids and bases (or what we would call ions) and not to worry too much about what form these acids and bases actually took in solution.[25]

A chemistry describing, explaining, and often advertising the virtues of a particular spring (or equally the dearth of active ingredients in a rival spring) had emerged as early as the fifteenth century.[26] Initially, an 'analysis' was little more than an examination of prominent physical characteristics of the water—odour, taste, temperature, colour—following suggestions of Pliny or Aristotle. During the Renaissance these observations were supplemented by manipulations and tests with reagents. The use of oak galls, which blacken iron-containing water, went back to Pliny; Gabriel Fallopius in 1564 put together a highly systematic collection of colour tests, as did the English chemist Robert Boyle over a century later.[27] Such indicator analysis probably developed from dyeing technology; the reagents used were frequently dyestuffs and their effects in water were as much an indication of the utility of the water for the dyer as they were of its chemical composition. In addition to testing waters with reagents, fourteenth century Italian physicians had begun the practice of characterizing the evaporative residue of a mineral water by its behaviour when heated, a practice widespread in Germany by the early fifteenth century.[28]

Bergman's Answer: The Triumph of Order

On these foundations was built an enormous medical and medico-chemical literature on the properties of various waters. As chemical knowledge grew and became increasingly better organized during the late eighteenth century, a number of attempts were made to systematize water analysis. The chemists responsible were the central figures in analytical chemistry during the period 1760 to 1830: Torbern Bergman, Richard Kirwan, M H Klaproth, A F de Fourcroy, and L N Vauquelin. The approach they took will be familiar to anyone who has taken an introductory inorganic analysis course; during this time chemists worked out many of the basic procedures for determining the identity and quantity of substances in aqueous solution.[29]

To many early nineteenth century water analysts it was the work done in the 1770s by Torbern Bergman (1735–84), professor of chemistry at Uppsala, that divided the asystematic and often incomprehensible analyses of earlier centuries from the rigour of their own science. He was their 'foundation,' as Thomas Thomson put it.[30] In 1778 Bergman published a large work on mineral water analysis which was translated into English along with a number of shorter essays on Swedish mineral springs, and on hot and cold artificial mineral waters. These appeared in the first volume of his *Physical and Chemical Essays* (1784).

The system of water analysis presented there utilized—and indeed canonized—three distinct approaches to understanding mineral waters: the examination of physical (including medicinal properties), a qualitative examination through the use of reagents, and a quantitative analysis of the evaporative residue (the salts that remained after all the water had evaporated) of a large quantity (perhaps several hundred pounds) of the water being tested. These stages, in this order, constituted a complete examination of a mineral water.

Each of these approaches had a long history. Distinguishing waters by physical characteristics went back to antiquity. A great deal could be learned from taste ('aerial acid' [carbonic acid] had a 'gentle sweetness or poignancy'; alum, a 'sweetish astringency'; natron or marine salt, a 'nauseous brackishness'); from texture (as in the sprightliness of carbonated waters); from strong odour (e.g. of hepatic waters, those containing H_2S); from the rate of heating and cooling, or even from sound—Lucas claimed that 'the purest water makes the greatest noise when poured from one vessel to another.' Colour was also important—iron or copper salts sometimes tinted water. Specific gravity showed how much dissolved matter a water contained.[31] Composition could also be inferred from medicinal effects, a curious form of argument since the ostensible purpose of analysis was to determine whether the water really had any medicinal properties.[32] While they maintained that this sort of physical characterization was an essential component of a complete mineral water analysis, Bergman and those who followed him rarely attached much importance to it in presenting their conclusions. Qualitative and quantitative evidence were far more important.

The second approach—qualitative examination through use of reagents or indicators—is often associated with Robert Boyle, though a long tradition of indicator analysis predated him.[33] Late eighteenth and early nineteenth century water chemists often studded their analyses with reagent tests, using not only colour indicators like galls, litmus paper, or tincture of cabbages, but also tests for specific substances, for example proving the presence of sulphur by the blackening of a silver spoon. Some tests utilized common replacement reactions (oxalic acid for the presence of lime, barium chloride for the presence of sulphuric acid).

Here too there was disagreement about the necessity of such procedures. Bergman, though insisting they be included in a complete analysis, saw them as secondary. They could be used to give a general notion of composition in cases in which one wished to avoid the trouble of a complete quantitative analysis. In the 1813 encyclopedia, *Pantologica*, of John Mason Good and Olinthus Gregory, reagent tests were portrayed as obsolete, being fast replaced by superior quantitative methods. Such tests were 'very uncertain ... their effects do not determine in an accurate manner the nature of the substances held in solution in waters ... [and] the cause of the changes which happen in fluids by their addition is often unknown.' The passage concluded that 'in the best works ... re-agents are only to be used as secondary means, which at most serve to indicate or afford a probable guess of the nature of the principles contained in waters.'[34] Others, notably the Irish analytical chemist Richard Kirwan and the French chemist A F de Fourcroy, saw qualitative analysis as a necessary precursor to quantitative analysis. In Kirwan's fabulously complicated and little used system of analysis there was no single quantitative approach that would work for all mineral waters. The protocol one chose for quantitative analysis depended on the set of salts one found on qualitative analysis since one had to make allowances for the presence of one substance interfering with the tests used for measuring another.[35] In fact, far from becoming peripheral or obsolete, such 'indirect methods' would come to be seen by the middle of the century as the only trustworthy approach to mineral water analysis.

It was the third approach, quantitative analysis of an evaporative residue, that Bergman and his followers emphasized. As early as the fifteenth century, analysis by fire—evaporation or distillation—had been a popular means for studying mineral waters. The hope had been to characterize a water by the crystals that formed during evaporation. The chief problem was one of separation. Under slow evaporation one could hope to separate the constituents since ideally each would crystallize and fall out of solution as the liquor reached supersaturation with respect to it. But too many salts crystallized too slowly and over too wide a range of concentrations to give good separations.[36] Another approach was to evaporate to dryness and characterize the residue as best one could: through the shapes and colours of crystals, their deliquescence, behaviour on heating, and so forth.[37] A few chemists, notably Robert Boyle, remained skeptical. They were concerned that a residue was a false or incomplete reflection of the water's true contents. They worried that in some unknown way evaporation changed the substance being evaporated. This concern, which we will see arising anew in the second decade of the nineteenth century, accounted in part for the attractiveness during the early eighteenth century of Boyle's reagent approach.[38]

What Bergman offered was a simple and systematic way of separating the salts in the residue. His approach was to evaporate the water to dryness, during which time any dissolved gases were collected for analysis. The residue was then weighed and treated with rectified alcohol which dissolved iron vitriol (sulphate) and chlorides and nitrates of lime, magnesia, and barium. These were separated from one another at a later stage. Residue insoluble in alcohol was treated

Evaporate

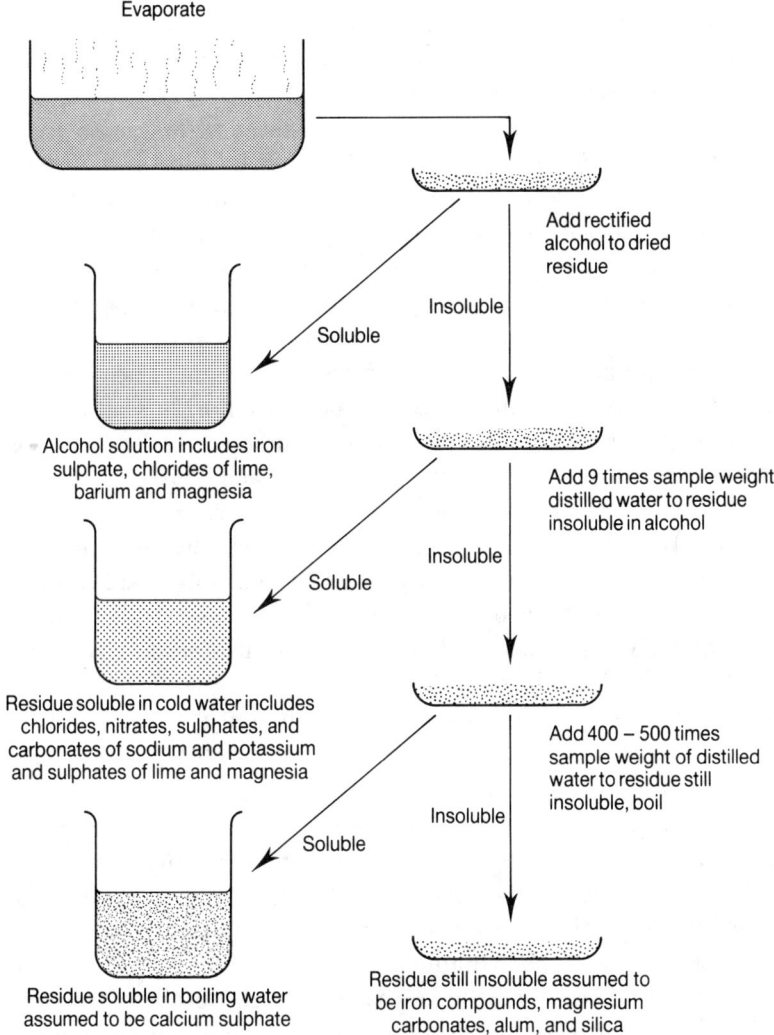

Add rectified
alcohol to dried
residue

Insoluble

Soluble

Alcohol solution includes iron
sulphate, chlorides of lime,
barium and magnesia

Add 9 times sample weight
distilled water to residue
insoluble in alcohol

Insoluble

Soluble

Residue soluble in cold water includes
chlorides, nitrates, sulphates, and
carbonates of sodium and potassium
and sulphates of lime and magnesia

Add 400 – 500 times
sample weight of distilled
water to residue still
insoluble, boil

Insoluble

Soluble

Residue soluble in boiling water
assumed to be calcium sulphate

Residue still insoluble assumed to
be iron compounds, magnesium
carbonates, alum, and silica

Figure 1.1 Bergman's protocol for mineral water analysis by evaporation. The diagram shows only the first level of separation. Individual salts could be later separated from each of the fractions.

with about nine times its weight of cold distilled water, which dissolved a great variety of alkaline and earthy salts, including nitrates, sulphates, carbonates, and chlorides of sodium and potassium, and sulphates of lime and magnesia. These too could be separated from one another at a later stage. Residue still insoluble after the first two treatments was boiled in 400–500 times its weight of distilled water, which would dissolve calcium sulphate. What remained was presumed to be a mixture of iron compounds, calcium and magnesium carbonates, alum, and silica. These too could be distinguished to a certain extent: iron by its colour, lime and magnesia by their reactions with sulphuric acid, alum by its solubility in hydrochloric acid and precipitation with potash, silica by its insolubility in the reagents that dissolved the others, and by the blowpipe.[39]

What made Bergman's system so attractive was that it made what had seemed chaotic, idiosyncratic, and ambiguous, appear orderly, uniform, and straightforward. Bergman's procedure was easy and it was imbued with Enlightenment optimism. The separations were believed to be complete—no calcium chloride was to remain undissolved in the alcohol. Its achievement seemed a triumph of eighteenth century rationality and it is no accident that Bergman, the achiever, was also the formulator of the most detailed affinity tables of the late eighteenth century on which the residue analysis was based. Both the water analysis and the affinity tables reflected the confidence that nature, at the level chemists studied her, was as intelligently organized and rigidly differentiated as it was at the organismic level studied by Bergman's colleague, Linnaeus. The comparison is not facile; at least one eighteenth century writer set out to produce a 'methodus aquarum, even as others have done of plants and animals.'[40]

This confidence in order and simplicity was to be short-lived. Bergman's protocol was neat, simple, and elegant, but it didn't always work. Beginning with A F de Fourcroy, and followed by Richard Kirwan, Martin Heinrich Klaproth, and Fourcroy's assistant and successor, Nicholas Vauquelin, almost every analytical chemist of stature found some reason to disagree with details of Bergman's system. Yet in most cases their criticisms did not threaten the basic approach of separating the salts in an evaporative residue. They complained that Bergman's separations failed to separate what he claimed they did or that his procedures failed to take into account substances newly found in mineral waters or failed to work where there were unusual combinations of ingredients. For example, Fourcroy pointed out that alcohol extraction dissolves some sodium chloride, which it is not supposed to, and fails to dissolve some calcium chloride, which it is supposed to. But these were problems that could be easily addressed by adding steps to the procedure, for example by isolating sodium chloride in both the alcohol fraction and the cold water fraction and combining the two quantities to get the total.[41] Extending some of Fourcroy's concerns, Kirwan recognized numerous cases in which the presence of one salt would interfere with tests for another. Hence in his system a qualitative analysis was essential so that one could know which interference problems would arise in quantitative analysis and make allowance for them.[42]

Despite criticisms, the Bergman protocol or a recognizable derivative remained the principal means of mineral water analysis used by British chemists well into the 1830s. As late as 1871 in a table of the composition of mineral waters, six of the eight analyses of European springs were by Bergman.[43] These were famous springs—Pyrmont, Spa, Seltzer—and had certainly been analysed by more modern techniques since Bergman's time, yet the author chose to list Bergman's analyses.

In the background, however, there remained the unresolved issue of whether there might be, as Meredith Gairdner put it, 'some lurking defect' in evaporation that altered the contents of a water. Most analysts, including Bergman, at least raised the issue. In commenting on the difficulty of water analysis, Bergman admitted that in part this was due to the fact that 'some of the principles ... are decomposed during the examination.'[44] The 1797 *Britannica* article on mineral waters noted in a similar context that 'the chemical operations used in the analysis of mineral waters, may sometimes occasion changes in the substances that are to be discovered.'[45] J M Good's *Pantologica* (1813) made the point even more emphatically: 'it cannot be doubted ... that the heat required to evaporate water, however gentle it may be, must produce sensible alterations to its principles, and change them in such a manner, as that their residues, examined by the different methods of chemistry, shall afford compounds differing from those which were originally held in solution.' The author concluded that it was a question 'whether the accurate results of the numerous modern writers afford any process for correcting the error which might arise from evaporation.'[46]

Prior to 1815 such concerns did not seriously threaten either the credibility of water analysis or the confidence and optimism of analysts. Such problems were simply part of what made water analysis tricky. There are a number of reasons why chemists were not much troubled by the possibility of analytical alteration. One was that by the end of the eighteenth century, most chemists felt that they understood the nature and extent of the more important alterations that occurred during analysis. The key here was the recognition of the role of carbonic acid in holding in solution earthy carbonates and iron. When water was heated, shaken, or simply allowed to stand, carbonic acid escaped as CO_2. From the quantity of CO_2, one could, however, work back to calculate the salts it must have held in solution, and thus correct for the alteration.[47] In fact, such carbonated or sparkling waters were one of the most important classes of medicinal mineral waters. To writers of the previous century, like Frederick Hoffmann, such effervescence had seemed to be the 'life' of the water, something mystical or vital that the chemist would never be able to capture.[48] Hence the unravelling of carbon dioxide chemistry was particularly satisfying: it demonstrated that even if analysis did alter the analysed water, chemists could still correct for that change.

Another reason for not worrying overmuch about accusations of analytical alterations may have been suspicion of the motives prompting such accusations. The claim that analysis irreparably altered what was being analysed could be a defence against the threat of exposure of a spring's impotence. A Sutherland, writing in the 1750s, regarded most claims of the inevitability of alteration in this way, as intended 'to promote use of particular springs mainly for financial reward.'[49] As we shall see in the next chapter, any sort of argument which held that the miraculous properties of mineral waters were for scientific reasons forever undiscoverable was useful to those wishing to uphold the uniqueness of the springs they were associated with. To admit the inadequacy of analysis would have been to give aid and comfort to such obscurantists.

A third reason not to worry was the availability of artificial synthesis of a mineral water as a way of checking the accuracy of analysis. Both Lucas and Bergman recommended this 'inverse method,' and Bergman went to considerable trouble trying to make it work, though with little success.[50] It turned out that one could not simply dump in the salts discovered on analysis and get a mixture with the potency of the original mineral water.[51] Often the salts found in the residue were hardly soluble in water. Yet despite the difficulties of emulating natural mineral waters, later writers continued to list synthesis as the ideal way to check the results of analysis.

Murray's Answer: The Triumph of Skepticism

In 1815 the Edinburgh physician and chemist Dr John Murray (to be distinguished only with difficulty from two other contemporary John Murrays who were also chemists) exploded this complacency by proposing that the salts obtained in analysis were not necessarily the salts that existed in solution.[52] Murray (d. 1820) is a surprisingly elusive figure. He took M.D. at St Andrews in 1814 and was an external lecturer in chemistry, materia medica, and pharmacy at Edinburgh. He was a fellow of the Royal Society of Edinburgh and the Geological Society of London, and an anti-Huttonian in geology. As a chemist, he was evidently a popular teacher and authored two texts, *Elements of Chemistry* and *A System of Chemistry*, each of which went through multiple editions. Murray was also one of the main British exponents of the unorthodox chemistry of C L Berthollet, and it was his interpretation and application of Berthollet's chemistry that posed so serious a threat to conventional mineral water analysis.

Murray's revolutionary proposals initially appeared as an afterthought in a paper on the 'Analysis of the Mineral Waters at Dunblane and Pitcaithly,' given to the Royal Society of Edinburgh in 1815. The Dunblane spring was newly discovered, and Murray was interested in demonstrating its medicinal properties. Using a modification of Bergman's process, he found the Dunblane water to contain sodium chloride, a small amount of calcium chloride, and calcium sulphate. But these results seemed inconsistent with the purgative effects claimed for the water. Sodium chloride and calcium sulphate were medically inactive; calcium chloride was active yet present in too small a concentration to explain the water's effects. Murray was not the first to recognize this sort of anomaly; others had proposed that the power of mineral waters lay in the great degree of comminution of salt particles in solution or had appealed to some mystical and unanalysable properties which gave springs their potency.[53]

Murray took the anomaly as an indication of the need for a reexamination of assumptions about the dissolved state. He wrote in the final section of the paper, subtitled 'Observations on the Composition of Saline Mineral Waters,'[54] that it was 'a question not unequivocally determined, and perhaps not capable of being determined, in what state the saline ingredients of a mineral water exist— whether the acids and bases are in ... binary combinations, ... or whether they exist in simultaneous combination, the whole acids being neutralized by the whole bases.'[55] The latter option (Berthollet's view) Murray regarded as unlikely: it was unable to account for the markedly distinct medicinal effects of different waters. Like most of his contemporaries Murray believed that in neutral solutions acids

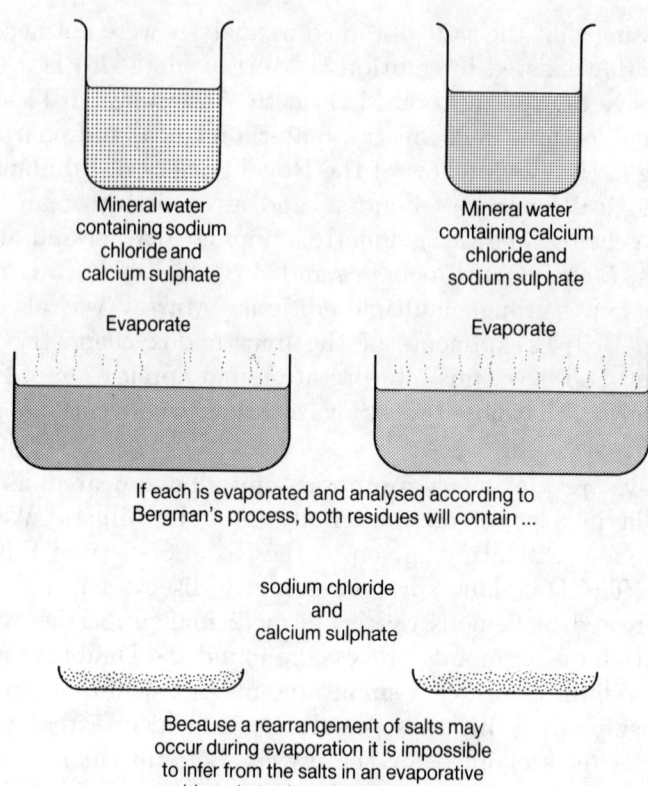

Mineral water
containing sodium
chloride and
calcium sulphate

Mineral water
containing calcium
chloride and
sodium sulphate

Evaporate

Evaporate

If each is evaporated and analysed according to
Bergman's process, both residues will contain ...

sodium chloride
and
calcium sulphate

Because a rearrangement of salts may
occur during evaporation it is impossible
to infer from the salts in an evaporative
residue what salts existed in the original
mineral water

Figure 1.2 The thrust of Murray's argument was that the finding of a
particular combination of salts, here sodium chloride and calcium sulphate,
in the residue of a sample which had been evaporated, was no proof that
those salts existed in the water. It was equally possible to believe that
recombination had taken place during evaporation, and that the original
water had contained calcium chloride and sodium sulphate. In Murray's
view there was no way to determine which was the case.

and bases existed as binary compounds—the question was which
ones. If one rearranged acids and bases, putting sodium preferen-
tially with sulphate, and calcium with chloride, the Dunblane water
suddenly showed a composition consistent with the pharmacological
activity claimed for it. But to do this flew in the face of the orthodox chemistry of affinities that had been developed throughout the
eighteenth century. That chemistry held that calcium would never
displace sodium from sodium chloride, because sodium and chlorine
had greater affinity for one another.

Murray was not denying that the inactive salts appeared in the residue; he was suggesting only that 'the state of combination ... [might be] modified by the analytic operations.' As evaporation took place the acids and bases might switch partners. Hence what patients drank might be quite different from the sort of mixture a chemist might try to synthesize using an analysis of residue as a guide. Unlike many of his early eighteenth century predecessors who had made this argument without having a clear idea of exactly what sorts of changes analysis was likely to produce, Murray was not without a theoretical framework which made these radical ideas not only plausible, but likely.

This was Berthollet's solution chemistry. As early as the 1810 edition of his elementary text, Murray had taken the view that in solutions an equilibrium of possible combinations of acids and bases developed, the equilibrium conditions being a function of mass, affinity, and all manner of physical conditions—temperature, cohesive states of the solutes, rates of change of temperature.[56] In complicated and dilute solutions these physical factors might totally overwhelm normal affinities and allow salts to be present that under normal circumstances would not coexist. In Murray's view the binary combinations that would develop in dilute solutions would be those which were most soluble. As evaporation proceeded, affinity—the force of cohesion—would become increasingly important and the strongest acids and strongest alkalis would crystallize into the salts that affinity chemistry predicted.[57] Conveniently, at least in the Dunblane case, the maximum solubility arrangement also gave Murray the set of salts most medically active.

But how to prove all this? If Murray were right that solutions were delicate equilibria, it followed that any analytical intervention, whether evaporation or the addition of a reagent, might alter the conditions of equilibrium. Thus there was no certain way of telling whether the Dunblane water really contained sodium sulphate and calcium chloride or sodium chloride and calcium sulphate. The fact that one found sodium chloride and calcium sulphate on evaporation meant nothing. There was, Murray wrote,

> perhaps no decisive experiment by which this question may be determined; for any method which would cause the *separation* of either substance as a binary compound, may also be conceived to operate by causing its *formation*. Thus, though sulphate of lime is obtained by evaporation, this is no proof of its prior existence, since the concentration of the solution might equally cause its formation .... Its

separation by a precipitant, by alcohol for example, ... is liable to
the same ambiguity; a certain degree of concentration of the watery
solution would be necessary for the effect, and the further operation
of the alcohol might be precisely on the same principle—diminishing
the solvent power of the water, and thus aiding the force of cohesion,
in determining the combination of the ingredients which form the
least soluble compound.[58]

Murray did do two experiments to test his hypothesis about the
double decomposition of sodium sulphate and calcium chloride in the
Dunblane water. First he added sodium sulphate to Dunblane water
with the expectation that if its presence in water were incompatible
with the calcium chloride shown by analysis, the two salts would
react to form calcium sulphate, which should, at the concentration
Murray was using, crystallize out of solution. No crystallization
took place; Murray concluded tentatively that the two medically active salts, sodium sulphate and calcium chloride, could coexist in
dilute solutions without undergoing double decomposition. Such a
conclusion was in violation of Richard Kirwan's rule (which many
other analysts took seriously, more so in fact than Kirwan) that
alkaline sulphates could not coexist with earthy nitrates or muri-
ates (chlorides).[59] Murray's second experiment was to add sodium
sulphate to the mineral water and then evaporate as in a normal
analysis. The sulphate appeared in the residue as calcium sulphate,
and Murray concluded that there was indeed a reaction between
calcium chloride and sodium sulphate during evaporation.[60]

Yet neither experiment was conclusive, Murray admitted, for precisely the reasons he had alluded to. In the first experiment, the fact
that no calcium sulphate crystallized was no proof that the added
sodium sulphate had not reacted with calcium chloride to form it.
Likewise, in the second experiment, because some of the sodium sulphate he had added reacted to form calcium sulphate was no proof
that sodium sulphate existed naturally in the water.[61]

What then could one conclude about the composition of mineral
waters? Very little. In Murray's view, any conclusion that what one
found in the residue of a mineral water was what was really in it
was 'merely oversight or prejudice.'

If it can be shown that the elements of these compounds may equally
exist in the water in a different state of combination, which the evaporation must change [which Murray felt he had demonstrated in his
two experiments], the conclusion that they do exist in such a state is
a priori as probable as the conclusion that they exist in the state in

which they are actually obtained. It is demonstrable that if muriate of lime [calcium chloride] and sulphate of soda exist together in a mineral water, ... they must by evaporation be obtained, as muriate of soda and sulphate of lime. The actual obtaining, therefore, of these latter compounds is no proof that they pre-existed ... in the water, to the exclusion of the opposite view.[62]

In fact, Murray was making three arguments in the Dunblane paper. The first, and most inclusive, was that it was impossible to prove that analytical intervention did not change the object being analysed. Taken seriously, the argument thoroughly undermined the possibility of any analytical chemistry and this Murray was unwilling to accept. The second argument, more narrow, was that the logical structure of evaporation analysis was at fault. On finding sodium chloride and calcium sulphate in a residue one could not infer that they had existed in the water which had produced the residue, especially when the opposite combinations, calcium chloride and sodium sulphate, could produce an identical residue. Murray's third argument was that the wrong sorts of evidence were being used in arriving at conclusions about the composition of mineral waters. In his view medicinal properties, the solubility of the salts, and perhaps even the likely geological sources of the salts, had to be taken into account in determining which of several possible arrangements of salts actually existed in the water.

Within two years Murray had developed an alternative approach, a new 'General Formula for the Analysis of Mineral Waters.'[63] Having demonstrated that one could never know what combination of salts existed in a mineral water, Murray maintained that analysis should only concern itself with the determination of individual acids and bases, leaving their combination a matter for speculation. Determining the quantities of these was reasonably straightforward: the bases and acids could be precipitated through replacement reactions.

Murray's most radical proposal was his suggestion for the presentation of analytical results. Each analysis was to have three sets of results. First the analyst was to list the quantities of acids and bases, the only claim that he could legitimately make. Second, one was to list acids and bases as binary compounds, 'inferred from the principle that the most soluble compounds are the ingredients.' This was Murray's view of the composition, the version that he believed best explained the medical properties the waters possessed. Third, the analyst was to present the acids and bases combined as they would

be in a conventional evaporative analysis, with strongest acids combined with strongest bases. This would allow comparison with other
analyses. Thus in Murray's view the first presentation represented
fact—all that was scientifically warranted; the second, truth (though
it could never be shown to be true); while the third was an heuristic
device for communication with other analysts.

Murray's compromise was not widely accepted, either by other
chemists or by their clients. A set of multiple options of what a water might contain was not what clients consulted chemists to obtain.
The only certain information that Murray offered them they weren't
interested in knowing: knowledge of the acids and bases in a mineral water was of little help to physicians convinced that salts were
the active ingredients. The first quarter of the nineteenth century
was a growth period for British spas, especially in the years when
war limited access to continental resorts. In their struggles against
established spas, proprietors of the new resorts looked to chemistry
for legitimation, not for unresolvable epistemic quandries. As for
chemists, while many acknowledged the cogency of Murray's criticisms and accepted that they could not go back to Bergman, they
were hesitant to adopt his alternatives. Had Murray simply been
wrong, it would have been much easier to deal with him. But at
least some of his points were convincing. He had a plausible explanation for the enormous and embarrassing variability in the results
analysts using different processes gave for the same water.[64] He had
a plausible explanation for the difficulty chemists had in synthe-
sizing imitations of the waters of famous springs.[65] And he had a
plausible explanation for the frequently commented upon inconsistency between a water's medical uses and the composition claimed
for it.[66] Yet whatever his achievements, Murray's attack on certainty in water analysis left chemical analysis far short of the degree
of definitiveness expected of it, by chemists as well as their clients.
Like Berthollet's chemistry on which it was based, Murray's water
analysis exposed the flaws of conventional views but was unable to
provide any satisfactory alternative.[67]

Hofmann and His Students: The Triumph of Pragmatism

The chemists' main response to Murray was the development of
a double standard—a willingness to provide for clients one set of
results, philosophically suspect perhaps, but done according to well-
known conventions, while admitting elsewhere, or as an aside, that

the water's actual composition might be quite different and remained a matter for speculation. For example, J J Berzelius, the foremost inorganic chemist of the period, noted in a paper on the springs at Carlsbad that even Murray had not recognized how fully Berthollet's chemistry exploded conventional ideas about solution. Berthollet's view was that all possible combinations of acids and bases would exist to some degree in solution, Berzelius noted. Hence the real composition of a water would be neither Murray's second, maximum-solubility version, nor his third, based on affinity chemistry. Having placed the matter in an even more uncertain light, Berzelius took the view that it made no difference. What was important was that results be comparable, not that they be correct.[68]

Such was the response to Murray well into the '60s. One might admit that Murray was right, yet see it as essential to present results according to convention. By the 1850s many analysts were acknowledging the unsoundness of residue analysis and were giving two versions of their results—acids and bases (determined according to Murray's procedures or variations of them) and their presumed combination as salts in solution. In general they did not adopt Murray's maximum solubility principle, however, but combined acids and bases 'as much as possible according to ... affinities, the strongest bases being supposed in combination with the strongest acids,' i.e., in the way Bergman would have arranged them.[69]

This convention was established in a series of analyses begun in the late 1840s by advanced students at the Royal College of Chemistry, under the direction of A W Hofmann, the Liebig student who headed the college. It had become common enough by the early '60s that Hofmann could say simply that an analysis had been done in 'the usual manner' without indicating what that was.[70] Those who chose not to follow it might be obliged to explain themselves. In an 1861 analysis Augustus Voeckler pointedly explained why he had 'deviated from the ordinary mode of uniting acids and bases' in listing carbonic acid as combined with potash rather than with an earth. Qualitative analysis showed that the water 'in its natural condition really contains alkaline carbonates,' Voeckler noted, and it would be wrong to pretend otherwise simply in order to hold to convention.[71] Those who stuck steadfastly to conventions could find themselves in embarrassing situations. The Hofmann student Edward Bennett, assigned the analysis of water from the Thames estuary at Greenwich, found, in effect, sea-water without salt. In a footnote he apologized:

1. ANALYSES OF THE NEW RIVER, EAST LONDON, KENT AND HAMPSTEAD WATERS.

	New River Water Company.	East London Water Company.	Kent Water Company.	Hampstead Water Company.
	Grains in an Imperial Gallon.			
Lime	5·7192	6·9034	7·82	2·9160
Magnesia	0·5280	0·7336	0·62	1·7098
Potassium	0·4972	0·5600	0·35	1·6471
Sodium	1·1634	0·9989	0·86	7·5761
Iron, alumina and phosphates .	trace	0·4760	trace	trace
Sulphuric acid (SO₃) . . .	3·2550	2·5830	1·86	9·1702
Chlorine	1·0500	1·0682	1·52	4·1230
Carbonic acid	11·1020	11·4527	11·56	10·9823
Silica	0·5005	0·6216	0·49	0·0728
Nitric acid	0·0150	0·4800	„	0·050
Ammonia	trace	trace	trace	trace

2. ANALYSES OF THE NEW RIVER, EAST LONDON, KENT AND HAMPSTEAD WATERS.

	New River Water Company.	East London Water Company.	Kent Water Company.	Hampstead Water Company.
	Grains in an Imperial Gallon.			
Carbonate of lime	7·82	10·16	11·64	4·95
Sulphate of lime . . .	3·23	2·33	3.16	„
Nitrate of lime	0·02	0·72	„	0·07
Carbonate of magnesia . .	1·09	1·51	1·28	3·53
Chloride of sodium . . .	1·73	1·76	2·24	6·79
Sulphate of soda . . .	1·49	0·94	„	15·14
Chloride of potassium . .	„	„	0·66	„
Sulphate of potassa . . .	1·11	1·25	„	1·40
Carbonate of potassa . . .	„	„	„	1·80
Silica	0·50	0·62	0·49	0·07
Iron, alumina and phosphates .	traces	0·47	traces	traces
Ammonia	traces	traces	traces	traces
Organic matter	2·79	4·12	2·61	1·84
Total	19·78	23·88	21·08	35·59
Solid residue obtained on evaporation . . . . . }	19·50	23·51	29·71	35·41

Figure 1.3 A W Hofmann, professor of chemistry at the Royal College of Chemistry, responded to Murray's dilemma by listing acids and bases separately in the first table followed in the second by their probable state of combination in the water (*J Chem Soc* 4 [1851]: 379).

the arrangement as above does not exhibit any chloride of sodium in the water, which no doubt must exist, as such, in the Thames at Greenwich, especially at high water .... But as the precise form or the proportions in which the individual constituents are distributed in a mixed solution is not known—for experiment proves they may be variously associated in solution according to the degree of concentration—every arrangement must be more or less hypothetical. We have therefore adopted the principle followed in the preceding analyses of combining the strongest acid with the strongest base as affording the best means of comparison.[72]

At issue in the development of conventions for representing results were two questions, one having to do mainly with chemistry itself, the other with the use of chemical knowledge. The first was the question of how far, and how significantly, the conventional representation of salts differed from the real composition of the water. The second was what the purpose of analysis was and what sort of representation of results best met that purpose. In an 1848 report on the Cheltenham waters, F A Abel and Thomas Rowney, two of Hofmann's students, took up these questions. With respect to the first question, they frankly admitted that the combinations they listed might not represent the actual composition of the water, but they did maintain that 'such an arrangement may be considered to furnish the nearest approach to the real distribution of the various constituents of the water.'

With respect to the second question, they argued that correspondence to reality was not the primary concern: there had been great 'want of principle and accordance in the arrangement of the results of … analytical investigation.' As a result it was 'impossible for the medical man to institute a correct comparison between different waters, and much of the interest and use of such analyses is lost, as it is this comparison of different waters that is most likely to lead to a knowledge of the manner in which individual constituents act.' They even went so far as to insist that 'a uniform and comparable arrangement of results, is of such primary importance, that the choice of the principle to be adopted is of less moment.'[73] Thus, while on the one hand Abel and Rowney claimed that they had finally gotten the composition pretty nearly right, on the other they admitted that even if they had not it made little difference since conventions were what mattered. Pursuing truth was all very well, but mineral water analysis was a practical matter. An expeditious and pragmatic system of reference, no matter how artificial, was of primary importance.

Thus the approach of Hofmann and those who shared his views was to admit the double standard and defend it too. Cataloguing and classifying waters took precedence over coming to an accurate understanding of aqueous solutions, and the two goals might not even be compatible. Yet was this admission really an honest and forthcoming one? Or was the double standard simply a way of legitimating error and ignorance by explaining them away as necessary convention? In 1856 J H Gladstone, whose research was slowly establishing rules for the behaviour of salts that coexisted in solution, criticized the Hofmann programme for promulgating an illusion of certainty in its analytical reports. The rule of combining acids and bases according to maximum affinities he regarded as 'utterly fallacious, ... purely empirical, and almost incapable of application, since our knowledge is very vague as to which [acid or base] is stronger and which weaker.' To be sure, Abel and Rowney had admitted the possibility that their system might not accurately reflect nature, yet Gladstone 'feared that the semi-scientific and the general public were deceived ... *and that chemists also often came to believe there was some truth in their own arbitrary mode of expressing the results of analysis.*'[74]

This last allegation is intriguing. As Linnaeus had found in trying to juggle artificial and natural classification systems, there was enormous and continual temptation for making the artificial and arbitrary into the real, for insisting that nature was simple, behaved in an orderly fashion and presented scientists with categories that conveniently corresponded to the means of investigation at their disposal. In this regard it is interesting to note the difficulties analysts working in the Hofmann tradition had in settling on an appropriate term to describe the logical relations between the acids and bases they had found and the salts into which these were presumably combined. Several claimed that the combination had been 'deduced'; a later writer, perhaps finding this too strong, used 'assumed'; others used a variety of other phrases of varying degrees of ambiguity.[75]

It is this set of issues—of whether analysis can ever hope to obtain an unambiguous indication of what really is in nature; of the necessity and yet the seduction of simplifications and conventions, of the confused relations among different kinds of evidence, physical, chemical, and medical—that makes Murray's problem so central in the history of mineral water analysis. Essentially the same problems preoccupied analysts in the second half of the century as chemists and bacteriologists groped to find some constant relation between the measurements they could make and the salubrity of the water they were analysing. Murray's problem is important then because it stuck, unresolved, with mineral water chemistry, and with the potable water chemistry that succeeded it.

I dwell on Murray's criticism and the response to it to suggest that it would be very wrong to think of water analysis in the nineteenth century as the rote application of standard methods. Instead, from the beginning of the century to the end, water analysts were wrestling with the great issues of the philosophy of science. But at the same time they were having to make their results useful to clients, whether these were proprietors of mineral waters or, increasingly after 1850, municipal authorities, public health officials, or water companies. The next chapter takes up the social transformation of practical chemistry in early nineteenth century Britain. During this period chemists developed markets for their services and these markets shaped the sort of chemistry that developed quite as much as did the unanswerable questions about the nature of solutions.

1   Charles Lucas, *An Essay on Waters*, II, p 63.
2   R S Surtees, *Handley Cross or Mr. Jorrocks's Hunt* (New York: Viking Press, 1930), pp 34–5. See also George Eliot, *Felix Holt, the Radical* (New York: Houghton Mifflin, 1909), v 1, pp 64–5.
3   Charles Perry, *An Account of an Analysis of the Stratford Mineral Water*, p 3; William Saunders, *Treatise on Mineral Waters*, p iii; Robert Bud, 'The Discipline of Chemistry,' pp 56–7.
4   A B Anderson and M D Anderson, *Vanishing Spas*; P J Neville Havins, *The Spas of England*; P J Waller, *Town, City, and Nation: England 1850–1914* (Oxford: Oxford University Press, 1983), pp 133–4; T B Dudley, *From Chaos to the Charter*; Saunders, *Treatise on Mineral Waters*, pp iv, 110–11, 209.
5   Roy Porter, *English Society in the Eighteenth Century* (London: Allen Lane/Penguin, 1982), pp 245–6; John Patten, *English Towns, 1500–1700* (London: Dawson/Archeon, 1978), pp 180–1; David Gadd, *Georgian Summer, Bath in the Eighteenth Century* (Bath: Adams and Dart, 1971); William Addison, *English Spas* (London: Batsford, 1951); J H Plumb, *Georgian Delights* (London: Weidenfeld and Nicholson, 1980).
6   A B Granville, *The Spas of England, II, The Midlands and South*, pp 290–9, 506–7. On Granville see Morris Berman, *Social Change and Scientific Organization*, p 116. But see R Phillips, 'Analyses of Two Sulphurous Springs near Plymouth' *Phil. Mag.* 3rd series, 3 (1833): 158–9.
7   Granville, *Spas*, II, p 161.

8   *Ibid*, pp 610, 105–6. See Also Diederick Wessel Linden, *A Treatise on Chalybeat Waters*, p 110; Bud, 'The Discipline of Chemistry,' pp 56–7.

9   Granville, *Spas, I, The North*, pp 163–4.

10 *Ibid*, II, pp 130–1, 274–5, 332, 405–9.

11 *Ibid*, I, p 165.

12 *Ibid*, II, pp 572–3, 161–2, 132; I, p 133. On Schweitzer see Bud, 'The Discipline of Chemistry,' p 279.

13 Saunders, *Treatise*, pp xii–xiii, 447–8.

14 Charles Lucas, *An Essay on Waters*, I, p 126. See also Jon B Eklund, 'Chemical Analysis and the Phlogiston Theory,' pp 252–6.

15 Lucas, *Essay*, I, p 164. See also Linden, *Treatise*, pp 1–2.

16 On Lucas and Bath see W J Williams and D M Stoddart, *Bath—Some Encounters with Science* (Bath: Kingsmead Press, 1978), pp 82–3; Noel G Coley, 'Physicians and the Chemical Analysis of Mineral Waters' 133–5.

17 Lucas, *Essay*, III, pp 246–7, II, p 103.

18 *Ibid*, III, pp 245–6.

19 Linden, *Treatise*, p 110.

20 Lucas, *Essay*, II, pp 13–16, 3–4, I, pp 135–6.

21 Linden, *Treatise*, p 6.

22 Coley, 'Physicians and the Chemical Analysis of Mineral Waters,' p 124. See also Eklund, 'Chemical Analysis,' p 231; J K Crellin, 'The Development of Chemistry in Britain through Pharmacy and Medicine,' pp 234–50.

23 Lucas, *Essay*, I, p 4; Frederick Slare, *An Account of Pyrmont Waters*, p 23; Saunders, *Treatise*, p 8. On medicine see D M Vess, *Medical Revolution in France, 1789–1796* (Gainesville: Florida State University Press, 1975), pp 10–18; and L M Beier, *Sufferers and Healers: the Experience of Illness in Seventeenth Century England* (London: Routledge and Kegan Paul, 1987), pp 8–32.

24 *Encyclopedia Britannica* (1797), s.v. 'mineral waters'; see also William Nicholson, *The British Encyclopedia, or Dictionary of Arts and Sciences* (1809), s.v. waters, mineral; John Mason Good and Olinthus Gregory, *Pantologica. A New Cyclopedia* (1813), s.v. mineral waters, vol 8; Daniel Gibbon, *A Compendium to the Chemical Chest* (London: T Hurst, 1837), p 93; Saunders, *Treatise*, p 23.

25 Harry C Jones, *The Nature of Solution* (New York: D Van Nostrand, 1917), pp 20–34; Saunders, *Treatise*, pp 16–17.

26 Allen Debus, 'Solution Analyses Prior to Robert Boyle,' *Chymia* 8 (1962): 41–61; E H Guitard, *Le Prestigieux Passé des Eaux Minerales*, pp 81–96; W Kirkby, *The Evolution of Artificial Mineral Waters*, pp 5–23.

27 Debus, 'Solution'; *idem*, 'Sir Thomas Browne and the Study of Colour Indicators,' *Ambix* 10 (1962): 30; Hermann Kopp, *Geschichte der*

Chemie, II, pp 58–9.

28 Debus, 'Solution,' pp 43–8; Slare, *Pyrmont*, p 24.

29 Ferenc Szabadvary, *History of Analytical Chemistry*, pp 29ff; Debus, 'Solution,' p 41; W T Brande, *Dictionary of Science, Literature, and Art* (New York: Harper Bros, 1870), s.v. chemistry, analysis.

30 Thomas Thomson, *The History of Chemistry*, 2 vols (London: H Colburn and R Bentley, 1831), II, p 41; see also Kopp, *Geschichte der Chemie*, II, p 64; Eklund, 'Chemical Analysis,' pp 235–40; Garnett, *Harrowgate*, p 15; Coley, 'Physicians and the Chemical Analysis of Mineral Waters,' pp 142–4.

31 John Elliott, *An Account of the Nature and Medicinal Virtues of the Principal Mineral Waters of Great Britain and Ireland and those in Repute on the Continent*, 2nd ed. corr. and enlarged (London: A Johnson, 1789), pp 102–4; Lucas, *Essay*, I, p 82; Kopp, *Geschichte der Chemie*, II, p 52. See also J A Fabricius, *Theologie de l'Eau* (Paris: Chaubert-Durand, 1743), pp 90–1.

32 Henri de Heers, *Spadacrene ou Dissertation Physique sur les Eaux de Spa*, nouvelle ed. (La Haye: P Paupie, 1739), pp 32–3.

33 Debus, 'Sir Thomas Browne and the Study of Colour Indicators,'* pp 29–31.

34 *Pantologica*, s.v. mineral waters.

35 W A Smeaton, *Fourcroy, Chemist and Revolutionary, 1755–1809* (Cambridge: W Heffer, 1962), pp 112–14; Nicholson, *British Encyclopedia* (1809), s.v. waters, mineral; Abraham Rees, *The Cyclopedia or Universal Dictionary of Arts, Sciences, and Literature* (1819), s.v. water; *Encyclopedia Britannica* (1810), s.v. chemistry, p 710; Partington, *A History of Chemistry*, III, p 669.

36 Torbern Bergman, 'Of the Analysis of Waters,' in his *Physical and Chemical Essays*, I, pp 158–59. See also Eklund, 'Chemical Analysis,' pp 240–1.

37 Debus, 'Solution,' pp 47–8.

38 Allen Debus, 'Fire Analysis and the Elements in the Sixteenth and Seventeenth Centuries,' *Annals of Science* 23 (1967): 127–47; Frederic L Holmes, 'Analysis by Fire and Solvent Extractions: The Metamorphosis of a Tradition,' *Isis* 62 (1971): 132.

39 Bergman, 'Of the Analysis of Waters,' pp 159–82; Szabadvary, *History of Analytical Chemistry*, pp 73–6. Chloride of lime is our calcium chloride, not bleaching powder (A Ure, *Dictionary of Chemistry* [1823], s.v. 'lime'.

40 John Rutty, *A Methodical Synopsis of Mineral Waters* (London: J Johnston, 1757) p vi. On Bergman and Linnaeus see J A Schufle, *Torbern Bergman: A Man Before his Time* (Lawrence, KS: Coronado Press, 1985), pp 119–25.

41 Smeaton, *Fourcroy, Chemist and Revolutionary, 1755–1809*,* pp 115–

6; Coley, 'Physicians and the Chemical Analysis of Mineral Waters,' pp 141–2; F L Holmes, 'From Elective Affinities to Chemical Equilibria,' 106–7. Cf Frederick Accum, 'Analysis of the Lately Discovered Mineral Waters at Cheltenham,' p 25, steps 8 and 16.

42 *Rees Cyclopedia*, s.v. water.

43 'Table exhibiting the composition of the principal mineral waters of Europe and the United States,' *J. of Materia Medica*. Supplement, 1871, pp 205–13.

44 Bergman, 'Of the Analysis of Waters,' p 109; Meredith Gairdner, *Essay on Mineral and Thermal Springs*, p 64.

45 *Encyclopedia Britannica* (1797), v 12, s.v. water, p 44.

46 *Pantologica*, v 8, s.v. mineral waters; Coley, 'Physicians and the Chemical Analysis of Mineral Waters,' pp 131, 139; John Rutty, *An Essay towards a Natural, Experimental, and Medical History of the Mineral Waters of Ireland*, p x.

47 Uno Boklund, 'Torbern Bergman as Pioneer in Mineral Waters,' pp 122–4; Partington, *A History of Chemistry*, III, pp 123–4; Eklund, 'Chemical Analysis,' pp 268–70; Guitard, *Prestigieux Passé*, pp 108–9; Saunders, *Treatise*, pp xiv–xv, 23.

48 Bergman, 'Of the Analysis of Waters,' p 101; Saunders, *Treatise*, p xiii, 4–5. Nevertheless Hoffmann himself tried to synthesize artificial waters (Noel G Coley, 'Preparation and Uses of Artificial Mineral Waters,' pp 34–5).

49 Quoted in Coley, 'Physicians and the Chemical Analysis of Mineral Waters,' p 131.

50 Lucas, *Essay*, II, p 98; Bergman, 'Of the Analysis of Waters,' pp 116–7.

51 Boklund, 'Torbern Bergman as Pioneer in Mineral Waters,' pp 119–20.

52 'Murray, John,' *DNB*; John Murray, Letter to the Editor, *Annals of Philosophy* 8 (1816): 471.

53 John Barker, *A Treatise on Cheltenham Water*; *British Cyclopedia of the Arts and Sciences* (1835), v II, s.v. water, p 892; Saunders, *Treatise*, p ix; *Rees Cyclopedia*, s.v. water.

54 John Murray, 'An Analysis of the Mineral Waters of Dunblane and Pitcaithly,' pp 347–63.

55 *Ibid*, pp 347–8.

56 John Murray, *Elements of Chemistry*, 2 vols (Edinburgh: Creech, 1810) I, pp 42–50.

57 Murray, 'An Analysis of the Mineral Waters of Dunblane and Pitcaithly,' p 350.

58 *Ibid*, p 348.

59 William Henry, *The Elements of Experimental Chemistry*, 4th American from the 7th London ed. (Philadelphia: James Webster, 1817), p xxxi; *Encyclopedia Britannica* (1810), s.v. chemistry, p 710; Richard Kirwan, *Essay on Mineral Waters*, pp 136–40, 164. For a more modern

treatment see Jones, *Solution*, pp 22–9, 142–3; Coley, 'Physicians and the Chemical Analysis of Mineral Waters,' pp 142–4.

60 Murray, 'An Analysis of the Mineral Waters of Dunblane and Pit-caithly,' p 349. For the lasting presence of this problem and similar experiments see A B Northcote, 'On the Water of the River Severn at Worcester,' *Phil. Mag.* 4th series 34 (1867): 262.

61 Murray, 'An Analysis of the Mineral Waters of Dunblane and Pit-caithly,' p 349.

62 *Ibid*, p 354.

63 John Murray, 'A General Formula for the Analysis of Mineral Waters,' 93–98, 169–77. See also his 'Analysis of Sea Water, with Observations on the Analysis of Salt Brines,' *Phil. Mag.* 1st series 51 (1818): 10–25, 91–103.

64 Murray, 'General Formula,' p 95.

65 Murray, 'An Analysis of the Mineral Waters of Dunblane and Pit-caithly,' p 358.

66 *Ibid*, p 353; *Rees Cyclopedia*, s.v. water, v 38; 'On Mineral Waters, Natural and Artificial,' *Lancet* 1828, ii, pp 251–2; Gairdner, *Essay*, pp 63, 70–3. An anonymous annotator noted in the margin of Edwin Godden Jones' article 'Chemical Analysis of the Mineral Waters of Spa' (*Trans. Medico-Chirurgical Assn.* 7 pt 1 [1816] 69, Michigan State University Library copy) that Murray's approach had finally resolved the inconsistency between medical effects and composition. See also Rutty, *A Methodical Synopsis of Mineral Waters*, p xiv.

67 F L Holmes has pointed to a similar treatment of Berthollet's views during the first half of the century. Without accurate quantitative knowledge of affinities, chemists were unable to predict how mass would affect reactions; they could only assert that traditional affinity relations did not work (Holmes, 'From Elective Affinities to Chemical Equilibria,' pp 111–25; Thomson, History of Chemistry,* II, p 223).

68 J Berzelius, 'Examen chimique des eaux de Carlsbad, de Toplitz, et de Konigswart,' *Annales de Chimie et de Physique*, 2nd series 28 (1825): 258–60. See also J F Comstock, *Elements of Chemistry* (Hartford: D F Robinson, 1831), pp 344–5; and Holmes, 'From Elective Affinities to Chemical Equilibria,' pp 122–3.

69 Abel and Rowney, 'On the Mineral Waters of Cheltenham,' p 194. For varied views and ways of combining acids and bases see 'Water Analysis: Statement of the Results,' *CN* 3 (1861): 285; C G B Daubeny, 'Report on Mineral and Thermal Waters,' pp 47–8; R Henderson, 'On the General Existence of Iodine in Spring Water,' *Phil. Mag.*, 2nd series 7 (1830): 11–12; Marie Bach, *Des Eaux Gazeuses Alcalines de Soultzmatt (Haut-Rhin), suivi d'une nouvelle analyses des eaux de Soultzmatt par M. Bechamp* (Paris: Balliere, 1853), p 256; James Barratt, 'Analysis of the Water of Holywell, North Wales,' *Q. J. Chem. Soc.* 12 (1859):

52–4; Thomas Graham, A W Hofmann, and W A Miller, 'Chemical report on the supply of water to the metropolis,' pp 378–9; J T Way in Royal Commission on Water Supply, 1868–9, App. F, p 36; L Playfair in *GBH*. Report on the Supply of Water to the Metropolis, App III, p 77.

70 A W Hofmann, 'Analysis of the Saline Water of Christian Malford near Chippenham,' *Q. J. Chem. Soc.* 13 (1860): 80–4; H M Noad, 'Analysis of the Saline Water of Purton near Swindon, North Wilts,' *Q. J. Chem. Soc.* 14 (1861): 43; Barratt, 'Analysis of the Water of Holywell, North Wales,'* pp 52–4; Northcote, 'On the Water of the River Severn at Worcester,'* p 255; Edward T Bennett, 'Analysis of the Thames water at Greenwich,' *Q. J. Chem. Soc.* 2 (1849): 199. The problem, of course, is that not all those using the phrase used the same 'usual' method (Cf R Phillips, 'Analyses of Two Sulphurous Springs near Plymouth,'* p 158). Fullest development of the Hofmann approach is George Merck and Robert Galloway, 'Analysis of the water of the Thermal Spring at Bath,' *Phil. Mag.* 3rd series 31 (1847): 56-67.

71 Augustus Voeckler, 'On the Composition of the Purton Saline Water,' *Q. J. Chem. Soc.* 14 (1861): 46–7. Cf Henry M Noad, 'Analysis of the Saline Water of Purton near Swindon, North Wilts,'* pp 43–6, who analysed the same water using the Hofmann conventions.

72 Bennett, 'Analysis of the Thames water at Greenwich,'* p 199.

73 Abel and Rowney, 'On the Mineral Waters of Cheltenham,' pp 193–4.

74 J H Gladstone, 'On the Salts actually present in the Cheltenham and other Mineral Waters,' *Proc. BAAS* for 1856 (London: John Murray, 1857), sections, pp 51–2 (italics mine); see also Partington, *A History of Chemistry*, IV, p 582; Holmes, 'From Elective Affinities to Chemical Equilibria,' pp 134–42.

75 On enlightenment attempts to classify waters, see Guitard, *Prestigieux Passé*, p 107. For 'deduced' see T J Herapath, 'Analysis of a Medicinal Water from the Neighbourhood of Bristol,' *Q. J. Chem. Soc.* 2 (1849): 205; J M Ashley, 'Analysis of Thames Water,' *Q. J. Chem. Soc.* 2 (1849): 77; F A Abel and Thos Rowney, 'Analysis of the Water of the Artesian Wells, Trafalgar Square,' *Q. J. Chem. Soc.* 1 (1848): 100. For other alternatives see Northcote, 'On the Water of the River Severn at Worcester,'* p 264 ('assumed salts'); Bennett, 'Analysis of the Thames water at Greenwich,'* p 199 ('ingredients ... assume the subjoined form'); W T Brande, 'Analysis of the Well-Water at the Royal Mint with Some Remarks on the Waters of the London Wells,' *Q. J. Chem. Soc.* 2 (1850): 349 ('Upon the whole, I am inclined to regard the following as a tolerably correct statement of the proximate saline constituents of this water').

2 Water Analysis and the Hegemony of Chemistry, 1800–40

It does not seem of much practical consequence of what our mineral waters are composed, for we have gone on with them hitherto, content with analyses of the most wretched kind, if the latest analyses of some of them be correct.[1]

Lancet

During the first part of the nineteenth century chemistry became in Britain the sort of profession ordinary people made their livings at and analytical disputes of the sort considered earlier increasingly took place among self-defined chemists, struggling to forge a new profession. The typical eighteenth century chemist had been a well-to-do eccentric (Cavendish), a dabbling clergyman (Hales), or an unschooled manufacturer. By contrast, by 1850 a large group of trained 'practical' chemists made their livings as analysts and were doing useful work in industry, commerce, government, law, and education. This transformation occurred first in Scotland. There, beginning about 1750, university chemists began to involve themselves in local industries: agriculture, the extraction of alkali from kelp, bleaching, and later sulphuric acid and dyeing. In the north of England in the early nineteenth century there developed a similar set of opportunities for chemists in the mining, metal working, and textile industries, and soon in the new artificial alkali industry.[2]

Around London chemists became involved in brewing, paper-making, and the new gas industry. In the metropolis there was also an active trade in scientific lecturing: to medical students, burdened with chemistry requirements by the Apothecaries Act of 1815; to artisans in the mechanics institutes; or to the genteel patrons of the Royal Institution or its imitators.[3] For water analysis London was the dominant locus. As potable water analyses became increasingly important in public health policy-making during the second half of the century, London chemists with close ties to the Local Government Board or with experience as expert witnesses secured greater credibility for their analyses and methods than did their northern rivals.

It is noteworthy that practical chemistry expanded during the years of the chemical revolution, at the same time that scientific chemistry was acquiring its familiar form of elements and compounds. Yet it would be wrong to think that the expansion of practical chemistry was only made possible by the progress of pure chemistry. Instead, as Bud and Roberts have shown, relations between practical and scientific chemistry were complicated, with the modern distinction between pure and applied science only emerging during the period.[4] The practical and the academic were not different breeds of chemist, nor was contributing to science unrelated to manufacturing, analysis, or any other form of commercial chemistry.

And the transformation of chemistry was not even wholly a matter of new knowledge, be it pure or applied. What chemists offered for sale was credibility, authority, and rationality, as much as it was new knowledge. Their success was made possible by the spreading belief that chemists and chemistry were the providers of profitable techniques and true answers. Morris Berman has argued that this credibility was not directly a function of competence, for in many cases early nineteenth century chemists claimed authority over technological territory which they were not competent to hold.[5] Instead, the hegemony of chemistry was a result both of the needs of industrial society for an authority which could settle social problems by technical means and of the ability of chemists to establish themselves as providers of that authority.[6] Water analysis illustrates Berman's point. By mid century chemists had wrested from medical men the final say both on the medicinal and the pathogenic potential of waters. Yet their achievement of this authority was not mainly a result of any specific pieces of knowledge that chemical analysis could demonstrably deliver. To be sure, there had been significant achievements in water chemistry; some medically important substances, notably iodine, had been discovered in a number of springs. But enormous and recognized areas of ambiguity and ignorance remained; that chemists held 'water' was a reflection of the fact that a combination of social needs and aggressive marketing had made chemistry one of the most important sources of social authority.

The Marketing of Chemistry: Brande and Taylor

To get a better idea of the context of water analysis in early nineteenth century Britain it will be helpful to review the careers of
two of the main promoters of London chemistry, William Thomas
Brande and Alfred Swaine Taylor. Neither figures very largely in
most histories of chemistry, nor was either primarily a water analyst. Yet between 1820 and 1850 these two, Brande overseeing a
Royal Institution–Royal Society nexus, Taylor ensconced as professor of chemistry and medical jurisprudence at Guy's Hospital, were
among the most important chemists in London. They were in demand as consultants and expert witnesses; were active as authors,
editors, lecturers, and publishers, and they influenced the careers of
numerous younger chemists.

Brande (1788–1873) has received more attention than Taylor, due
to his involvement with the Royal Institution during the Davy–
Faraday era. Brande came from a family of German apothecaries
who had accompanied George I to England and had continued to
serve the Hanoverians for the rest of the eighteenth century. The
Brandes were thus well-off and oriented toward chemistry. During
the first decade of the nineteenth century, Brande, in his teens, studied medicine and chemistry at Hunter's Windmill Street School and
at St George's Hospital. By 1808, at age 20, he was teaching a
variety of subjects at both schools and at the Cork Street Medical School. He was elected F.R.S. in 1809 at 21, made a professor
at Apothecaries Hall in 1812, and at the Royal Institution in 1813.
Subsequently he became master of Apothecaries Hall, coinage officer
to the Royal Mint, and an examiner at the University of London.
His *Manual of Chemistry* was one of the most respected texts of the
day and he advocated his views as editor of the Royal Institution's
Quarterly Journal of Science, Literature, and the Arts.

Brande was not without genuine scientific talent, but he made
little attempt to apply himself to original research. Instead his business was selling chemistry. To Brande almost any technical problem was being trifled with if it was not informed by chemistry. As
Berman has noted, the Royal Institution became for him the proper
forum for any 'controversy involving cost–benefit analysis, technical consultancy, litigation and patents, government testimony, and
problems of pollution.'[7] That contemporary chemistry was often incapable of providing useful guidance did not matter. Brande, like
others we shall see, discoursed with an air of authority, especially to

Parliamentary select committees on issues he knew little about. On occasion his ignorance was exposed by a cross-examining barrister but in the long haul Brande's programme of making chemistry an indispensable authority succeeded amazingly well.[8]

While Brande ruled over general and technical chemistry, Alfred Swaine Taylor (1806–80) was the foremost toxicologist and forensic chemist of the day. Son of an East India captain, Taylor studied medicine at St Thomas's and Guy's Hospitals in the mid 1820s, working with the famous surgeon Astley Cooper. In 1828 he went to Paris where he studied chemistry with Gay-Lussac and toxicology with Orfila. He became the professor of medical jurisprudence at Guy's in 1831; in 1832 he began sharing the Guy's chemistry lectureship with Arthur Aikin. He was elected F.R.S. in 1845 and F.R.C.P. in 1853, wrote the authoritative text on medical jurisprudence, edited the *London Medical Gazette*, and co-authored with Brande an elementary text, *On Chemistry* (1863).[9]

Like Brande's, Taylor's scientific contributions were modest, empirical investigations, dealing with the detection of poisons and other phenomena of forensic interest. For neither man are the usual indicators of scientific achievement satisfactory; both came to hold numerous positions of trust, appeared regularly in courts of law and before select committees of Parliament, and influenced the careers of others without being prominent knowledge-producers themselves. Most importantly, their careers typified those of a great many practical chemists—with decent laboratory skills, passing familiarity with the contents of the journals, tolerable lecturing talents, good connections, and untouchable confidence one could make a decent living in London as a practical chemist. A few, like Davy and Faraday, tried to keep a distance between their serious science and practical work that subsidized it, and looked with scorn upon the Brandes of the world; yet there were many more Brandes and Taylors.[10]

The Contexts of Mineral Water Analysis

Mineral water analysis embodied these more general tensions in British chemistry. They were the source of the double standard discussed in the last chapter. Clients who commissioned water analyses wanted results listed as salts. Chemists complied, despite their inability to confirm that the salts they listed were actually in the water. But to understand fully how water analysis fit into the growing profession we need to consider more exactly what were the purposes of analyses and the big issues of water chemistry.

Why analyse water at all? By the mid eighteenth century a complete water analysis had become a lengthy and complicated matter. In one sense mineral water analysis seems a science without purpose. Supposedly mineral springs were to be known by their effects, not their constituents. Many had proven their worth long before there were chemical means either of distinguishing waters from one another or of determining their composition. Analysis was rarely the means by which new mineral waters were discovered nor was it instrumental in protecting patients from ingesting poisonous salts—two of the main uses one might expect.[11]

Yet analyses were done, published in books, papers, and tracts, and made much of. It would be wrong to impose rigid categories on these works, but one can recognize three distinct contexts in which mineral water analyses were done. First, many were advertisements. A chemical analysis was scientific legitimation for the medical claims being made. Quantitative analysis could show, more convincingly than any testimonial, how closely the contents of an unknown spring resembled those of well-known resorts. (Of course, defenders of the well-known spring might counter with analyses done with improved techniques and revealing new uniquenesses.)

A second reason to analyse mineral waters was to establish a basis for imitating them. There had been attempts at artificial mineral water manufacture in the mid seventeenth century; by the nineteenth imitating famous waters had become an important industry. An understanding of carbon dioxide chemistry brought with it the ability to manufacture artificial carbonated waters, still known in some places as 'minerals.' Here too there were bitter controversies. Upholders of the old spas denied natural waters could be imitated. Entrepreneurial chemists argued that they could not only imitate natural waters but improve upon them by adding active ingredients and leaving out deleterious salts.

The third class is a catch-all category. It includes works concerned mainly with increasing natural knowledge and analyses undertaken as lab projects by chemists-in-training such as those done by Hofmann's students. The main scientific rationale for mineral water analyses was geological; the soluble substances that impregnated ground water were clues to the composition of the earth's interior and the sources of subterranean heat. Among British analysts, Richard Kirwan exemplified this perspective. He saw mineral water analysis as the key to 'sound notions of universal geology, a science intimately connected with the principles of morality and religion .... Arising from unknown depths, ... [mineral waters] *alone* announce to us ... the awful operations therein transacted.'[12]

It is tempting to see the first two contexts as applications of the third, which would represent a core of pure science. Yet a split between pure and applied science was only emerging during the period, and to impose such a pattern on early nineteenth century water analysts would be to impose an unfamiliar and unacceptable set of distinctions. To appreciate how these contexts fit together and how the most brazen advertising might be accepted as a scientific contribution, we need to look more closely at these contexts themselves and consider how mineral water analysis reflected aspects of the Baconian ethos of early nineteenth century British science.

Analysis as Travelogue: Frederick Accum, 1808, 1819

Two striking examples of the way mineral water analysis might be used to advertise a particular spa are the reports on the mineral waters at Cheltenham (1808) and Thetford (1819) by the Anglo–German chemist, entrepreneur, and food-adulteration crusader Frederick Accum. Accum (1769–1839), an émigré German apothecary, exemplified the sort of chemistry Brande was practising and advocating. He spent most of his career in a wide-ranging private practice that included analysing whatever was brought in to be analysed, as well as a good deal of lecturing and laboratory instruction. He also wrote a great deal: texts on chemistry and mineralogy, treatises on gas lighting and manufacture, food adulteration, dietetics, brewing, bread making, wine making, and building materials. He also consulted and served as an expert witness, built and sold experimental apparatus, and promoted speculative technical ventures such as supplying London with gas.[13] To Humphrey Davy, defender of the honour of science, Accum was 'a cheat and a quack,'[14] yet his accomplishments in food analysis and gas manufacture were real enough.

Though the reports on Cheltenham and Thetford were published in the *Philosophical Magazine*, one of the main British forums for research in physics and chemistry, portions of them read like travel brochures.[15] Accum used stirring accounts of scenery, salubrity, and sociability to introduce the details of analysis. Cheltenham was one of those 'choice and suitable spots ... particularly favourable to the curative effects of mineral waters.' It had 'uncommon fertility ... romantic scenery ... [which would] present a picture dear to the man of taste as well as to the invalid.' The rain there 'seldom prevent[ed] ... walking or riding for any length of time,' the houses 'in point of taste and elegance may vie with any modern buildings whatever.' The food was good and the natives known for their longevity.[16] Eleven years later when Accum reported on Thetford (not a leading spa, but trying to become one), he reused the Cheltenham hyperbole. Here too were the 'houses, which in point of taste and elegance may vie with any modern buildings,' the 'uncommon fertility,' 'romantic scenery,' the 'picture dear,' the rain that never precluded outdoor exercise and so on.[17]

Accum was candid about the promotional potential of his analyses. The Thetford spring had failed as a resort in the mid eighteenth century and had been closed, 'till [now] a happier spirit of research seems once more likely to liberate it ... and to diffuse those benefits it is so well calculated to ensure.'[18] As his biographer Browne makes clear, Accum was expected in such cases to act as a publicist.[19] He insisted that in praising Cheltenham, he intended 'no invidious comparisons ... with other springs.' He was just telling truths that needed to be told: 'unbiassed as I stand, a humble labourer in the field of chemical science, it is merely my wish to furnish a clear idea of the nature and composition of those fountains of health, so as to present truth in a simple form, and to establish it upon legitimate foundations; in order to enable the medical practitioner to select in a judicious manner the springs so bountifully given to the spot by the hand of Nature, and to apply them with advantage in the routine of his profession.'[20]

Implicit in Accum's apology were two claims for the utility of analysis. First, analysis was to demonstrate that Cheltenham's water really did have medicinal properties; it could not be legitimately dismissed. Second, Accum implied that guided by chemistry, Cheltenham's physicians might now rationalize their physic. This latter idea had been central for Bergman: once chemists had correlated composition with medical effects, physicians would be able to prescribe with great specificity certain doses of certain waters to certain patients. As Thomas Garnett claimed, chemistry 'emboldens the practitioner to make trials of the efficacy of mineral waters, in cases in which a person ignorant of chemistry would never think of, and which it would be rash to attempt without previous knowledge of the properties and composition.' Mere experience would never make a physician, he added.[21] In fact, while chemists often gave lip service to the idea, this second rationale was mostly pretence, a way of augmenting the reputation both of the analyst and the spa analysed. There are two reasons for thinking that this was the case.

First, physicians had made good livings prescribing the waters long before Accum. They made their reputations matching regimen to constitution. So, while qualitative analysis might have a small utility in matching a particular class of patients—say those with skin diseases—with the right sort of waters (sulphurous), a quantitative analysis represented a degree of precision that would be unlikely to have much of a bearing on the doses the doctor prescribed. As John Barker put it with regard to the Cheltenham water, 'it may indeed be ... proved, that iron enters into its composition, which may be known as well by the taste. But what purpose will it answer to calculate that there are about four grains in every quart?'[22] And given the notorious discrepancies in methods and results of analysis, it would have been an exceptionally naive doctor who would have committed his practice to the pronouncements of a chemist. Linden complained that misbegotten chemistry had led to claims of opposite diseases (constipation and diarrhoea) being cured by the same water.[23]

Second, analysts themselves were often perversely vague when it came to explaining what medicinal uses their analyses indicated. Many (Accum included) went directly from listing their analytical operations or giving tables of constituents to statements of medical effects without explaining how the tests led to the statements (or for that matter to the composition). To an extent, of course, such knowledge would be expected of physicians and wealthy invalids, yet the contrast between details of reagent tests and reticence about routes of inference suggests that it was the appearance of thoroughness that was to impress the reader.[24] Analyses indicated the relative goods the rival springs promised to deliver and symbolized too that someone knew what was going on, that the medicinal environment one was to encounter was comprehended and would be applied in a precise and rational way.

While Accum is unusual in so openly making his sales pitch in an ostensibly scientific article, he was by no means unique. A Dr Evans, writing in the *Philosophical Magazine* in 1805 on 'Sutton Spa, near Shrewsbury,' alluded to the splendid scenery and low prices of this yet-undeveloped spa.[25] Others mentioned that their analyses had been undertaken at the request of the proprietor of the waters, and that their purpose in writing was to ensure that the particular spring received its due share of attention.[26] In some cases, portions of the analyst's report were excerpted for incorporation into promotional materials, as was the case with A W Hofmann's 1854 report on Harrogate.[27]

Glorifications of bucolic scenery were also standard fare, even among more moderate writers like Saunders, who spoke of the Bristol Hotwell as 'one of those choice and favoured spots that are peculiarly calculated for the pleasure and comfort of the invalid ... the whole adjacent country abounds with beautiful scenery and romantic prospects.' Because environment was held to contribute to cure, and because not all beneficially situated springs had been sufficiently utilized, such brazen advertising could almost be viewed as disinterestedly in the service of the public good. Saunders wrote: 'it is merely advantage of situation or accidental causes that have given some of these [springs] a superior reputation over the rest; and where this is owing to beauty of site or local conveniences, it is well merited, as these circumstances have no small share in the general plan of cure, by enabling the invalid to employ daily exercise, and giving that irresistible charm to the spirits, which the sight of a beautiful or romantic country almost always excites.'[28]

To the modern mind it may seem strange to find such undisguised commercialism in scientific journals. While journal standards varied, Accum's excesses do raise questions about the consulting scientist's relationship with his client. Early nineteenth century sensibilities on such issues differ from those prevalent today, and even then there was no consensus on these matters. Some, like Davy, were scornful of such performances, but such men were condescending toward commercial chemistry in general. In part, the ethos which legitimated articles like Accum's was a Baconian ethos, but before considering the ways in which mineral water analysis was Baconian, we need to consider the second context of mineral water chemistry, the use of analysis to guide the synthesis of artificial waters.

Torbern Bergman and the Synthesis of Artificial Mineral Waters

One component of Torbern Bergman's systematic approach to mineral water analysis was the suggestion that each analysis be checked by dissolving the constituents found in the analysis in distilled water. If the synthetic water possessed the medicinal qualities of the original, the analysis could be assumed accurate; synthesis was the check. As mentioned in the last chapter, such syntheses were difficult and few, if any, chemists customarily used synthesis to verify their analyses. But Bergman had another interest in synthesis. If a mineral water could be successfully synthesized, it could be produced on a large scale and its benefits made much more widely available.

This question of whether one could expect analysis to guide synthesis was thus not only one of the most philosophically problematic but also one of the most socially significant questions water chemists faced. An industry that could manufacture artificial mineral waters would threaten the exotic and expensive springs. Already by the later eighteenth century some of the best known continental springs—Seydschutz, Seltzer, Spa, Pyrmont—were exporting bottled water in sealed bottles to ensure authenticity.[29] Never of strong constitution, Bergman was himself a consumer of these bottled mineral waters and it was in part his dissatisfaction with their expense, frequent unavailability, and variable quality that led to his interest in synthesis.[30] By the mid 1770s he was using his own imitations of several continental waters to treat his own illnesses and those of a few patients.

Bergman was realistic about the opposition a large-scale program of synthesis would arouse.

> From the very nature of the thing it must be obvious, that an invention of this kind, however useful, cannot possibly be universally pleasing.—Many who are incapable of ascertaining or judging of the truth, will distrust it, not without reason, on account of its novelty;—many contend, that to imitate nature is impossible, without considering, that when the component parts are thoroughly known, the success of the process cannot in any degree depend upon the hand which combines them. Some who prescribe, and others who sell the foreign waters, condemn the artificial, for obvious reason; and not a few are urged by motives too trivial to be detailed.[31]

To Bergman this resistance was resistance to progress. The advantages of a free market in mineral water manufacture seemed straightforward. In Sweden such an industry would make the waters available year around (they were unavailable in winter and spring); it would ensure better quality, a lower price, and stop the flow of money 'out of the kingdom.'[32] But such resistance was also resistance to science. Those who insisted on the inimitability of natural springs on such grounds as that their properties resulted from 'a certain degree of fermentation, as they are pleased to call it,' were

Figure 2.1 However rigorous the attempts to certify their quality, as with official seals pictured here, bottled mineral waters were of variable quality and expensive to boot. Progressive chemists like Torbern Bergman envisioned a synthetic mineral water industry to remedy these shortcomings (D W Linden, *A Treatise on Chalybeat Waters*, pp xix–xx).

clinging to an obsolete alchemical perspective. Such views were held. John Barker of Cheltenham, for example, held that 'there are specific properties in almost every mineral water, wherein it differs from every other of the same class. Nay, there are qualities in the water, and even in the spirit of every common spring, whereby it is peculiarly different, in many respects, from all others.' The belief that chemistry could be used to 'elucidate things of so high a nature, and enable us to imitate them' was in Barker's view 'a gross mistake, the crude conception and immature production of a deluded mind.' In rejecting this view Bergman was advocating a chemistry fully representative of the spirit of the enlightenment: the materials of the earth were compounds, lawfully combined, of simple substances and whether they were put together in 'the bowels of the earth ... [or] artificially added ... can make no difference in the result.'[33]

In making the link between the progress of society and the advance of science Bergman was placing great trust in analytical chemistry. Not only was he assuming that he could isolate and identify all the constituents of the waters he analysed, but that he knew which were pharmacologically active. Indeed, so confident was Bergman that he saw no need to duplicate natural waters exactly. It was quite proper to leave out inactive constituents and salts that might be harmful.[34]

British proponents of artificial mineral waters enlarged on Bergman's arguments. Synthetic waters would be more accessible, affordable, and effective.[35] Rather than sticking to mere imitations, chemists could improve on natural waters by leaving out harmful ingredients and increasing concentrations of active ingredients. One might even go so far as 'to form new and valuable compounds [in artificial mineral waters] which are no where to be met with in a natural state.'[36] Ironically, the very chemistry that cast doubt on the accuracy of analysis—Murray's—would resolve one of the major problems of the artificial waters industry. Murray's argument that a rearrangement of salts occurred as a water evaporated, and therefore that the salts found in a residue were not necessarily the salts in the water, implied that one could supply constituents in highly soluble forms, rather than struggling to dissolve the insoluble compounds found in residues.[37]

Yet the conception of social progress so central to Bergman's justification of artificials turned out to be double-edged, for there was as much opportunity for quackery in the artificial water industry as in the natural, and rather than ending the oligopolistic control of mineral waters, the rise of the synthetic industry simply provided an opportunity for a new group of chemist–entrepreneurs. There was thus ample room for quarrels, for attacks of synthesizers on anti-synthesizers and vice versa.[38] The most important British centre of artificial mineral water manufacture was F A Struve's Royal German Spa at Brighton, where continental mineral waters were imitated.[39] Successful and well respected in Britain, Struve was accused by a continental chemist of ignoring the latest analyses and hence trading under false colours in representing his concoctions as imitations. Again there were financial interests involved; Struve was competing with continental spas. The Oxford chemistry professor Charles Daubeny defended Struve, noting that emulation was less important than effectiveness, and that the criticism was 'scarcely candid.'[40]

The Logic of Mineral Water Analysis

Ultimately, as the more sober commentators on the imitation issue pointed out, the question was one of the adequacy and completeness of analysis. To be confident that one had imitated a mineral water one had to be confident that analytical chemistry in its current state was good enough to detect the active constituents in the water. Yet the abilities of analysts were continually changing and improving: Daubeny (1795–1867), who served Oxford as professor of chemistry, rural economy, and botany (not all at the same time), wrote in 1836 of 'chemists, [who] in the pride of half knowledge ... [had] smiled at the faith reposed' in a spring which their analysis had showed to be devoid of medicinal properties, yet which had later turned out to have active components not yet discovered at the time of the analysis.[41] Daubeny was writing in the wake of the discovery of iodine and bromine in many springs, and while iodine was not yet clearly recognized as the cure for goitre, it did seem to him that these elements—and other only recently detected components of mineral waters such as manganese, zinc, strontium, potassium, lithium, and phosphoric and fluoric acids—might account for 'the unexplained virtues attributed to certain mineral waters.'[42] Given this history of discoveries it was unwise for chemists to think that they finally knew all the components of a mineral water.

In principle Daubeny had a resolution for this kind of problem: one ought to insist on a correlation between medical effects and chemical composition. 'To refuse credence to the reports given by medical men with respect to the salutary or injurious effects of a particular water, merely because the chemist can discover in it no active principle, would seem a proceeding not less unphilosophical, than ... treating as fabulous the accounts given of stones that had fallen from the sky, because ... [we] did not understand how such ponderous masses could have continued suspended in it,' he wrote. Likewise, 'granting that a spring possesses peculiar virtues, we must suppose that it differs, either in its mechanical, or chemical properties, from the rest.'[43]

In fact this sort of correlation was both correlation and explanation and that made it problematic. As Daubeny posed it, the problem was one of inductive reasoning. One started with an empirical finding, medical effects. The validity of a theory explaining these effects, i.e. what chemical analysis supplied, was determined by the consistency of that theory with empirical findings. If an inconsistency between theory and observation arose one was to discard theory: medical truths were to drive out chemical truths. Once generalizations had been obtained, they would permit recognition of anomalies—waters whose effects could not be explained by their composition—and lead to progress. If a water had different effects than others of like composition, one simply looked for new ways in which it differed chemically and physically. In this way new medicinal substances might be discovered.

To many chemists this was an unacceptable way of posing the problem, for it represented not the correlation of medicine and chemistry but the subjugation of chemistry to medicine. Indeed, this was precisely the argument being made. 'Chemistry,' wrote Barker of Cheltenham, was a 'good servant to physic, though a very bad master.' It provided 'imperfect knowledge ... apt to mislead weak minds.'[44] Where Daubeny had treated medical effects as empirical findings capable of being unambiguously demonstrated, most chemists saw chemical composition as the only thing that could be empirically determined. Claims of medical effects were unproved assertions—unsound theories explaining why invalids got well from diseases they probably either didn't have or would have recovered from without the waters. To them mineral water chemistry was predicated on the idea that medicinal properties had to be deduced from chemical composition. There were good grounds for this view. Evidence of the medicinal virtues of springs was almost entirely in the form of testimonials; there were no controlled clinical experiments demonstrating the efficacy of mineral waters in certain conditions. Already it was beginning to be admitted that a great deal of the healthfulness of spas was due to relaxation, regimen, and climate, hence it was quite defensible to refuse to accept any claims for medicinal virtues which could not be confirmed by analysis.[45] Chemists could cite older writers (and moderns like Barker) who had regarded waters of a certain spring as irreducibly unique, 'exquisitely formed by the hand of nature, produc[ing] effects very different from those of any other mineral waters about this place.' In place of such obscurantism they could point out that 'analysis shews however, that this water must possess less active and stimulant powers than any of the others.'[46]

There was no good resolution for this problem. Meredith Gairdner, according to Daubeny the best (certainly one of the more sober) of modern writers on the subject, recognized both horns of the dilemma. He regarded Murray's chemistry, for example, as having shown 'the great, and in many cases dangerous, errors into which the physician might fall, who, *a priori*, judged entirely of the medicinal effects of a mineral spring from the results of chemical analysis' and admitted that medical effects were often 'the reverse of what we should expect from ... composition.' He called for 'impartial experience,' but admitted 'although I assume experience to be our principal guide in judging of the real effects of any spring, ... let it not be supposed that I undervalue chemical analysis, or am of the number of those who regard them [mineral waters] as specifics prepared by the Hand of Nature for the cure of the more obstinate maladies with which human nature is afflicted. This would be to render the whole a system of mystical empiricism, and to place an insurmountable barrier to the acquirement of any true theory of their action, totally incompatible with the present state of medical science.' A new era of mineral water chemistry, founded in Murray's ideas of the pharmacological activity of acids and bases, was the only way out of this dilemma.[47]

The Foundations of Authority: Baconians and the Ideology of Progress

Neither the utility of chemistry for advertising mineral waters nor the possibility of making new and stronger mineral water medicines explains the authority mineral water chemists achieved as the determiners of the medicinal potency of waters. These contexts of water chemistry only reflect that authority or, at most, contributed to it in a small way. Nor was that authority a function of the degree of certainty mineral water chemistry had attained, for while it is undeniable that the capabilities of analytical chemistry were improving during the period, by the late 1840s chemists were still likely to come up with quite different compositions for water from the same spring. In part that authority derived from the authority chemistry in general was acquiring, but it also reflected chemists' success in placing mineral water analysis within particular traditions of scientific progress. By making it clear how far chemists had come, how worthy were the programs of investigation they were pursuing, and how useful had been their results so far, these traditions made it possible to accept an authority that was not truly authoritative.

To the modern mind the contrast between the seriousness analytical chemists attached to their own pronouncements and the catalogue of contradictory results they produced is amazing and appalling. One wonders how a community of scientists could go on, year after year, contradicting one another without seeing something as dreadfully wrong and determining either to insist on a standard method of analysis, or to exclude incompetent analysts from the profession, or to decide, like Murray, that unwarranted inferences were completely undermining their practice. So rapidly were the contents claimed for various springs changing that Brande's *Quarterly Journal of Science* observed, 'It does not seem of much practical consequence of what our mineral waters are composed, for we have gone on with them hitherto, content with analyses of the most wretched kind, if the latest analyses of some of them be correct.'[48] In one sense the writer was correct: whether chemists were correct mattered little; what mattered was that patients trusted chemistry and went to the spas and got themselves cured (or got themselves to believe they were cured). But most chemists were not so cynical and the observation sheds no light on how they saw their endeavour.

To understand how analyses could be taken seriously we need to go beyond the social circumstances in which analysis was done and understand the motifs or ideologies of analysis, the throwaway rhetoric of context and significance with which analysts began the reports they published in scientific journals. In early nineteenth century Britain analysts' reports reflect two prominent motifs which may be labelled 'Baconianism' and 'Enlightenment.'

Articles on mineral water analysis reflected the resurgence of Baconianism in the early nineteenth century.[49] Mineral water analysis was Baconian in a number of respects. First it relied on an army of fact-gatherers, whose contributions, when sorted and organized, would lead to accurate knowledge. These contributors were not coordinated as well as they would have been at Salomon's House (or under the auspices of the British Association) and there were great problems with comparability and completeness. Kirwan complained of 'a labyrinth of particular facts, betwixt which we can trace no connection, nor consequently apply to no useful purpose' and believed that 'to select, ... compare, repeat, and correct where need should be found, and occasionally add to these, ... [could] be undertaken and properly executed only by a society of skilful and well-informed persons, instituted for that particular purpose.'[50] Gairdner admitted in 1832 that 'it cannot be denied that the subject is involved in much obscurity ... but this should be a stimulus to renewed exertion; observing without intermission, and applying the Baconian philosophy at every step, we may arrive at results that ... now perhaps would be rejected as visionary.'[51]

Second, mineral water chemists saw their enterprise as Baconian inasmuch as they equated scientific progress with more facts, to be gained from continued study of known springs or discovery of new springs. No conceptual rearrangement was foreseen. When Daubeny wrote about the progress of analysis he wrote of the new substances discovered in springs, not of the theoretical changes that had given chemists the new entities to look for or the new means to find them.[52]

Third, water analysis was Baconian in that water analysts saw themselves as building a foundation for technical and medical progress. When bromine and iodine were discovered in a few mineral waters, Daubeny undertook a nationwide survey of their presence. Even though medical effects had not yet been demonstrated for these substances, he regarded the project as worthwhile on the grounds that their existence might explain hitherto unexplained medical properties of some springs. Hence Daubeny's research was Baconian not in the sense that it promised direct benefits, but in the sense that he deemed it worthwhile to collect facts that might have social utility in the future. Even those whose analyses were essentially advertisements presented their findings as new knowledge which added to the common good.[53]

Fourth, like the histories of the trades undertaken during the early years of the Royal Society, the mineral water analyses were intended to contribute knowledge that both described—the analyst was contributing to a catalogue of what nature offered the invalid—and provided a basis for generalization—once the relations between composition and medical effects had been discovered, medical treatment could be carried on in a rational way and discoverers of new springs could know their medical properties without having to conduct lengthy and inconclusive clinical experiments.[54]

Finally, and again in common with the early years of the Royal Society, investigators of mineral waters found themselves forced to tolerate a great range of competence (and motive) among their colleagues. A part of the Baconian ethos was the idea that all information was grist for the mill, though of course much of it might need to be sifted out. Writers like Granville and Daubeny recognized that a great deal of the mineral water literature came from persons not competent in chemistry and was 'manifestly dictated by selfish motives,' yet they did not dismiss it on those grounds.[55] Instead they assumed that as chemistry progressed what truths there were in these works would become apparent while their falsehoods would disappear.

'Enlightenment,' the second motif, was more specific to mineral water chemistry, though it is certainly connected with broader intellectual currents of the age. In introducing their analyses mineral water chemists sometimes gave a short history of the discovery and progress of the medicinal use of mineral waters. In its most developed form, this history had four stages. First came discovery: at some point in the distant past ordinary folk, acting perhaps through instinct, had discovered the beneficial effects of the water and become accustomed to using it to cure their ills. A stage of primitive explanation followed. To explain why some springs had such special powers they had relied on pantheistic (or papist) superstitions: the powers of springs came from saints or spirits. The third stage was skepticism: with the coming of the Reformation (or some other religious authoritarianism) such pantheism had been condemned, and the springs, deprived of an animistic rationale, yet lacking scientific warrant owing to the primitive state of analytical chemistry, had fallen into unwarranted neglect. These were 'the days of intellectual bondage.' Yet by instilling a spirit of independent inquiry the Reformation had also given birth to the fourth stage of the enlightened scientific present when analytical chemistry, finally matured, had redeemed these forgotten springs.[56]

It will be apparent how nicely this history fit into the larger history of protestant and industrial Britain in the early nineteenth century. It was the triumph of enlightenment over superstition, a scientific confirmation of the common sense of good, plain folk. This version of history was particularly attractive to those trying to develop unknown springs into important resorts. It suggested that there were far more medicinal springs than currently recognized. The ideology also elevated chemistry above medicine: through the centuries of neglect the doctors had either failed to recognize these springs or been unable to persuade others of their powers. While medicine could not provide proof, chemistry, by contrast, was progressive: one could, or would soon be able to, offer a final answer to the question of what, if any, medicinal properties a spring possessed.[57] To opponents of chemistry the same prospect was a threat. Barker confessed himself 'so old fashioned [as] to think, that their uses [of mineral waters] ... were much better known ... in the last, and even some preceding generations, than the present.'[58]

Together, these ideologies made it possible to take seriously Accum's hyperbole about standing 'unbiassed ... a humble labourer in the field of chemical science.' Accum made it clear that his purpose was to enlighten ('to furnish a clear idea of the nature and composition of these fountains of health'); to convince by unimpeachable evidence ('to present truth in a simple form'); and to legitimate medical use of the Cheltenham waters ('to establish it [the composition] on legitimate foundations').[59] In his Thetford analysis, he represented himself as coming to put an end to long-standing doubts about the efficacy of the waters there. He noted that a mid eighteenth century analysis by Dr Matthew Manning, a local physician, had been well done for its time, 'but the science of chemistry at the time ... was not sufficient to enable him [Manning] to trace their [the constituents of the water] true combinations,' and hence the spa had not succeeded. Since that time however, in no other part of analytical chemistry had there been 'greater acquisition, in point of real matter of fact,' than mineral water chemistry and now 'modern chemistry [was able to give] ... us clear and accurate information as to the nature and qualities of all the foreign matters.'[60]

Accum's representations are ironic in two senses. First, the 'happier spirit of research' in which he placed himself and understood his abilities was rapidly disappearing in the face of Murray's criticisms. Second, Manning, the object of Accum's condescension, had made a similar offer of enlightenment three quarters of a century earlier, in his explanation of what his analysis would do for Thetford. Manning's analysis was

> to establish its [the Thetford Spring's] virtues on the principles of sound science, that no one should, henceforth, presume to refuse them his assent. This analysis has, happily, succeeded beyond my utmost expectations; having most clearly proved these waters to *abound* in all those mineral substances required for the cure of *chronic complaints*.[61]

Ducking Disagreement; Avoiding Anomalies

The juxtaposition of the claims of Manning and Accum, both supremely confident that enlightenment had finally arrived, helps make sense of one of the most remarkable characteristics of early nineteenth century British mineral water analysis: the toleration of discordant results. One of the diagrams in figure 2.2 is taken from an 1847 paper by Merck and Galloway, two of Hofmann's protégés, on the waters at Bath. The paper included their new analysis of the Bath waters (done by Hofmann's 'usual method') and this comparison of their results with those obtained by five earlier analysts, going back to Richard Phillips' analysis in 1806. As the authors noted previous analysts had disagreed significantly as to what was in the Bath waters: 'Besides great differences in the quantitative analysis, we find discrepancies even in regard to the presence and absence of certain constituents.'[62] Indeed it was these discrepancies that seemed to warrant a new analysis.

We can understand the tolerance of this range of results only in terms of the optimism that pervaded chemistry in the early nineteenth century. At each stage practitioners felt they finally had gotten the chemistry right, just as we do now; hence each subsequent analysis was to make clear which if any of the past efforts had been pretty nearly right and which wrong. If there was anything ironic it was the longevity of this optimism. The first of the authors Merck and Galloway considered, Richard Phillips, had raised the same issues in his report, wondering how the analysts of the eighteenth century could have come up with anywhere from 17 to 34 grains/quart of solid matters in the Bath waters.[63] Likewise, there was widespread recognition among chemists that incompetent and fraudulent analyses were sometimes done, and that there was need for more uniformity in technique and more stringency in qualification. Yet it was always another who was incompetent and fraudulent. Not until the mid '70s when these concerns finally led to the formation of the Institute of Chemistry and the Society of Public Analysts would the community of chemists find sufficient organization to tackle these problems.[64]

There was also a wholly benign explanation for the discrepancy, the expectation that nature varied enormously. Often chemists explained discrepant analyses of a particular spring by attributing these differences to variations in the waters themselves. While this excuse could plausibly be stretched only so far, it did have the advantages of deflecting public criticism from chemists' competence and techniques, of dissipating incipient intra-professional conflict, and of suggesting the need for more analyses so that the full range of variability could be determined.[65]

These perspectives carry over into the potable water analysis of the second half of the century. There too we find a continuing strong

Notices respecting New Books. 67

	Phillips.	Scudamore*.	Walker.	Noad.
Carbonate of lime	7·680	5·280	10·667	
Carbonate of oxide of iron	0·274	0·200	0·243	0·521
Carbonate of soda	......	......	......	5·760
Sulphate of lime	86·400	98·320	81·624	96·240
Sulphate of potassa	......	......	2·927	
Sulphate of soda	14·400	1·520	19·371	
Chloride of sodium	31·680	12·240	15·122	27·456
Chloride of magnesium ...	......	15·360	13·339	7·142
Alumina	......	......	0·150	
Silicic acid	1·960	1·920	3·233	3·360
	142·394	134·840	146·676	140·479
Quantity directly observed	144·125	......	147·622	149·72
Carbonic acid...............	11·52 cub. in.	......	7·60 cub. in.	

TABLE

GIVING THE COMPOSITION PER GALLON OF THE WATERS OF THE DEAD SEA, AS SHOWN BY THE ANALYSES

OF DIFFERENT CHEMISTS.

	MM. Lavoisier, Macquier, and Sage.	Dr. Marcet.	Prof. Klaproth.	M. Gay-Lussac.	Prof. Gmelin.	Dr. Apjohn.	Messrs. Herapath.
Specific gravity	1·2403	1·2110	1·2450	1·2283	1·2120	1·1530	1·17205
Boiling-point	undetermined.	undetermined.	undetermined.	undetermined.	undetermined.	221°	221°·75
Chloride of calcium	} 33·122·2115	{ 3·221·2600 / 8·561·7700	9·237·900	3·439 2400	2·726·8424	1·967·7098	2·014·2129
„ „ magnesium			21·090·300	13·163·6911	9 988·5526	5·948·3270	6·417·6780
Bromide of magnesium					372·7022	162·2272	206·0708
Iodide of magnesium ?							very minute traces.
Chloride of potassium		5·426·3125	9·050·0452	6·170·220	5·975·6795	1·420 0519	687·6492	998·7570
„ „ sodium						6·004·7206	6 326·8569	9·935·2036
„ „ ammonium						6·1084		4·9226
„ „ aluminum						76·0167		45·8975
„ „ iron								2·2304
„ „ manganese						179·6063	4·0355	4·9216
Organic matters								50·6510
Nitric acid								doubtful traces.
Carbonate of lime								traces.
Sulphate of lime			45·77588				44·7107	60·5325	55 6800
Silica and bitumen								traces
Fixed salts	38·548·5240	20·878·8510	36·498·420	22·578·6106	20·819·3118	15·157 3380	19·736·5254
Water	48 272·4760	63·891·1490	50·651·580	63·402·3894	64·020·6882	65·552·6620	62·306·9690
	86·821·0000	84·770·0000	87·150·000	85·981·0000	84·840·0000	80·710·0000	82·043·4944

ON THE WATERS OF THE DEAD SEA.

344

Figure 2.2 A troubling question for the early nineteenth century mineral water analysts was how closely should independent analyses of the same water agree. Note the figure for fixed salts in the Herapath and Herapath table of analyses of the Dead Sea, ranging from 15.15 to 38.5 grains. On the Merck and Galloway table note Noad's listing of the carbonate as carbonate of soda while the other analysts list it as carbonate of lime. John Murray's insights explained such discrepancies (*Phil Mag* 3rd series 31 [1847]: 67, *J Chem Soc* 2 [1849]: 344).

belief that a society in which chemistry informed decisions in matters of health and industry was vastly superior to one in which it did not. There too there was a willingness to accept as legitimate, analyses done under circumstances where those funding the analyses had a direct financial stake in the outcome of the analyses. There too one finds discrepant results (and more importantly, discrepant interpretations). Finally, there too one sees the conviction that the troublesome problems in assessing water quality lay in nature rather than in chemistry, and would be resolved by more chemistry.

1   *Lancet*, 1828–9, ii, p 110.
2   Clow and Clow, *The Chemical Revolution*, ch 25; Bud and Roberts, *Science versus Practice*, pp 19–45; J K Crellin, 'The Development of Chemistry'.
3   M Berman, *Social Change and Scientific Organization*, p 134; Robert Bud, 'The Discipline of Chemistry,' pp 35–56, 173; J N Hays, 'The London Lecturing Empire, 1800–1850,' in I Inkster and J Morrell, eds, *Metropolis and Province, Science in British Culture, 1780–1850* (London: Hutchinson, 1983), pp 91–119; Charles Newman, *The Evolution of Medical Education in the Nineteenth Century* (London: Oxford University Press, 1957), pp 98–9.
4   Bud and Roberts, *Science versus Practice*.
5   Berman, *Social Change*, p 153; C H Spiers, 'William Thomas Brande, Leather Expert,' *Annals of Science* 25 (1969): 179–201. See also June Z Fullmer, 'Technology, Chemistry and the Law,' *Technology and Culture* 21 (1980): 1–28; A Chaston Chapman, *The Growth of the Profession of Chemistry*, p 5.
6   M Berman, *Social Change*, p 151–4.
7   *Ibid*, p 151.
8   On Brande see Partington, *A History of Chemistry*, IV, pp 75–6; Berman, *Social Change*, pp 130–6, 151–5; Spiers, 'William Thomas Brande'*; E S Scott, 'Brande, William Thomas,' *DSB* 2, p 420; Aubrey A Tulley, 'The Chemical Studies of William Thomas Brande,' MSc Thesis, Univ. of London, 1970.
9   On Taylor see *DNB*, v 19, p 403; and George T Bettany, *Eminent Doctors: Their Lives and Work*, 2 vols (1885; rpt. Freeport, New York: Books for Libraries Press, 1972), II, pp 291–4.
10   J Z Fullmer, 'Davy's Sketches of his Contemporaries,' *Chymia* 12 (1967): 134; Berman, *Social Change*, pp 154–5; L Pearce Williams, *Michael Faraday: A Biography* (New York: Clarion/Simon and Schuster, 1971), pp 106, 322; W V Farrar, 'Andrew Ure FRS and the Phi-

losophy of Manufactures,' *Notes and Records of the RSL* 27 (1973): 299–324. For concerns see Thomas Thomson, *The History of Chemistry*, 2 vols (London: H Colburn and R Bentley, 1831), II, p 231; *idem*, 'The History and Present State of Chemical Science,' *Edinburgh Review* 50 (1829): 275–6.

11 William Saunders, *Treatise on Mineral Waters*, p iii, but see p 197.

12 R Kirwan, *Essay on Mineral Waters*, pp 2–3; C G B Daubeny, 'Report on the Present State,' pp 20–9, 56–75, 80–93; Guitard, *Le Prestigieux Passé*, pp 179–83.

13 C A Browne, 'The Life and Chemical Services of Frederick Accum,' *J Chem Ed* 2 (1925): 829–51, 1008–35, 1140–8; Partington, *A History of Chemistry*, III, p 827, IV, p 75; Berman, *Social Change*, p 60.

14 Quoted in Fullmer, 'Davy's Sketches of his Contemporaries,'* p 134.

15 On the circumstances of these analyses see Browne, 'Accum,'* p 850.

16 F Accum, 'Analysis of the lately discovered mineral waters at Cheltenham; and also of the medicinal springs in its Neighbourhood,' *Phil Mag* 31 (1808): 17.

17 F Accum, 'Analysis of the Chalybeate Spring at Thetford,' *Phil Mag* 53 (1819): 359–60. Concern about rain drying quickly was widespread (Saunders, *Treatise*, p 133; [A Hofmann], *Synopsis of the Analyses of the Mineral Springs of Harrogate extracted from Dr Hofmann's report, with Practical Remarks by the Medical Section of the Water Committee* [n p, 1854], pp 9–10).

18 Accum, 'Thetford,'* p 360.

19 Browne, 'Accum,'* p 1016.

20 Accum, 'Cheltenham,'* pp 15–16.

21 Garnett, *A Treatise on the Mineral Waters of Harrogate* (1792), p viii, quoted by Crellin, 'The Development of Chemistry in Britain,' p 279.

22 J Barker, *A Treatise on Cheltenham Water*, p 23.

23 Linden, *Treatise on Chalybeat Waters*, pp 1–2.

24 John Elliott, *An Account of the Nature and Medicinal Virtues of the Principal Mineral Waters of Great Britain and Ireland*, p 80. By the mid nineteenth century few analytical reports had anything to say about the medical significance of the results, and this perhaps signals the declining popularity of mineral waters (A B and M D Anderson, *Vanishing Spas*, p 8). See also Eklund, 'Chemical Analysis and the Phlogiston Theory,' p 234.

25 Dr Evans, 'An Account of Sutton Spa, near Shrewsbury,' *Phil Mag* 22 (1805): 61–8.

26 T J Herapath, 'Analysis of a Medicinal Water from the Neighbourhood of Bristol,' *J Chem Soc* 2 (1849): 200–5; A Voeckler, 'On the Composition of the Purton Saline Water,' *Q J Chem Soc* 14 (1861): 46–7; Mr Howell, 'Analysis of the Carbonated Chalybeate Well, lately discovered at Middleton Hall, the Seat of Sir William Paxton Kt., near

Llanarthney in Carmarthenshire,' *Phil Mag* 35 (1810): 179–80. See also Chapman, *The Profession of Chemistry*, p 23.

27 G W Pigott, *On the Harrogate Spas and Change of Air: Exhibiting a medical Commentary on the Waters founded on Professor Hofmann's Analysis*, p 280.

28 Saunders, *Treatise on Mineral Waters*, pp 114, 254.

29 Torbern Bergman, 'Of the Artificial Preparation of Cold Medicated Waters,' in his *Physical and Chemical Essays*, I, p 232. English waters were also being bottled (Gwen Hart, *A History of Cheltenham*, p 124). See also Linden, *Treatise*, pp xxi–xxv; E H Guitard, *Le Prestigieux Passé*, pp 152–75.

30 Bergman, 'Of the Artificial Preparation,'* p 275.

31 *Ibid*, p 276.

32 *Ibid*, p 233. See also John Rutty, *An Essay towards a Natural History of the Mineral Waters of Ireland*, pp iii, ix; Noel G Coley, 'The Preparation and Uses of Artificial Mineral Waters,' p 32. Bergman made his case directly to the king, Gustaf III, in his retirement address as president of the Academy of Science (J A Schufle, *Tobern Bergman. A Man Before his Time* [Lawrence, KS: Coronado Press, 1985], pp 372–3).

33 Bergman, 'Of the Artificial Preparation,'* pp 263–4; Barker, *A Treatise on Cheltenham Water*, pp 6–7; J B Gough, 'Lavoisier and the Fulfillment of the Stahlian Revolution,' *Osiris* 2nd series 4 (1988): 15–33.

34 Bergman, 'Of the Artificial Preparation,'* pp 271, 277.

35 *Pantologica*, s.v. mineral waters.

36 *Rees Cyclopedia*, s.v. water; 'On Mineral Waters, Natural and Artificial,' *Lancet* 1827–8, ii, pp 251–2; Wm Kirkby, *The Evolution of Artificial Mineral Waters*, pp 77–80. There was similar concern with quality and purity of pharmaceuticals generally (Crellin, 'The Development of Chemistry in Britain,' p 139).

37 *Rees Cyclopedia*, s.v. water.

38 R Phillips, 'An Analysis of the Salts prepared by Mr Henry Thompson from the Cheltenham Waters,' *Ann Phil* 11 (1818): 28–31. See also W T Brande and S Parkes, 'A descriptive Account of Mr Thompson's Laboratory at Cheltenham, for the Preparation of the Cheltenham Salts; with a Chemical Analysis of the Waters whence they are produced,' *Q J Science, Literature and the Arts* 3 (1817): 54–71; Hart, *History of Cheltenham*, p 125.

39 N G Coley, 'Preparation and Uses of Artificial Mineral Waters,' pp 42–4; A B Granville, *The Spas of England*, II, pp 572–3.

40 Daubeny, 'Present State,' pp 54–5.

41 *Ibid*, p 46. Cf P J Macquer cited in Eklund, 'Chemical Analysis,' p 242.

42 Daubeny, 'Present State,' pp 15–19; *idem*, 'Memoir on the Occurrence of Iodine and Bromine in Certain Mineral Waters of South Britain,' *Phil Trans Royal Society of London* 120 (1830): 224; R Henderson, 'On

the General Existence of Iodine in Spring water,' *Phil Mag* 2nd series 7 (1830): 11; Saunders, *Treatise on Mineral Waters*, pp 143–4.

43 Daubeny, 'Present State,' p 44. See also Eklund, 'Chemical Analysis,' pp 234–5; Guitard, *Le Prestigieux Passé*, pp 96–107.

44 Barker, *A Treatise on Cheltenham Water*, p 4.

45 E.g. Edwin Godden Jones, 'Chemical analysis of the Mineral Waters of Spa,' p 68; Granville, *The Spas of England*, I, p xxxiv; Saunders, *Treatise*, pp 92, 95; T B Dudley, *From Chaos to the Charter: A Complete History of Royal Leamington Spa*, p 352.

46 Jones, 'Chemical Analysis,'* p 56.

47 Meredith Gairdner, *Essay on Mineral and Thermal Springs*, pp 356–7.

48 Cited in *Lancet*, 1828–9, pt 2, p 110.

49 Jack Morrell and Arnold Thackray, *Gentlemen of Science: Early Years of the British Association for the Advancement of Science* (Oxford: Clarendon Press, 1981), pp 268–9. See also Eklund, 'Chemical Analysis,' p 234; Richard Yeo, 'An Idol in the Market Place—Baconianism in 19th Century Britain,' *History of Science* 23 (1985): 258.

50 R Kirwan, *Essay on Mineral Waters*, pp 6–7; Abel and Rowney, 'Cheltenham,' pp 193–4; Dr Evans, 'An Account of Sutton Spa, near Shrewsbury,'* p 67; Amicus, 'Analysis of the Mineral Waters of Caversham, Berkshire,' *Annals of Phil* 8 (1816): 123; Linden, *Treatise on Chalybeat Waters*, pp 132–3. See also Eklund, 'Chemical Analysis,' p 235.

51 Gairdner, *Essay on Mineral and Thermal Springs*, pp 2–3.

52 Daubeny, 'Present State,' pp 15–9. See also Crellin, 'The Development of Chemistry,' pp 140–7.

53 Daubeny, 'Memoir on Iodine,'* p 223.

54 Torbern Bergman, 'Treatise on Bitter, Seltzer, Spa and Pyrmont Waters and their Synthetical Preparation,' p 32.

55 Granville, *Spas of England*, I, pp xvi–xvii.

56 Saunders, *Treatise*, pp iii–vi; Linden, *Treatise*, pp 125–6, 136; Pigott, *Harrogate*, pp 1–13.

57 Howell, 'Analysis of the Carbonated Chalybeate Water,'* p 179.

58 Barker, *A Treatise on Cheltenham Water*, p 79.

59 Accum, 'Cheltenham,'* p 16.

60 Accum, 'Thetford,'* p 361.

61 *Ibid*, pp 360–1.

62 George Merck and Robert Galloway, 'Analysis of the Water of the Thermal Spring at Bath,' *Phil Mag* 3rd series 31 (1857): 56, 67.

63 R Phillips, 'Analysis of the Hot Springs at Bath,' *Phil Mag* 24 (1806): 342, 355–6. See also Henry W Freeman, *The Thermal Baths of Bath: their History, Literature, Medical and Surgical Uses and Effects*, pp 209–10. On conflicting results see also Gairdner, *Essay on Mineral and Thermal Springs*, pp 62–80.

64 Saunders, *Treatise on Mineral Waters*, pp 139–40; Colin Russell, N
   G Coley, and G K Roberts, *Chemists by Profession*; Bernard C Dyer
   and C Ainsworth Mitchell, *The Society of Public Analysts and Other
   Analytical Chemists*; Chapman, *The Profession of Chemistry*, pp 5–19.
65 R Phillips, 'Bath,'* pp 342–3; *Encyclopedia Britannica*, 1797, s.v. 'Mineral Waters,' v 12, p 44. See also Gairdner, *Essay on Mineral and
   Thermal Springs*, pp 2–3.

3   London's Water: The Dress Rehearsal of 1828

It is vain therefore to say, that where nothing is discovered there is nothing wrong.[1]

William Lambe

In 1828 Dr William Lambe, fellow and censor of the Royal College of Physicians, graduate and fellow of St John's College, Cambridge (fourth wrangler, 1786), and 'one of the most elegant medical writers of his day,' asserted that the ordinary water a great many people habitually drank was deadly: 'I believe the evil to be deep and serious; not merely injurious to cleanliness and comfort but ... a mischief which saps the foundations of life, and brings multitudes to a premature grave.'[2] In particular, Lambe believed that ordinary drinking water, particularly that supplied to Londoners from the Thames, was laden with organic matter in a state of decomposition, which it was 'generally agreed' (Lambe gave no references) had an injurious effect on health.[3] Sometimes these decaying organic matters were perceptible—water would be noticeably foul—and 'in this case, the common feeling of disgust induces men to reject it as unfit for human use.'[4]

Yet all too frequently, contamination was not only imperceptible to the senses but undetectable by water analysts, who had almost exclusively (and scandalously, in Lambe's view) focused on the spa waters consumed by the wealthy and ignored the potable water that everybody drank. The vast bulk of organic contamination, Lambe was convinced, existed as 'vaporous substances which taint the whole body of the fluid, and of whose influence on the body we are for the most part ignorant.'[5]

Many of Lambe's ideas—that organic contaminants were more serious than dissolved salts; that decomposing organic matter, or something about it, was particularly dangerous, and that ordinary analytical chemistry was ill-suited to detect these decomposing matters—will seem to us both prudent and prescient. They seem to presage John Snow's recognition of water-borne diseases of 1849–54 and the arguments of Edward Frankland and Benjamin Brodie the younger a decade later that with respect to water-borne disease it was wiser to follow common sense (and epidemiology) than analysis. Yet Lambe's ideas were not taken seriously in the late 1820s; as he put it himself, his views were 'met with an exceedingly cold reception from the mass of mankind.'[6] Lambe is thus a good transitional figure, one who recognized the importance the purity of public water supplies would come to have, yet stuck with the mineral water chemistry of the past. The occasion of his remarks is also significant: the 1828 inquiry on London's water of the Royal Commission on Water Supply has been seen by some as marking the beginning of the public health movement.[7] That assessment is likely to raise as many questions as it answers, but it does seem clear that the inquiry was the first significant discussion in Britain of what standards of quality ought to be expected of a public water supply. It also marks the laying out of the essential issues and arguments that would characterize six subsequent major extra-parliamentary investigations of London's water during the nineteenth century as well as numerous select-committee investigations on the metropolitan water supply as well as the water supplies of provincial cities.

For this reason I label these hearings a 'dress rehearsal,' but the language of the theatre is also apt for other reasons.[8] First, the Commission's hearings were a spectacle. Expert scientists, including Lambe, brought forth proofs that London's water was either beyond reproach or continually sapping the health of the citizenry. The drama of learned contradiction was a new mode of interaction between water analysts and the public. As we have seen, mineral water analysts had tried to circumvent open conflict. They had been reluctant to make accusations of incompetence to explain discrepant results.

In public hearings (or court cases) such delicacy was not possible. The questioning was in the hands of commissioners or members of parliament or barristers, who insisted that answers be relevant to policy problems. Unpleasantly for scientists their disagreement unavoidably spilled out to be soaked up by the press. A *Times* leader on a later outbreak of water controversy made light of the degree to which the whole business had become a show: before the select committee would be brought

> long-necked phials from all the ponds within fifty miles, with analyses by Professor Brande and contrary analyses by Faraday; there will be repentent offers from the London Waterworks and extravagant promises from the New River Company ... . Recriminations will be rife of animal and vegetable and mineral pollutions; one sample will be accused of a greenish tinge; one company will attribute a superfluity of lime to its opponent's fluid, and the second will retaliate upon the first the presence of some noxious animalcule. We shall learn ... the evils to which water is heir—carbonates and sulphates, iron and lime, will be pitted against one another as conducive to bone or detrimental to fibre; nay, the very substances will be produced in court, and one advocate will triumphantly point to the plaster cast, and another to the horse shoe, which his clients subtracted from the reservoirs of Vauxhall or the cisterns of Belgravia.[9]

Such hearings were also dramatic in that they reflected deep conflicts on public issues of enormous importance. Unlike the conflicts of mineral water chemists over the merits of rival springs or the fabrication of imitation waters, conflicts in potable water chemistry involved central issues of social philosophy: what rights did people have to good water, who should control water supplies, how safe did water have to be, and how were inter-regional disputes about water use to be settled. All were subjects of debate for most of the century. Hence when analysts took conflicting positions on a water's contents and their significance, they were often taking positions on these social issues as well.

Finally, in light of the frustrating and protracted debates on London's water that were to occupy the next 75 years, these hearings are appropriately regarded as a dress rehearsal of a long-running controversy. They made public a pattern of outrage, inquiry, and inconsequential public response that persisted: in 1850–52, 1866–69, 1880–84, 1892–93, and 1898–99 chemists, medical men, engineers, geologists, meteorologists, economists, accountants, and public administrators would make essentially the same arguments, only changing details as their sciences progressed. Throughout this period the players in the drama remained more or less constant: there were the private water companies that supplied London, the advocates of various (and sometimes conflicting) forms of public control, and the groups of engineers and investors seeking to replace the existing supply (drawn mainly from the polluted Lea and Thames) with their own purer alternatives. Scientists usually appeared in these forums as paid representatives of some one of these interests, and such conditions of engagement tarnished the good name of science, wrote a *Fraser's* author in 1834: 'it is curious and interesting to observe the manner in which palpable truths are frittered down, and become lost to the public, by the ingenuity and disingenuousness of scientific men, when called upon to support particular interests. They deal with axioms and facts as if they were hypothetical, and with hypotheses as if they were facts, just as in turns it serves their purpose.'[10]

To be sure, in some respects calling the 1828 inquiry a rehearsal is anachronistic, for it would be wrong to think the contending parties were only preparing for a long run, or that at least some of the witnesses and consultants did not passionately believe that their words and deeds were of utmost immediate importance. (Others may in fact have regarded their testimony as so much forensic exercise.)[11] Yet in retrospect we can see the hearings as the trying out and adjustment of arguments that would be raised repeatedly for the rest of the century. And certainly many mid-Victorian activists were candid in regarding investigating commissions as theatre: inquiries were the circuses with which governments dodged their obligation to act.[12]

Ordinary Waters

The 1828 hearings are important here because they mark the establishment of the context in which a great deal of the discussion about water analysis took place. They do not however mark the beginning of the discussion of the proper qualities of domestic water supplies. Works on mineral waters, on technical or manufacturing chemistry, and encyclopedia articles on water frequently dealt with potable water quality. For example, Charles Lucas gave much space to domestic water, and particularly to London waters, in his 1756 *Essay on Waters*. He analysed several London well waters and found most suitable 'for the ordinary purposes of families.'[13] Yet Lucas' purpose was to show that common waters contained many of the medicinal properties that others saw as exclusively properties of mineral waters, and his approach was not typical.

To gain perspective on the water quality discussions that took place in 1828–30 it is necessary to recognize that the term 'water' itself had distinctly different connotations than it now does. Though the composition of water as an hydrogen–oxygen compound had been established in the late eighteenth century, older ideas persisted of water as a principle, a set of watery characteristics to which various ethereal spirits might annex themselves. In such a perspective it made sense to think not so much of waters of greater or lesser purity but of a variety of waters with various qualities and uses. One characteristic Lucas did share with other writers was an interest in classifying waters, following a tradition that went back to Pliny and Hippocrates. Waters from dew, rain, snowmelt, ponds, lakes, muddy-bottomed rivers, and rocky-bottomed rivers could all be distinguished by their medical effects, suitability for various industrial uses, and their purity, understood as their tendency not to become foul.

With respect to medical effects, water from melted snow, for example, caused goitre, while hard spring waters were believed to cause bladder stones.[14] As for industrial uses, dew and rain water were soft, and hence useful for cooking, bread making, tanning, paper-making, bleaching and other uses, but also tended to putrefy. Hard water was good for making mortar.[15] As for purity, a common scheme held that after distilled water, rain (and dew) water were purest, followed by snowmelt, spring water, river water, and stagnant pond water.[16]

At the heart of a popular concept of the purity of domestic waters was the issue of fermentation and putridity. Almost all waters were recognized as having some potential to putrefy or ferment. Indeed, water itself was closely linked with these processes in two important senses. First, both fermentation and putrefaction had long been intimately linked with water, since it had been early observed that dry things did not rot. Water was thus a 'putrefacient': 'all putrefactions ... are ... performed by means of water alone; and without it there would be no such effect in all nature.'[17] Second, the relentlessness with which water dissolved things made it clear that no water would remain pure for long. In the pre-stainless steel world the phenomena of organic decomposition were pervasive in a way they no longer are, and it was futile to think water would long remain uncontaminated by organic matter.

The concept of water contaminated with putrefying matter was thus not new with Lambe. Nor was Lambe original in recognizing its danger. Pliny had advised against fetid or slimy water. The 1810 edition of the *Encyclopedia Britannica* warned that 'putrid water' was 'in the highest degree pernicious to the human frame, and capable of bringing on mortal diseases even by its smell.'[18] In 1828, then, the question was not whether there was a threat from organically contaminated water, but how serious it was, and the answer hinged on the twin issues of detection and purification.

While most of Lambe's contemporaries accepted that water was frequently contaminated with fermenting or putrefying matter, they were also confident that this contamination disappeared eventually through processes of self-purification, or could be made to disappear through storage, filtration, precipitation, or settling and decantation. Putrefaction was after all a transitory state. It might be true, Lucas pointed out, that the Thames was 'tainted with an infinite variety of adventitious bodies from the streets and sewers of our capital,' yet once its water had fermented and the impurity transformed into an inflammable air or a solid deposit (or 'after all the queer stuff has sunk to the bottom'), the water would again be pure.[19] Some such process was said to occur in the casks of river water mariners took to sea. If the casks were opened too quickly the water would be found 'so black and offensive as scarcely to be borne,' yet if left to complete its fermentation such water would be excellent.[20]

One can draw several tentative conclusions from these conceptions of foulness. The first is that the people of pre-industrial Britain would probably have been familiar with a greater variety of waters, possessing differing degrees and kinds of impurities, than we are today. They would appreciate that water grossly impure for drinking purposes (stagnant pond water, for example), might be excellent for dyeing, bleaching, or tanning.[21] Few of them would expect to encounter 'pure' water; instead purity would be a matter of degree. As *Rees Cyclopedia* observed, since pure water was virtually non-existent it was 'probably never intended as an article of drink for mankind; certainly, at least, not as one absolutely necessary for their existence, or even healthy condition.'[22]

Second, it is likely that determining water quality would have been a judgement within the competence of a layman (or for that matter a beast). Recognizing foul water required no expert or analysis, but only eyes, a nose, and a sense of taste. Again, *Rees Cyclopedia* put it plainly: the classification of waters 'according to their sensible properties, coincides likewise, as well perhaps as the present state of the subject will admit, with their chemical and medicinal properties.'[23]

Third, purification was also within a layman's capacity. There are numerous pre-nineteenth century examples of filtration and chemical purification schemes for use on all scales: domestic, industrial, or municipal.[24] An anti-Lambe article in the *Westminster Review* asserted that there was not a 'boor peasant in England' who did not know 'that if he wants to keep a covered spring or well pure for use, he must put a frog into it if he does not find one there,' since the frog would eat the smaller crawling things. The same article complained that a great deal of the fuss about purity came from those who weren't keeping their cisterns clean, and that no one could prevent water from depositing its clay in cisterns, where it would become a breeding ground for all manner of 'flying seeds.'[25] To be sure such arguments were especially useful to those defending the sale of water taken from a polluted river in shifting blame from water companies to their customers. Yet it is also true that one issue of the 1828 controversy was who bore responsibility for purity.

Hence it would be wrong to think that Londoners were oblivious to the quality of their water before Lambe brought it to their attention. One might acknowledge unpleasantnesses in the water without classifying it as a problem resulting from a particular cause, or seeing a practicable alternative, or believing that someone was culpable.[26]

William Lambe and the Menace of Putrefaction

Lambe's much stronger conclusions reflect the fact that he approached the issue from quite a different direction than did most writers on waters. For him organic contamination was a medical issue first, and only secondarily a matter of the bio-geo-chemical cycles of the world. While practicing in Leamington in the 1790s, Lambe had become concerned about lead poisoning from use of lead pipes for conducting water. This led him to a concern about organic contamination and eventually to 'a perfect revolution on the subject of common water.' Rather than seeing water as 'the source of health,' Lambe came to see it as 'the prolific source of disease.'[27] By 1810 he had developed a theory that vegetarianism along with the consumption of very pure (distilled) water was the key to health and a specific against cancer as well.[28]

On these foundations, Lambe adopted in his 1828 pamphlet a part of the 'ubiquity-of-putrefaction' motif considered above. He accepted the idea of ubiquitous putrefaction, was skeptical toward the notion of automatic purification. Thus he was scornful of Thomas Thomson's analysis of the waters of the Clyde as containing nothing more than a modest dose of magnesium and sodium chlorides and sulphates. If this were 'the whole truth,' wrote Lambe,

> it would be perfect childishness to suspect any noxious influence from the use of these waters. But before forming this conclusion, we ought to inquire, does the water of the Clyde preserve, when kept, its sweetness and freshness? Does it never become offensive to the smell and the taste during the heat of the summer? Are there no fish in the Clyde which perish and rot, and which contaminate the stream with excrementitious matter? Are not the streams which, finally collecting, swell its ample bed, originally the washings of tens of thousands of acres, over which are strewed, and in the substance of which are embedded, the decayed and decaying remains of myriads of animals and vegetables, in every stage of decomposition and putrefaction?[29]

(One can almost imagine a wealthy Londoner finishing Lambe's pamphlet and expostulating, 'Well, if you put it like that!') As Lambe points out, we cannot avoid answering 'yes' to his questions. When we do we are faced with the further question of effects: 'these matters being constantly applied to human bodies, the whole species is concerned to know, whether they are useful to the animal frame, ... noxious, ... or inert.'[30] Lambe presented indirect evidence that they were harmful but he was more concerned that chemistry was 'unable to throw the smallest glimmering on the subject.'[31] Nor were lay judgements helpful, he pointed out. It was well known that water that was 'perfectly pellucid, void of odour, and agreeable to the palate' would putrefy if left standing. Unknowingly one might be continually poisoning oneself by drinking tainted water.[32]

Unlike others, who were confident that they could recognize bad water and purify it, Lambe painted a picture of a malevolent nature in which danger was omnipresent and invisible. The only way to ensure survival was to retreat into the haven of distilled water and vegetables. His ideas may seem far-fetched and his fear of the world paranoid, yet the themes he set out were taken up by chemists and reformers in the '50s, '60s, and '70s. The possibility that nature was indeed malign, or at best neutral, became more plausible in the wake of the visitations of cholera that began in 1832. Later writers agreed that there was enormous potential for dangerous impurity in nature, that analysis was useless (or even worse than useless) for the detection of this impurity, and that a water policy founded on caution was wiser than one based on analytical demonstration.

The 'Dolphin' and its Aftermath

What triggered the 1828 inquiry and Lambe's pamphlet on 'Thames Water' was the appearance in the spring of 1827 of a scurrilous pamphlet, 'The Dolphin,' by one John Wright, a journalist. The 'Dolphin' of Wright's title was the intake pipe of the Grand Junction water company, one of eight which served greater London. Like most of the other companies, the Grand Junction took its water from the tidal Thames. Usually the companies purified their river water by storing it in subsidence reservoirs, but it often emerged still turbid or discoloured. What outraged Wright about the Grand Junction company in particular was the proximity of its intake to the outlet of the Ranelagh sewer, one of London's great old sewers; in such a situation sewage contamination was inevitable. Wright's case was not based solely on disgust; he submitted testimonials from medical men who believed their patients to have been adversely affected by the London waters, and included a lengthy discussion of the theory of the pathological action of organically contaminated water based on Lambe's ideas.[33]

Wright's pamphlet ignited widespread dissatisfaction with the water supply in general, not merely with the Grand Junction supply. Heavily attended public meetings in April 1827 passed resolutions condemning the companies. The matter was discussed in newspaper leaders and learned reviews, petitions were circulated, and finally in July a Royal Commission on the Metropolitan Water Supply was established.

Much of the outrage was not focused on the Dolphin itself or even on the poor quality of the water. Wright's pamphlet had provided an outlet for a growing frustration with the high cost and poor service of the companies' supplies that had begun with fundamental changes in the companies' business practices starting in 1817. Prior to 1817 the eight companies—The Grand Junction, Chelsea, New River, West Middlesex, and East London serving districts north of the river, the Lambeth, Southwark and Vauxhall, and Kent serving the south side—had acted independently of one another, either serving different areas of the metropolis or, as was especially the case after 1810, competing to serve customers in a single area. Beginning in 1817 however, they had agreed to divide the metropolitan market by allotting to each a monopoly for service in a given area. With monopoly came large increases in water rates—some witnesses to the 1828 commission reported increases of as much as 400 per cent between 1818 and 1821, and rates had been high even before the agreement.[34] At the same time, service was deteriorating, partly because the companies no longer had an incentive to give good service and partly because the metropolitan Thames and its tributaries were themselves growing filthier owing to refuse from the new gas industry and increased flushings of the sewers themselves, a paradoxical result of increased water supply. Indeed, it was about 1820 that the Thames ceased to be a salmon river, salmon being unable to cross a de-oxygenated zone around London to reach their upstream breeding grounds.[35]

The fact to which Wright drew attention, that the water of one company was disgusting and probably dangerous, was not therefore the main issue. Indeed, throughout the controversy, the quality issue was bound up with more general dissatisfaction with service: too many people felt they were paying too much for a supply that was too often dirty, unreliable, and insufficient in quantity. Of five petitions calling for water reform, only one objected explicitly to the Dolphin.[36] In light of the cholera epidemic that would come four years later and in light of later epidemiological work that would clearly link cholera and typhoid to polluted water, it is hard to realize how marginal health issues were to this controversy. Almost every witness the Commission heard expressed disgust at the water, but only a few worried that such a supply was a major threat to health or the cause of acute disease, though others believed such water probably unsettled the constitution and was certainly unlikely to improve health.

To understand why the drinking of sewaged water was not taken more seriously we need to look more closely at three closely associated problems. The first concerns defining what sort of a public problem water supply was to be. Whose responsibility was it and what kinds of knowledge and action were necessary to solve it? What were scientists and other experts to contribute? The second had to do with deciding what process of evaluation was appropriate for judging water quality. What tests would be used, and how would their results figure in making policy? The third concerned determining what harm the consumption of sewage-polluted water was likely to do. For each there were conflicting answers.

To Advise or to Act: Working out the Expert's Role

It is striking that when faced with what was basically a call for renegotiation of the terms on which consumers purchased water, the government appointed a compact commission of experts, thereby treating the issue as a technical problem. The commissioners were the chemist W T Brande, a physiology professor, Peter M Roget, both of the Royal Institution, and the eminent but ageing civil engineer Thomas Telford. Initially they took an expansive view of their task: they saw their job as devising a rational scheme of water supply to replace the ill-coordinated arrangements that had arisen over previous centuries. They asked that their gentleman-secretary be replaced with an engineer capable of undertaking the necessary surveys. There was a change in government during the winter of 1828, but both Home Secretaries involved, first Lord Wharncliffe and then Sir Robert Peel, had in mind a much briefer and narrower inquiry, aimed at documenting the condition of the supply, with further action to be left to Parliament. In March 1828 Peel was urging the Commission to get on with analysis of the water, in his view one of its key tasks. Peel's view prevailed, but the Commissioners did manage to make clear their view of the need for a more rational water supply by appending to their report both the correspondence on the scope of the inquiry and several proposals of alternative sources of supply they had received.[37]

At issue here were conflicting notions of how science was to contribute to public policy. The Commission's perspective was technocratic. In its view supplying a city with water and even administering that supply were as much technical problems as building a road or canal. That the government was the client in this case was of little moment to Telford. A water supply was needed; as engineer his job was to come up with a plan for getting a sufficient quantity of suitable water to the metropolis.

The government was not unsympathetic to the need for better water. The call for reform came not from the unrepresented masses but from the politically powerful: from middle class merchants, hoteliers, and publicans, and from wealthy and often aristocratic householders in fashionable neighbourhoods. Lord Wharncliffe, one of those Home Secretaries who clipped the Commission's wings, had been one of the anti-monopoly petitioners of 1827 (as had Brande, for that matter). But for both Home Secretaries water policy remained a legislative matter. The evidence science provided might contribute to a legislative resolution but could not be allowed to displace other considerations. This was exactly the philosophy Parliament was developing for dealing with other technical issues such as the building of canals, and the same approach was later used with the railways: technical initiatives had to be made compatible with demands of equity and tradition. At issue then were both the proper limits of technical means of solving problems—to what extent were complaints about a public utility a matter of engineering, chemistry, or physiology—and what standing did the findings of the experts have in the resolution of the problem, i.e., were the experts actually to solve the problem or simply to discover facts relevant to its solution?

Chemistry and the Burden of Proof

In fact, even the narrower mission of discovering the quality of London's water gave the Commission trouble. The Commissioners reported that they had begun their inquiry by 'examining such analyses of the Water as had already been made, and were communicated by the companies, as well as by several individuals of high authority on these matters.' These were found 'to be so far at variance with each other as to prevent our drawing from them satisfactory conclusions.'[38] They did not mention whose analyses they meant, but they did publish in appendices analyses by Dr George Pearson and John Gardner (for the Grand Junction Company of its sources of supply), by Richard Phillips (for the New River Company of its water), and by Lambe (of the metropolitan Thames, from which most of the companies took their supplies). Pearson and Gardner, and Phillips found the water satisfactory, while Lambe found it appallingly polluted.[39] In their report the Commissioners implied that the most likely source of these opposite conclusions was nature herself—the variability of the Thames.[40] To resolve the issue they commissioned Dr John Bostock, professor of physiology at Guy's Hospital, to analyse a set of nearly 40 samples taken from a number of locations under precisely stated conditions of weather and tide. Bostock's work did little to resolve the issue, however: he found some of the samples dreadful, others quite acceptable for a public water supply.[41]

Sampling variability may have had an impact on the quantities of suspended and dissolved matters the various analysts discovered, but their opposition was grounded in differing systems of analysis, which were based on different and incompatible conceptions of where responsibility for purity lay—with company or consumer. The main problem was that the most nearly relevant body of analytical procedures, mineral water chemistry, simply did not treat the issue of potability. Instead, judging potability was, as we have seen, a matter of common sense. It might be admitted that common sense was not always trustworthy, yet there were still no conventions as to the chemical definition of safe water, and as no conventions existed chemists were pretty much free to adapt and interpret mineral water analysis as they saw fit. They did so.

Pearson and Gardner treated the Grand Junction water as an ordinary mineral water, neglecting to entertain the possibility that determining potability might entail an entirely different set of standards and methods. Their approach was a modification of Bergman's protocol: evaporating to dryness and treating the residue first with alcohol, then with repeated washings of distilled water, and finally with nitric acid. They found the water to contain chlorides of magnesium and sodium, calcium sulphate and carbonates, and silica, none of these in unusually large quantities. These, however, were only the dissolved constituents, for Pearson and Gardner had allowed their samples to settle and only analysed the supernatant liquid. In a separate analysis they did work up the suspended matter but refused to take it into account in arriving at a conclusion about the water. Their rationale was that whatever was only in mechanical suspension was 'adventitious' with respect to water quality. Consumers need never drink such matter because it would settle out in a reservoir, cistern, or tumbler. They concluded that the water was 'as perfectly harmless as any spring water of the purest kind used in common life.'[42]

Such a statement is a defensible interpretation of analytical findings, but it does not state the assumptions Pearson and Gardner had made. First and most importantly, they were assuming that negative results were meaningful. Accepting a negative result of an analysis—saying that analysis detects nothing harmful—implies that one has used procedures that would have given a positive result had anything harmful been in the water. In failing to identify either what components or what quantities of components they would judge injurious, Pearson and Gardner were in principle free to match any conclusion to any analytical result. In practice, of course, their analysis could identify some presumably deleterious constituents—any large quantity of a medically active salt would be grounds for judging a water unsuitable for domestic use—but their analysis did not warrant the claim that they had looked for all the dangers which some physicians (Lambe for example) were beginning to think bad water might manifest.

Second, Pearson and Gardner were assuming (as the companies' defenders would assume in later inquiries) that water users held a great deal of the responsibility for water purity. This assumption is evident in the decision to disregard 'adventitious' suspended matters since these would settle out. Consumers were responsible for keeping cisterns clean; hence it was their job both to keep such matter from becoming re-suspended and to prevent contamination of the cistern by airborne filth.

In contrast with Pearson and Gardner, Lambe's analysis was directed toward discovery of an impurity defined so broadly that it could hardly be missed, and which no one could ever claim to have removed totally. He maintained that 'in an examination of the properties of a water applied to dietetic uses, the principal object of inquiry should be, whether the water be impregnated with organic matter in a state of decomposition.'[43] It made no difference whether this was animal or vegetable organic matter and in any case it was impossible to identify it by origin. He pointed out repeatedly that mineral water chemistry was useless for this task. Organic matter undergoing decomposition would interfere with the usual affinity relations on which mineral water analysis depended, and even for mineral substances the results would be meaningless.[44]

Yet some form of analysis was necessary since the senses could not be trusted to recognize bad water, and Lambe too used processes adapted from mineral water chemistry. Decomposing organic matters were to be discovered through addition of lead chloride or proto-nitrate of mercury, both of which precipitated what Lambe called 'charcoal,' by which he apparently meant something akin to what later writers called organic carbon.[45] He regarded the finding of this substance, along with the presence of hydrogen sulphide, methane, and nitrous acid, as proof that the Thames contained decomposing organic matter.

Where Pearson and Gardner had employed an analytical scheme almost guaranteed to give negative results, Lambe's approach could hardly fail to yield positive findings. That the Thames drained farmland, towns, and the growing metropolis was obvious without analysis. To discover some indication of organic decomposition, whether of the process itself or of its products, would not have been a startling conclusion. Even if one could not find actual traces of it one could still infer, as Lambe did, that putrefaction was taking place.[46]

As with the Pearson/Gardner analysis, what was controversial about Lambe's interpretation were his assumptions. Just as Pearson and Gardner had assumed that their chemistry was adequate to discover the noxious ingredients in water if any existed, so Lambe was assuming, indeed representing as the consensus of medical thought, that whatever his reagents precipitated was the harmful matter in water. The main issues of contention were thus at the level of unvoiced assumptions about what the significant impurities in water were, what tests to use to discover them, how to interpret the results, what actions the results necessitated, and ultimately perhaps how bountiful nature was in cleaning up after humans. Sampling variability hardly began to account for the differences.

John Bostock's infusion of empiricism and neutrality did nothing to resolve these issues. Bostock had analysed samples from all the companies' intakes taken at different stages of tide. He had used a third analytical approach and his conclusions were coloured by his own assumptions about how chemistry could measure potability. He began by classifying the samples into three classes—good, bad, and in between—according to taste, smell, and especially appearance. He attached great importance to this classification: 'by simple inspection of the specimens I think myself fully warranted in asserting, without any reference to experiment, that in all the waters except those of the first class [the best waters], the quantity of extraneous matter was so considerable as to render them improper to be employed in diet.'[47]

To use visual inspection as grounds for rejecting waters was not radical. Physical examination had a central place in traditional evaluation procedures for both mineral and potable waters. The senses did indeed provide good grounds for making some classifications; few would argue that the waters Bostock rejected—visibly tinted, swarming with animalcules, smelling and tasting foul—could be desirable for a public water supply. But Bostock went further. For him a visual inspection was sufficient basis for accepting as appropriate for domestic use those waters which appeared free of contamination. He wrote of his first class of waters that some of them, 'although not without extraneous matter, may be styled sufficiently free for ordinary purposes.'[48]

Bostock had thus arrived at his main conclusions without doing any chemistry at all and what analysis he did was anticlimactic. He did quantitative analyses on only two of the samples, those judged best and worst of the thirty-odd initial samples. He found both to contain calcium carbonate (the main ingredient), calcium sulphate, chlorides of magnesium and calcium, and organic matter, the latter determined by the loss in weight when the residue was ignited. With the exception of the organic matter, the two waters were almost identical, but the worst, taken at high tide from the Lambeth Company's intake, had slightly over 14 grains/gallon of organic matter, while the best, from the intake of the West Middlesex Company at low tide, had only 0.49 grains.[49] Bostock's conclusion was a bit lame—the water was sometimes satisfactory, sometimes dreadful.

With regard to the history of analytical chemistry, Bostock's performance may seem both timid and primitive—'a trumpery examination which any chemical student could have made,' sneered the *Westminster Review*.[50] Yet whatever its lack of sophistication, Bostock's analysis is an important one, for he raised and offered a resolution to the fundamental problem that nineteenth century water analysts faced, that of the relationship between knowledge derived from analytical operations and that derived from other sources, such as knowledge of a water's origins, of the physiological effects it was held to have or, in Bostock's case, of its physical characteristics. In using physical characteristics as the basis of his first (and only) evaluation of quality, Bostock was treating these characteristics, not composition, as the primary index of quality. That is, he was not merely trying to correlate composition with appearance, odour, and the like, nor suggesting that water had to be both physically attractive and chemically pure. Instead, he was finding out how the most important qualities, those evident to the senses, were manifested chemically.[51]

Just as the mineral water literature reflects the tension over whether physicians' experiences in treating patients or chemists' analyses of composition provided the true answer to the question of whether a mineral water 'worked,' the literature of potable water analysis would be stuck with this problem of explaining just exactly what degree of authority chemical analysis had and why it had it for the rest of the century. Bostock's resolution, that chemistry could illuminate ordinary observation but not displace it, reinforced an understanding of water quality that went back to the writings of Pliny and Vitruvius: the idea that in classifying waters for domestic use, physical characteristics were sufficient. The long-standing classification of water which stank, or was murky, swarming with life, or in any other way disgusting, was a prominent part of this taxonomy, and in undergirding disgust with analysis Bostock was reinforcing it, rather than modifying or rebutting it. Even Brande, champion of the chemists, was uncertain how to approach this issue.

He wrote in 1850 of a water high in nitrates and organic matter, 'yet ... [which was] bright and colourless, has no unpleasant taste, and is abundantly resorted to ... by a very populous neighbourhood' and was unsure what to conclude.[52]

A consequence of this role for chemistry was to leave analysis in a secondary and peripheral role in the evaluation of water quality. With the exception of eccentrics like Lambe who worried about the imperceptible evils that could get into water, most witnesses who testified on this issue before the Commission took the view that judging water quality was a matter of common sense. The views of William Somerville, a physician, and Telford's assistant James Mills, make this clear. It seemed to Somerville

> that the question of the purity of the water has been placed on a very erroneous footing by many, who say that there is no ingredient in the water ... to produce disease, and that chemical analysis detects nothing deleterious in the mixture; this reasoning would equally apply to water taken from the pan of a water-closet. The very idea of impurity is, in my estimation, sufficient ground for rejecting water that flows from a foul source.[53]

Mills was equally plain:

> A slender portion of common sense ... authorizes me to affirm, that a stream which receives daily the evacuations of a million human beings, of many thousand animals, with all the filth and refuse of various offensive manufactories, ... cannot require to be analyzed, except by a lunatic, to determine whether it ought to be pumped up as a beverage for the inhabitants of the Metropolis of the British empire.[54]

Analysis did not begin to be accepted as having an independent (and ultimately a superior) warrant in the determination of potable water quality until it was accepted as contradicting verdicts based on common sense. As the observations of Somerville and Mills suggest, initially this contradiction would come as the call for accepting negative conclusions, for accepting a water as pure because analysis found nothing harmful in the water, even though it might be condemned by common sense. As we shall see during the next round of London's water controversies in the 1850s, chemical analysis was mainly used by the defenders of the existing river-water supply in precisely this way: to show that the water was harmless despite the complaints that were made against it and despite its undeniable contamination with sewage. Only in the late 1860s did chemistry begin to be used to provide a warrant for condemning as impure, water which common sense or physical examination would accept as pure. Thus one of the great tasks (and one of the great successes) of water analysis in the second half of the century was the overthrow of the common sense standard, the convincing of the public that appearance was not a reliable index of salubrity. Prior to the 1870s, however, Bostock's correlation would have been accepted by many as the proper standard of potable water quality.

Constitutional Medicine and the Concept of Bad Water

Ultimately what Pearson and Gardner, Lambe and Bostock disagreed about was not so much what was in the water Londoners drank, but what the water's composition signified, i.e. what effects such water must have on the health of those who drank it owing to its composition. To understand how they interpreted their results we need to understand the contexts of contemporary physiology and pathology on which their conclusions were based. The medical theories of the day supplied a framework that differed in two important ways from the framework within which diseases like cholera or typhoid fever would later be linked to polluted water. The differences lay first with regard to what it meant to speak of a specific disease entity, and second, with regard to what sorts of initial assumptions a physician started with in ascertaining the causes of a case of disease. These differences were tied to a different conception of the relations between doctors and patients, especially with regard to what a doctor was expected to do for a patient. What is important about this framework is that it failed to supply a rationale for social action— no matter how much one might deplore the polluted condition of the water, the dominant medical theories themselves lessened the importance of water as a cause of disease and undermined the immediacy of water reform. This context of medical theories left even Lambe, the harshest and most astute of the critics, without strong arguments with which to oppose use of polluted river water.

What may appear to us as a hesitancy, ambivalence, or an equivocal attitude toward the drinking of polluted water is unmistakable in the medical testimony of the Commission's report. The Commission itself interviewed only a few doctors with most of the medical testimony coming in John Wright's 'Memoir,' an answer to criticisms of his original 'Dolphin' pamphlet which the Commission printed in full in an appendix. Wright had sent samples of Grand Junction water to prominent London physicians and surgeons asking for opinions of its quality and likely effect on the health of those who drank it. Almost all (at least of those he quoted) expressed revulsion. The samples were 'filthy,' 'impure,' 'most disgusting to the imagination.' When it came to ascribing ill-health to such water, however, his respondents were cautious, ambiguous, or evasive (or, more properly, they seem so in hindsight). H Leigh Thomas accepted that such water must be 'prejudicial to the health,' Robert Keate called it 'injurious to the health of thousands,' Robert Bree urged that it 'must be deleterious to the health,' James Johnson, that it 'cannot be salubrious,' J R Hume that its continued use was 'capable of producing deleterious effects,' though he did not know whether it had. Thomas Turner and Henry Halford, while rejecting the water as unsuitable, would not commit themselves as to its effect on health.[55]

The strongest statement came from R Hooper, who wrote:

that the daily use of impure water has a tendency to produce, or is a cause of many diseases, there cannot be any doubt; and it is a question of much importance, whether such matters in the stomach do not greatly contribute to the production of that state of faulty digestion, and impurity of blood, of which the inhabitants of this and other large cities are constantly complaining.[56]

Wright also cited published medical authorities, from the ancient (Hippocrates) to the modern (Cabanis, Lambe, James Johnson) on the general effects of water on health. They too saw impure water as a medical problem but one no different in kind from poor diet or lack of exercise.

The vagueness of many of these statements does not mean that these medical men were taking water quality less seriously than their successors who understood how polluted water could transmit cholera and typhoid. But such statements were as strong as the prevailing constitutional medicine of the early nineteenth century allowed one to get.

This 'constitutional medicine' has been best described in several works by Lester King.[57] It was the medicine of learned physicians in the eighteenth century and for centuries earlier. It was not so much a particular theory of pathology or therapy as a broad perspective toward medical practice, and it has been so thoroughly repudiated by modern scientific medicine (and by historians who have consecrated modern medicine) that it is hard for us to regard it with sympathy. The key difference between the two is that where modern medicine is concerned with the curing of diseases, understood as distinct microbial or physiological entities, constitutional medicine was concerned with maintaining or restoring health—ensuring that the mind was at peace, the stomach settled, the bowels in order, that undue lassitude or frenetic activity did not distress the patient. As Lucas explained, 'the body of man consists of various heterogenous parts, and it is of a very lax and fragile texture. It is liable to an infinite variety of disasters, and hardly ever to be pronounced in a perfect state of health.' The concern was with counteracting (or perhaps facilitating) the patient's way of living. Mineral waters, for example, were especially useful for 'gross and corpulent persons, that have fed fully and foully, used little or none exercise, and whose bowels are so furred, as to have lossed [*sic*] their natural sensibility.'[58]

This was a very old medical perspective, one present in the humoral theory of Galen and Hippocrates in which health was balance or *krasis*. It had adapted to changes in physiological theory through the centuries, being equally compatible with chemical and mechanical theories of the body. With its emphasis on health, constitutional medicine was centrally concerned with the relations of habits to the environment, and in particular with Galen's 'six things non-natural': diet, air, activity and rest, excretory habits, quality of sleep, and passions.[59] All these could be affected to some degree by conscious changes in the patient's way of living and his surroundings; all to some degree affected one another, and all clearly affected health. Upon entering practice in an area, the physician was to discern the qualities of the environment, including the properties of the water, as these were among the non-naturals.[60] This orientation toward habit and balance had implications for the concept of what diseases were. Classification was based on symptoms, or essences, not on the presumption of a unique cause, microbial or otherwise, for each disease. A variety of combinations of causes might produce a single disease.

Here the issue is how this perspective affected the evaluation of the effects of Thames water on health in 1828. Waters, both ordinary and mineral, were of central importance in constitutional medicine. Water cooled, neutralized, dissolved, diluted.[61] It was important as an article of diet, or through its effect on the air, or through the effect of its temperature when applied externally or internally. The medicine of the spa made sense in this context. One did not take mineral waters because they acted as specific remedies for certain diseases, but because of the effects they had on the constitution.

Consider, for example, typical accounts of the constitutional effects of a chalybeate water: it 'stimulates the fibres of the stomach and bowels, increases the quickness and strength of the pulse, promotes different secretions in the more remote parts, and represses inordinate discharges into the intestinal canal; the pale emaciated countenance, from its use, assumes a healthy florid colour, and the alvine, renal, and cuticular excretions are increased,' or according to another writer, it helped in 'hypochondriack Diseases, where pent up vapours and flatulent Humours rack and lacerate the tender Fibres and Membranes of the Stomach and Bowels.'[62]

Much the same sort of thing could be said for water contaminated with decomposing organic matter. According to Cabanis, for example, 'water loaded with putrid vegetable matters, or with earthy substances' tended to predispose one to 'cold and slow diseases' and to 'blunt the sensibility, enervate the muscular force.' Thomas Percival wrote that water 'if impure, will gradually produce some morbid changes on the body. On the robust indeed, its action may perhaps be slow, and imperceptible; but the tender and valetudinary will find themselves sooner and more sensibly affected by it.' He saw water as a cause of many endemic diseases, especially chronic conditions of children.[63] Such effects were due to the composition of the water, but also to its temperature, the time and circumstances of its consumption, and the state of the individual who drank it, and they might as easily be harmful as beneficial. Together, a host of causes determined what result would ensue from drinking water of a particular chemical composition and there was no justification for treating any single one of these causes as more important than any of the others.[64]

This then is the context of the testimonials Wright assembled, presumably the strongest he could find to condemn the water. Contemporary medical thinking, even of those medical men most opposed to the existing water supply, led directly away from the attribution of dramatic ill effects to water exclusively, no matter how foul it might be. For much of the century the effects of the ingestion of polluted waters would be admitted as predisposing causes of disease, conditions of environment that so weakened the victim's resistance that he took sick. By the end of the century the hunt for the microbes responsible for specific disease would make the question of predisposition moot—the state of the patient's constitution, predisposed or otherwise, was irrelevant since if one got rid of the exciting, microbial cause there would be nothing to worry about.[65] Predisposition was also dismissed as inaccessible to experiment, since there might be so many possible predisposing causes as to prohibit their recognition through controlled experiment. Yet in the context of pathologies of balance and without clear understanding of the constant relations of lesions and symptoms, these same 'predisposing causes,' water and all the rest, were nothing less than the causes—the sufficient conditions—of illness.[66]

The consequence of all this was that the Commissioners did not have even an argument, much less epidemiological evidence, that would permit them to single out bad water as being responsible for a certain amount of excess death and disease. They admitted as much:

> There must always be considerable difficulty in obtaining decisive evidence of an influence, which although actually operating to a certain extent as a cause of constitutional derangement, may yet not be sufficiently powerful to produce immediate and obvious injury. It cannot be denied that the continued use of a noxious ingredient in diet may create a tendency to disorders, which do not actually break out until fostered by the concurrence of other causes; for we unquestionably find an influence of the same kind exerted by other agents, which occasion merely a certain predisposition to disease, and of which the immediate operation must therefore be extremely insidious and difficult to trace.[67]

So weak in fact was the medical case against the water, that defenders of the supply found it unnecessary to refute it with their own medical experts. The *Westminster Review* simply quoted from the testimonials Wright had supplied and concluded that 'of such stuff is what is called physic, and the philosophy of physic, and physical writing ... but will any body answer how it was, and is, and will be, that all the physicians which were muster in London on this question could not give an opinion about it, except a foolish, or an ignorant, or a neutral one, or—none at all? And when water, too, is the universal medicine, and the universal cause of disease moreover. Oh, ye doctors, ye shall not doctor us when we are sick. But enough of ye all.'[68]

The case for reform of the London water supply thus depended on the authority of chemists, who arrived at opposite, ambiguous, and somewhat dubious conclusions (but did not hesitate to attach the authority of exact science to them), and of medical men, who uniformly regarded impure water as dangerous but were unable to ascribe particular illnesses to particular waters and did not possess the requisite methods of statistics, pathology, bacteriology, nor the philosophical perspectives about causation that would enable them to do so. Even though the Commission's report was backed up in a report by a Commons Select Committee which met a few months later in the summer of 1828 and looked into the business arrangements of the water companies, it did not lead even to the tabling of a bill for a reformed water supply. Telford continued to investigate alternative sources of supply and select committees met in 1834 and 1840 to consider his recommendations.[69] Yet the furor over the Dolphin had spent itself, and reform of the London water supply would not again be seriously considered until Edwin Chadwick and some independent entrepreneurs re-raised the issue in the late 1840s. The inquiry of 1828 did, however, lead some of the water companies (most notably the Chelsea, with its famous engineer, James Simpson) to establish filtration works for treating some of their water.[70]

1   Lambe, *An Investigation of the Properties of Thames Water* (London: Butcher and Underwood, 1828), p 52.

2   *DNB* 11, p 497; T B Dudley, *From Chaos to Charter*, p 348; Lambe, *Properties*, p iv.

3   Lambe, *Properties*, p 8.

4   *Ibid*, p 8.

5   *Ibid*, p 52.

6   *Ibid*, p viii.

7   D Lipschutz, 'The Water Question in London,' pp 510–26. Cf A Hardy, 'Water and the Search for Public Health,' pp 259–65.

8   See Steven Shapin, 'The Audience for Science in 18th Century Edinburgh,' *History of Science* 12 (1974): 95–7; A Mazur, 'Disputes between experts,' *Minerva* 11 (1973): 243–7; Brian L Campbell, 'Uncertainty as Symbolic Action in Disputes among Experts,' *Social Studies in Science* 15 (1985): 429–53; and Brian Wynne, *Rationality and Ritual: The Windscale Inquiry and Nuclear Decisions in Britain* (Chalfont St Giles, BSHS, 1982).

9   *Times*, 31 Dec 1850, 4e.

10   [Wall], 'Metropolis water supply,' *Fraser's Magazine* 10 (1834): 563. See also Viscount Ebrington in debate on the 1851 water bill (Hansard, 3rd series, v 122, c 857).

11   A S Taylor, *Principles and Practice of Medical Jurisprudence* 2nd edn (Philadelphia: Henry C Lea, 1873), pp 31–41; C Hamlin, 'Expert Witnessing and Scientific Method,' pp 485–513.

12 Morris Berman, *Social Change and Scientific Organization*, pp 104–5, 154, 173.

13 Lucas, *An Essay on Waters*, I, pp 144–5. See also A Hardy, 'Water and the Search for Public Health,' pp 254–9.

14 Lucas, *Essay on Waters*, I, pp 29, 150; William Saunders, *A Treatise on Mineral Waters*, pp 4–8, 66–89; John Barker, *A Treatise on Cheltenham Water*, pp 12–17.

15 *The British Cyclopedia of the Arts and Sciences*, v 2, s.v. water, pp 885–6; Lucas, *Essay on Waters*, I, pp 2–7, 24–8, 39, 57.

16 Pliny the Elder, *Natural History* trans. WHS Jones (Cambridge: Harvard University Press, 1963), Book 31, c. 21; Baker, *Quest for Pure Water*, 3–4; *Pantologica*, v. 12, s.v. water.

17 E Chambers, *Cyclopedia or Universal Dictionary of Arts and Sciences*, 1741, s.v. water; Barker, *A Treatise on Cheltenham Water*, pp 12–17.

18 Pliny, *Natural History**, Bk 31, c 22–37; *Encyclopedia Britannica* 1810, s.v. 'putrid water,' v 20, p 647.

19 [Richard Horne], 'Father Thames,' *Household Words* 2 (1851): 446; Lucas, *Essay on Waters*, I, p 35. See also 144–5; Baker, *Quest for Pure Water*, pp 20–7, 302, 362–6, 392; *Encyclopedia Britannica*, 1797, s.v. water, v 18, p 811.

20 Saunders, *Treatise on Mineral Waters*, 78–9, 216; *Encyclopedia Britannica*, 1810, s.v. water, v 20, p 647; *Pantologica*, s.v. waters, v 12.

21 Saunders, *Treatise on Mineral Waters*, p 81; Lucas, *Essay on Waters*, I, pp 141–2.

22 *Rees Cyclopedia*, s.v. water.

23 *Rees Cyclopedia*, s.v. water, v 38; Lucas, *Essay on Waters*, II, p 2.

24 Baker, *Quest for Pure Water*, pp 19–25, 65–77; Hartley, *Water in England*, pp 198–200.

25 'The Thames Water Question,' *Westminster Review* 12 (1830): 37–8; Thomas Percival, *Experiments and Observations on Water*, pp 6–7, 72–3.

26 [Wall]. 'Metropolis water supply,' p 566.

27 Lambe, *Properties*, pp vii–viii.

28 Lambe, *Properties*, pp viii, 56–61; *Rees Cyclopedia*, s.v. water; H Saxe Wyndham, *William Lambe MD, A Pioneer of Reformed Diet* (London: The Vegetarian Society, 1940).

29 Lambe, *Properties*, p 3.

30 *Ibid*, p 4.

31 *Ibid*, p 5.

32 *Ibid*, p 8.

33 Daniel Lipschutz, 'The Water Question in London,' pp 510–26; Hardy, 'Water and the Search for Public Health,' pp 260–3.

34 A physician who had paid £2/year for water before 1818 paid nearly £7/year thereafter (evidence of Robert Kerrison, *RCMWS*, p 44. For

comparison, this was a period when a middle class artisan might expect an annual income of about £100 (E J Hobsbawm, 'The British Standard of Living, 1790–1850' in A J Taylor, ed, *The Standard of Living in Britain in the Industrial Revolution* [London: Methuen, 1975], p 75).

35 R S R Fitter, *London's Natural History* (London: Collins, 1945), p 94; Pentelow, *River Purification, A Legal and Scientific Review*, p 2.

36 *RCMWS*, pp 114–16.

37 *RCMWS*, pp 117–23. See also Brande, 'The Supply of Water to the Metropolis,' pp 350–6; [Wall], 'Metropolis water supply,' p 563.

38 *RCMWS*, p 8.

39 *Ibid*, pp 84–99.

40 *Ibid*, p 8.

41 *Ibid*, pp 77–83. On Bostock see *DNB* 2, p 885.

42 *RCMWS*, p 97. On Pearson see *DNB* 15, pp 610–11.

43 *RCMWS*, p 84.

44 Lambe, *Properties*, pp 2, 6–7.

45 These were not the regularly used reagents for this purpose. According to Lambe he was the first to use proto-nitrate of mercury for such a purpose. Lead salts, though usually lead acetate, were sometimes used for determining organic matter (Lambe, *Properties*, p 9). The standard precipitation test for vegetable extractive was a brown precipitate formed when silver nitrate was added to a water already freed of chlorides and sulphates (*Encyclopedia Britannica*, 1810, s.v. chemistry, v 4, p 710). According to Fresenius, mercurous nitrate will react with highly oxidizable organic acids with formation of metallic mercury (pp 82–3). Lead chloride reacts with ammonia to produce an insoluble white precipitate of lead oxychloride (p 154). K R Fresenius, *Manual of Qualitative Chemical Analysis*, translated into the 'New System' and newly edited by Samuel W Johnson (New York: John Wiley, 1879).

46 Lambe, *Properties*, pp 43–4.

47 *RCMWS*, p 78.

48 *Ibid*.

49 *RCMWS*, p 80.

50 'Thames Water Question,' p 35.

51 *Encyclopedia Britannica*, 1797 edition, s.v. mineral waters, v 12, p 44.

52 W T Brande, 'Analysis of the Well-Water at the Royal Mint,' p 350.

53 *RCMWS*, pp 52–3.

54 *RCMWS*, pp 62–3.

55 *RCMWS*, pp 137–40.

56 *RCMWS*, p 137.

57 Lester King, *The Philosophy of Medicine: The Early Eighteenth Century* (Cambridge: Harvard University Press, 1978); *idem, Medical Thinking: A Historical Preface* (Chicago: University of Chicago Press,

1982); Roy Porter, *Disease, Medicine, and Society in England 1550–1860* (London: MacMillan, 1987), pp 25–7.

58 Lucas, *Essay on Waters*, I, p 155, II, pp 266–7.

59 King, *Phil. Med.*, p 217, pp 228–30; Guitard, *Le Prestigieux Passé*, p 102.

60 Lucas, *Essay on Waters*, I, p 126.

61 J Barker, *A Treatise on Cheltenham Water*, pp 27–31; Saunders, *Treatise on Mineral Waters*, pp 12–3, 19, 364–70; Lucas, *Essay on Waters*, I, pp 159–65.

62 Michael Ryan, *Remarks on the Supply of Water to the Metropolis, with an Account of the Natural History of Water in its simple and combined states; and of the chemical composition and medical uses of all known mineral waters, ...* (London: Longmans, 1828), p 32; Frederick Slare, *An Account of Pyrmont Waters*, pp 27–9.

63 *RCMWS*, p 129; Percival, *Experiments and Observations on Water*, p 2.

64 Chapters on causation in early nineteenth century texts on medical practice are fascinating in this regard. They deal in a highly sophisticated manner with a great many different kinds of causes and recognize the error of simplistic mono-causal explanations. Cf Charles Williams, FRS, *Principles of Medicine comprising General Pathology and Therapeutics*, ed with additions by Meredith Clymer (Philadelphia: Lea and Blanchard, 1848), pp 18–36; George B Wood, *A Treatise on the Practice of Medicine* 2nd edn (Philadelphia: Grigg, Elliott, and Co, 1849), pp 125–38; Thomas Watson, *Lectures on the Principles and Practice of Physic, delivered at King's College London* (Philadelphia: Lea and Blanchard, 1844), pp 49–69; Wm A Guy, *Hooper's Physician's Vade Mecum: or Manual of the Principles and Practice of Physic*, 6th edn (London: Renshaw, 1858), pp 2–24, 310–11. See also W F Bynum, 'Cullen and the Study of Fevers in Britain, 1760–1820,' in W F Bynum and V Nutton (eds), *Theories of Fever from Antiquity to the Enlightenment, Medical History* supplement no 1 (London: Wellcome Institute, 1981), p 139.

65 Phyllis A Richmond, 'The Germ Theory of Disease,' in A Lilienfeld ed, *Times, Places, and Persons: Aspects of the History of Epidemiology* (Baltimore: Johns Hopkins University Press, 1980), pp 84–93.

66 *British Cyclopedia of the Arts and Sciences*, s.v. water, vol 2, pp 889–92; [Wall], 'Metropolis water supply,' 567.

67 *RCMWS*, pp 10–11.

68 'Thames Water Question,'* p 34.

69 S C on Metropolis Water, 1834; S C House of Lords on the Supply of Water to the Metropolis, 1840.

70 Baker, *Quest for Pure Water*, pp 99–112.

4 The 'Hard Water and Animalculae Sellers': Analysis and Politics in London, 1849–52

Your Stomach would turn because your mind turned first.[1]

Sir Edmund Beckett

In 1849 the question of how London was to be supplied with water again became a burning issue. The next three years saw investigations of new (and old) alternative supplies, a new series of indignant editorials, new attempts to squeeze from science definitive justification of opposing programmes. With the 1849–52 controversy we have clearly crossed a crucial watershed—we are now in familiar historical territory, in the great age of sanitary reform. Cholera had come and gone once and was just ending its second visit. Edwin Chadwick had defined sanitary progress, and during the controversy John Snow would advance his famous hypothesis of the specific link between cholera and the consumption of water befouled by the excreta of a cholera victim.

It may seem that these factors—the definition of public sanitary responsibility, an incipient scientific understanding of the terrible effects of bad water, and the shock of epidemic—would give the 1849–52 controversy a sharply different character from its predecessor. Yet the continuities outweigh the discontinuities. Again consumer issues were prominent: people objected to heavy charges for an intermittent supply of bad water. The conflicts were not the class conflicts of cholera epidemics, but conflicts over political and financial control of the water supply, and the scientific debates on water quality had less to do with working out epidemiological relations between water and cholera than with seeking a politically serviceable concept of impurity and determining which branch of professionals would henceforth rule over water quality.

Three distinct concepts of impurity surfaced during the controversy, each with its own methods for testing water and implications for policy. Two of these were new: the concepts of purity as softness and as freedom from microscopic life. Both new approaches were worked out by reformers trying to replace both the water companies and the polluted waters which they supplied. The companies' defenders, on the other hand, continued to rely mainly on mineral water chemistry to demonstrate that the existing supply was safe.

The John Wright of the 1849–52 controversy was Edwin Chadwick. Following the lead of his patron Bentham, Chadwick preached the creed of rationalized, efficient, and coordinated public services. With respect to water this meant complete urban drainage and plenty of water to flush all wastes from cities to sewage farms. This perspective emerged in his great and controversial report *On the Sanitary Condition of the Labouring Population of Great Britain* in 1842, and was embodied in the 1848 Public Health Act, with Chadwick himself a member of the General Board of Health that would administer the Act.[2] The Act did not apply to London however, and Chadwick's involvement with London's water was part of his campaign to bring to the metropolis the sanitary benefits the provinces were to receive. He saw need for three reforms: replacement of the polluted Thames water with pure water, elimination of the water companies, and transfer of the administration of the water works to a board of responsible professional administrators. To that end he prepared (and the GBH published in 1850) a two volume report on the Metropolis Water Supply.[3] The controversy, however, was well along when report appeared; it was neither the source nor the resolution.

Chadwick had, in fact, little influence on the structure of the public works industry. Water supplies were already in existence in many places; they had been developed piecemeal as groups of investors or far-sighted municipalities had succeeded in getting through Parliament the private bills that would give them a monopoly to supply a certain area. Public agitation, the gearing up of rate payers or investors for the expense and uncertainty of the Parliamentary ordeal, was an essential part of this process. In potential railway investors promoters planted the prospect of profit; in water consumers they contrasted the deplorable present with the delectable future.[4] Judging from the reports of breakdowns in the waterworks, of visible swimming things in the water, and of imperious responses from the companies to complaints about service or charges, London's water consumers had good reason to be dissatisfied. Yet their dissatisfaction was not spontaneous. As in 1828 public outrage became organized political activity through the efforts of engineers, speculators, and sanitarians who had various interests in seeing changes in the existing arrangements. Such groups commissioned the scientific studies that proved the existing supply to be impure, loosed the pamphleteers, organized the meetings that produced the petitions. A great deal more was at issue than the quality of the water and the welfare of those who drank it; at stake were the values of the land from which the new supply would come, the values of the companies' shares, the careers of engineers who planned or built new supplies, and the future powers of existing governing bodies, such as the London vestries. One count listed 25 alternative schemes which had been proposed for supplying London with water and the water quality issue was as much a means of struggle among these rivals as an end in itself.[5]

Already in December 1848 a Bishopsgate public meeting had attacked the water monopoly[6] and at a spate of meetings during the fall of 1849 and the winter of 1850 the companies were attacked for their unresponsiveness during the recent cholera epidemic. On October 11, 1849, a *Times* leader announced that it was time for the water monopoly to end. It was occasioned by the New River Company's 'business-as-usual' attitude when requested to supply extra water to areas heavily struck by cholera.[7] The editorial reflected Chadwick's views: that it was high time to dissolve the companies and that they had no right to compensation based on future revenues since under common law water belonged to the public.[8] Beginning on Christmas Day and continuing for the better part of a week, the *Times* made good its threat by publishing 'The Water Monopoly and the Sanitary Movement,' a record of the companies' highhandedness over more than 30 years.[9]

Meanwhile agitation was brewing in other quarters. At an October 23rd meeting sponsored by the Metropolitan Water Supply Association (associated with James Taberner's scheme for a well water supply), several parish medical officers voiced their concern about the poor quality of the supply.[10] From mid December to March a series of meetings took place in the parishes, sponsored either by Taberner's organization or by the London (Watford) Spring Water Company, promoter of Telford's Brushy Meadows supply. In part

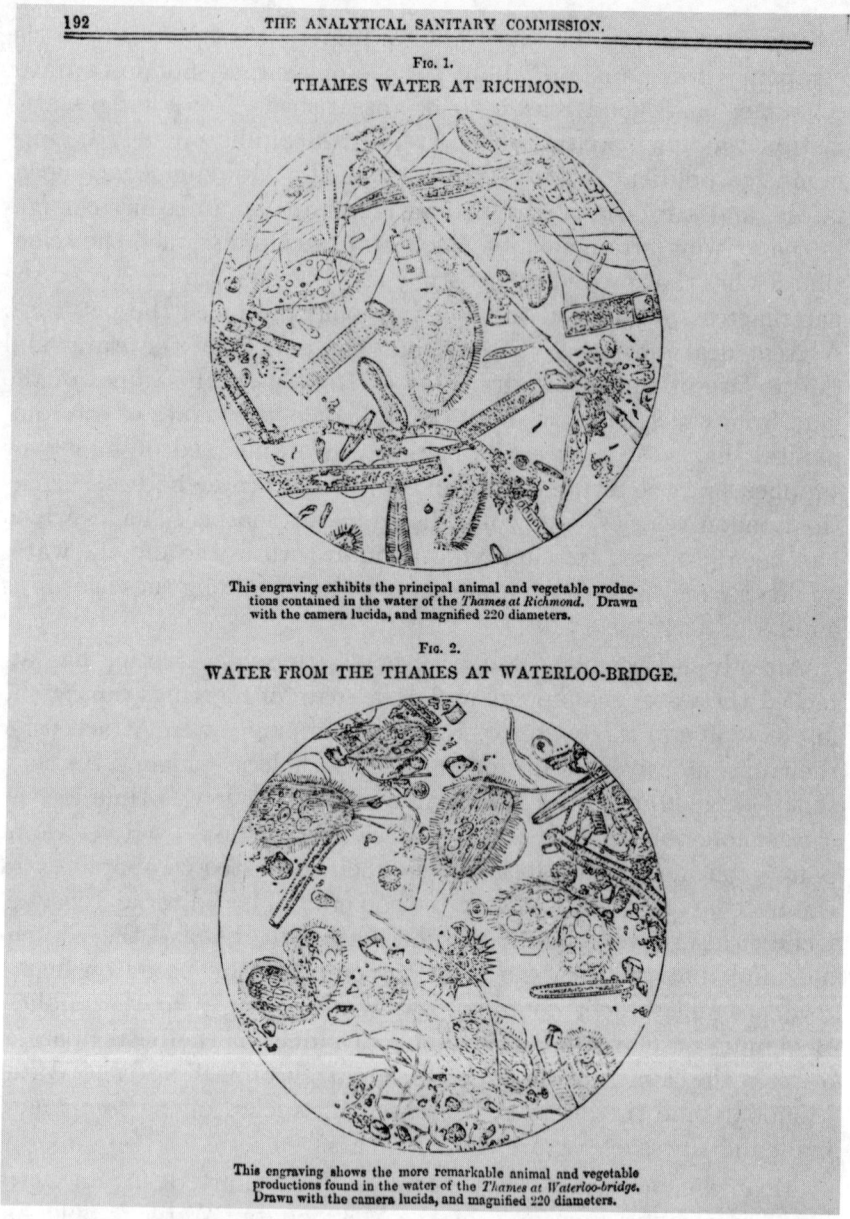

Fig. 1.
THAMES WATER AT RICHMOND.

This engraving exhibits the principal animal and vegetable productions contained in the water of the *Thames at Richmond*.   Drawn
with the camera lucida, and magnified 220 diameters.

Fig. 2.
WATER FROM THE THAMES AT WATERLOO-BRIDGE.

This engraving shows the more remarkable animal and vegetable
productions found in the water of the *Thames at Waterloo-bridge*.
Drawn with the camera lucida, and magnified 220 diameters.

Figure 4.1 A H Hassall's drawings of living things in the London waters
excited alarm during the 1849–52 controversy. Hassall admitted that the
drawings were composites; the microscopic field wasn't quite this crowded
(*Lancet*, i, 1851, pp 192–3).

the enthusiastic response came from the fact that these proposals

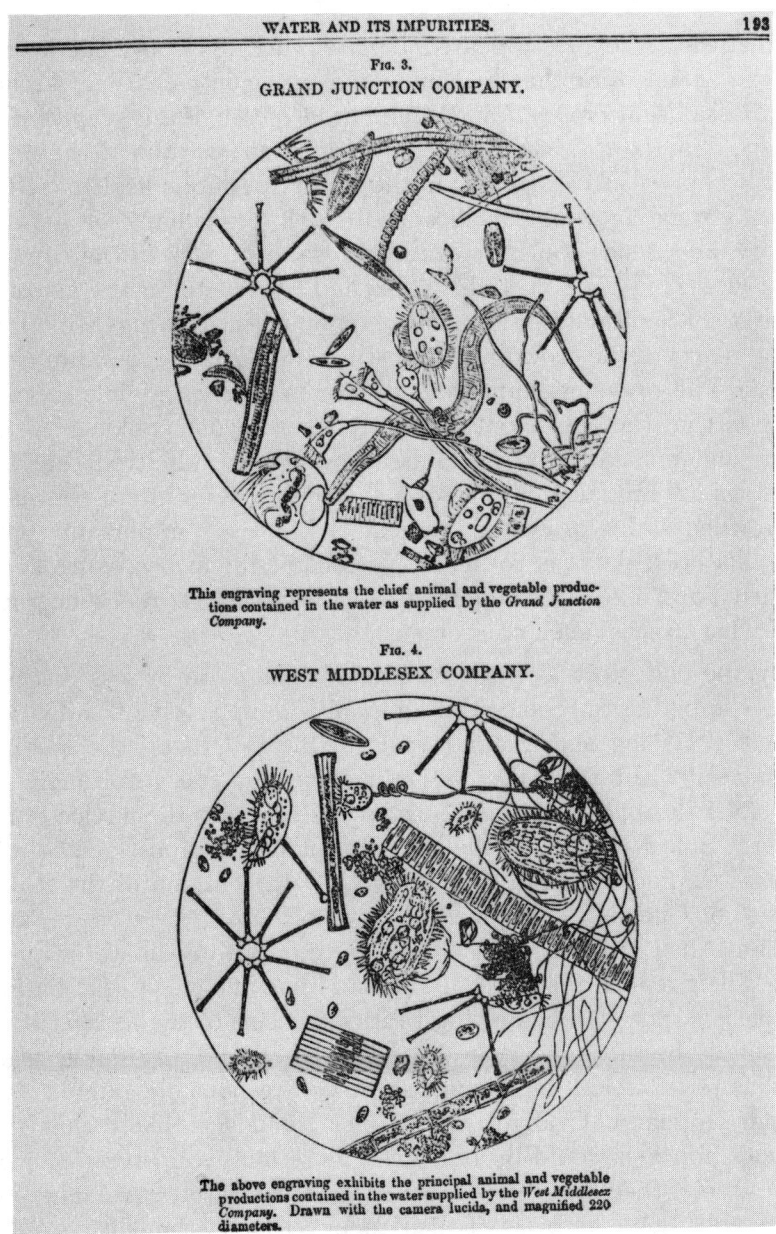

Fig. 3.

GRAND JUNCTION COMPANY.

This engraving represents the chief animal and vegetable productions contained in the water as supplied by the *Grand Junction Company*.

Fig. 4.

WEST MIDDLESEX COMPANY.

The above engraving exhibits the principal animal and vegetable productions contained in the water supplied by the *West Middlesex Company*. Drawn with the camera lucida, and magnified 220 diameters.

Figure 4.1   (continued)

posed no threat to London's system of parochial government. Unlike Chadwick, who preferred expertise to inefficient vestry democracy, Taberner and Samuel Homersham of the Watford Company were willing to work within the vestry system.[11]

Chadwick's report appeared in late May 1850, overshadowed, however, by a work that had appeared two months earlier, Arthur Hill Hassall's *Microscopical Examination of the Water supplied to the Inhabitants of London and Suburban Districts*, one of the most effective appeals to sensibility in the history of public health.[12] Hassall, a struggling young medical man with a penchant for natural history, had microscopically examined water of each of the companies that served London. In his book, and a year later as the *Lancet*'s 'Analytical Sanitary Commission,' he published drawings (coloured in the book) of the crowds of disgusting organisms he found in each water. The drawings appeared in circular frames, giving the impression that they represented exactly what anyone could see in his water under magnification. (Hassall eventually admitted that his drawings were in fact composites and that the waters were not quite so crowded as his drawings implied.[13]) The most important thing Hassall's book did was to make microscopic life a new category of impurity, and a great deal of debate in 1851 and 1852 was concerned with what exactly such creatures signified.[14]

By the end of 1850 there were still eight more or less distinct plans competing to replace the companies' supply, with Chadwick's scheme (utilizing springs and wells in Surrey) being the leading candidate.[15] But the water bill introduced by the government in May 1851 disappointed almost everyone. Ignoring the advice of its General Board of Health, the government had commissioned three of London's most prominent chemists, A W Hofmann of the Royal College of Chemistry, Thomas Graham of University College, and William Allen Miller of King's College, to report on the water question. Their report appeared in June and had the effect of deflating enthusiasm, not so much owing to its criticisms of the alternatives, but simply by raising so many unanswered and unanswerable questions as to make any big change seem an irresponsible gamble. Although Hofmann, Graham, and Miller thought it likely that the Thames and Lea would become even more foul as upstream towns built sewers, they found no evidence that the water was dangerous at present.[16] On the basis of their report, the government chose to make no change in the sources of supply. Its bill called for merger of the companies into a single company with dividends restricted to five per cent. Hardly anyone liked the bill and it disappeared in committee.[17] In 1852 the government presented another bill, one focusing on quality. The Thames companies were required to move their intakes upstream, to filter their water, to cover reservoirs, and to supply water on constant service. In exchange for these concessions they retained control of a profitable monopoly. Passage of this bill effectively relieved the companies of pressure to change the source of supply until 1866.[18]

Cholera and the Concept of Impurity—What Does Bad Water Do?

With this sketch we can turn to the arguments about quality and analysis that surfaced during the controversy. In the clamorous public meetings of 1849–50 there were numerous allusions to impurities in the water—to its 'organic matter,' to the fact that the Lambeth Company served 'larger and fatter animalculae,' to 'sulphate and carbonate of lime, and large quantities of saline matter, more or less injurious to the human system,' or simply to the 'impure matter,' 40,000 tons of which went into the river daily, to be 'stirred up by tides and steamers, and then … served up … for breakfast the next morning.'[19]

The water was thus found guilty of anything it could be guilty of. But such accusations were products of heated meetings, expressions in the idiom of the public health movement of the anger many people felt toward the companies. Consumers knew that what came from the tap was foul and that the sources of supply were foul. They might not know exactly which of the multiple sorts of impurity was most reprehensible but that made little difference. Any concept would do.

In the months after cholera had killed over 14,000 people in greater London, it is hardly surprising to find that disease too getting listed among the sins of the water supply. The physician and microscopist Edwin Lankester asserted that the areas worst hit by cholera were those with the worst water. He could not find anything that seemed directly responsible in the water, but its impurity was such that it was 'no surprise that disease was produced.'[20] Yet even Lankester did not suggest that the water was transmitting cholera, but only that it had likely 'greatly aggravated cholera mortality,' i.e. by acting as a predisposing cause or producer of debility. The chairman of the Bermondsey Board of Guardians went further, insisting that the cholera 'had peculiarly chosen for its ravages those districts south of the Thames supplied with its waters, and, as if to indicate its deleterious influence, had literally stayed at the point where that supply ceased.'[21] The chairman at a February 1850 meeting in Southwark spoke of the 'duty to remove, if not one of the causes, certainly one of the most powerful agents in the extension' of cholera.[22] In the spring of 1851 the *Times* itself spoke of 'impure water' containing 'the seeds of death' as certain fact. The GBH report likewise attributed cholera to water polluted with animal and vegetable contamination.[23]

It would be easy to see in these statements the beginnings of a modern concept of water as the vector of diseases like cholera even while recognizing that ideas about the nature of the morbid agents in the water remained various and vague. It is true that the 1849 cholera epidemic was the occasion both for the first edition of John Snow's famous work 'On the mode of the Communication of Cholera,' which correctly recognized water supply as the major vector, and for William Budd's announcement of a water-borne cholera germ, the cholera fungus, discovered by his associate, the Bristol microscopist Frederick Brittan. Yet the first edition of Snow's work was indistinguishable from the stack of contemporary speculations on cholera; Snow's convincing epidemiological analyses of the Broad Street outbreak and of cholera distribution in areas supplied by the different south London water companies were done during the 1854 epidemic and only appeared in the second edition. Likewise, Budd's claims that a fungus unique to cholera victims had been discovered and that it was *the cause* of cholera were rejected in short order, the first on grounds of faulty observation, the second on grounds of unsound inference.[24]

It is also striking that the water reformers made less use of the epidemic than they might have done. When their statements are closely examined, it is clear that Lankester and the others were not attributing cholera directly to the water. Instead their comments reflect a medical theory dominant in the early public health movement, sometimes misleadingly called 'miasmatism' but bearing a closer resemblance to the constitutional theories considered in the last chapter.[25] Health and disease were seen as consequences of the total environment. The conditions of city life—the stale and vitiated air, the uncleanliness, crowdedness, alcoholism, poor food, and foul water—acted collectively to undermine health: the combined effect of all these 'predisposing causes' was virtually the disease itself.[26]

Such was the context of water quality discussion in 1849–50. Lankester's assertion that bad water 'greatly aggravated' cholera mortality treated bad water as a predisposing cause which weakened those who caught cholera.[27] Others were concerned about the predisposing effects of dead sparrows or extract of churchyard in their water.[28] Chadwick's associate F O Ward wrote that 'the dilute impurities of even the clearest-looking Thames water, when introduced day after day into the blood, must produce a certain effect ... of a more or less injurious kind.'[29] The body was like a sponge, according to another: membranes in the lungs and stomach absorbing everything and passing all on to the blood. Even Hassall's animalcules were probably predisposers. These were 'liable to disturb the bowels, especially during such an epidemic as cholera; and in this way ... would act as a predisposing cause of the disease.'[30] These ideas were fully exploited in Chadwick's water report. Here cholera was seen as an unusually virulent form of the diarrhoea normally caused by bad water. The source of virulence was a decaying matter in the atmosphere, which, being soluble, could be transferred to (and thus by) water.[31]

This manner of viewing illnesses as products of the totality of predisposing causes had great utility for sanitary reformers. It provided the rationale for their concern with all physical, social, and moral components of the environment and suggested that any reform, no matter how small, produced real improvement. Yet when it came to the reform of a single factor, such as water supply, the same perspective was a handicap. Seeing each case of illness as an outcome of a long history of debilitating circumstances did not sanction the targeting of a single cause. One of the arguments against the Budd–Brittan cholera fungus theory (and against Snow, even after 1854) was that it was simplistic and illusory to think of 'the cause of cholera,' since like any other event it had many causes.[32] No matter how much critics might despise the existing water supply they did not have a set of concepts that would allow them to single it out as bearing the responsibility for cholera. Like Wright, they lacked a theory that would lend immediacy to their concerns.

Soft Water Becomes an Issue

Thus the 1849 cholera, which to us would provide the best reason for getting better water, had relatively little importance in the reformers' arguments. We can now look at the arguments they did use. Chadwick's main complaint was that London's water was too hard. This was not an important medical issue, though some eighteenth

century physicians had seen hard water as responsible for various illnesses, particularly bladder stones. Hardness had also been of little
interest to most of the authors on mineral waters prior to William
Saunders' *Treatise on Mineral Waters* (1800), where its importance
to the growing textile industry was emphasized.[33]

To a significant degree Chadwick himself was responsible for making the hardness issue so central. The foundations of his concern are
evident in the famous *Sanitary Report* of 1842. Already he had
decided that the prime cause of disease was atmospheric impurity
and that water could help by 'cleansing and removing solid refuse
and impurities.'[34] But too often there wasn't enough water to flush
the cities and scour the new narrow bore pipe sewers. As for water quality, Chadwick virtually ignored it in the report, noting only
that some of the provincial informants had suggested a relationship
between bad water and illness.[35] And even then he noted that while
the public usually found water with visible animal pollution most
objectionable, that containing mineral salts was actually more dangerous. The main brief against hard water was not directly medical
at all: it was unsuitable for removing urban filth because it wasted
an enormous quantity of soap. The report's final word on water
was that 'the formation of all habits of cleanliness is obstructed by
defective supplies of water.'[36]

In the 1850 report on London's water Chadwick discussed three
types of impurity: the organic contaminations from sewage which
presumably contributed to cholera; the microscopic (and occasionally macroscopic) organisms that lived in the water; and the hardness. Hardness was the most significant.[37] He argued that hard
water was unhealthy (it hindered solution and hence digestion but
did not, he thought, cause bladder stones). He cited the view of
the famous chef Alexis Soyer that soft water was better for cooking, and claimed that soft water made more (and better) tea from
the same quantity of leaves.[38] He devoted ten pages to the enormous amount of soap wasted each year in the metropolis—£630,000
worth, according to one estimate, all due to hard water.[39]

Most other water-supply reformers shared Chadwick's emphasis
on hardness and for most of a decade the issue dominated discussion of water quality. Thomas Clark of Aberdeen called hardness
the most important qualitative issue. Lyon Playfair lectured at the
Museum of Practical Geology that hard water caused disease and all
manner of industrial problems. He insisted, as did the Chadwickian
publicist F O Ward, that animals (and humans) instinctively chose

soft water over hard.[40]

Especially in the wake of a catastrophic cholera epidemic which Chadwick himself attributed in part to the water supply, matters of tea-steeping and soap-wasting might not seem to provide the most compelling reasons for a change in water supply. In some ways, however, the hardness argument was stronger than one founded on a putative link between cholera and water. Its advantages were twofold: first, it was impractical to do anything about it. While water softening processes existed they were not economical on a large scale. As Chadwick and his followers repeatedly pointed out, no matter how effectively the Thames might be freed from sewage, its water would still be hard.[41] Second, in contrast to such concepts of impurity as dissolved decomposing matter, hardness was unambiguous. Chemists might disagree about its significance, but there was little room for disagreement as to the level itself—the procedures for measuring hardness (either Clark's test based on the precipitate formed with soap, or the older method of driving off CO_2 which would precipitate most of the carbonates) were simple and well-accepted. The 1828 campaign had foundered on chemists' inability to agree on what sort of impurity they should be concerned with. Focusing on hardness avoided this problem, though of course it did nothing to resolve that vexed issue of just what it was in polluted water that Londoners ought to be worrying about.

The Vexed Problem of Life

Chadwick's other objection to the quality of London's water (in addition to its hardness and dissolved atmospheric impurity) was that water swarming with microscopic life could not be good to drink. The propriety of relying on waters populated with invisible creatures had been a minor issue in 1828, and not so much a medical question of whether such creatures might harm health as a question of sensibility. Witnesses had reported 'shrimp-like creatures,' fish, periwinkles, and 'little round black things' in their water; microscopic life had been regarded as disgusting and improper in a public water supply.[42]

The reformers of 1849–52 made a great deal more use of microscopical evidence. They hoped to show that one's instinctive revulsion at such water rested on an instinctive understanding that such creatures were either harmful themselves or infallible indicators of an otherwise undetectable danger.

Chadwick's chief sources on this issue were the Manchester chemist and sanitarian Robert Angus Smith, who had been active in sanitary science for several years, and Hassall, a newcomer to sanitary matters. Smith had investigated London's water for the Metropolitan Sanitary Commission and his report 'On the Air and Water of Towns' had been published by the British Association.[43] He had surveyed the length of the Thames, collecting data on chemical composition (including hardness, of course) as well as observing the changing flora and fauna in the river. It seemed to Smith that changes in the species of microscopic life might prove a sensitive indicator of changes that were indistinguishable by chemical means. He counselled water analysts to allow their samples to deposit their sediment, for the organisms in this deposit would be the best characteristic of the quality of the water.[44]

Yet though he would continue to advocate this procedure for the next two decades, Smith was never able to determine precisely which species corresponded to which conditions of pollution. In practice his indices of impurity were based not so much on species, but on the number of individuals, their motility, size, and whether they were animal or vegetable. Even aesthetic factors had some standing: Smith noted that samples from heavily polluted reaches of the inner London Thames 'contained animalcules larger, fatter, and uglier than any preceding.'[45]

Arthur Hill Hassall's handling of the issue was more detailed and sophisticated than Smith's, but he too came up against the same problem of assigning significances to the organisms living in various waters. Hassall had been born into a medical family in the Thames-side town of Teddington. While studying medicine in Dublin in the early '40s he had turned to natural history and become an authority on the microscopic marine life of the Wicklow coast. Having returned to London in the late '40s, Hassall became involved in the sanitary problems of the north London parish in which he was living. His microscopic studies of the water his patients drank grew into an investigation of the water supplied to other parts of the metropolis. Following the lead of Frederick Accum, he also began to look into food adulteration, and these activities brought him to the attention of Thomas Wakley, the reforming editor of the *Lancet*, who published his exposés, and of Chadwick and the GBH, which employed him as an analytical microscopist for most of the remainder of the decade.[46] Hassall examined water from the Thames itself, taken from the reaches where the companies got their supplies, waters from the outlets of sewers flowing into the Thames, and finally samples of the water the companies distributed, taken either from their reservoirs or from standpipes. Like Smith, he justified microscopy on the grounds that there were no chemical means to distinguish harmful from safe organic matter.[47]

Hassall found that all the waters contained microscopic life but his hope was to discover characteristic floras and faunas (and other distinct microscopical characteristics) of waters taken from different aquatic environments or treated in different ways. In a few cases he had some success. Water from the inner London Thames had lots of paramecia. Sewer water had worms, black carbonaceous matter, and wheat husks and other materials able to pass unscathed through the digestive tract. Examination of the companies' waters produced no surprises. Several contained traces of sewage yet Hassall was not able to recognise a distinct flora and fauna for each company as he had hoped to. On the contrary, there was great variation in the living things found in different pipes served by the same company.[48]

On the assumption that what worked for Chadwick and the General Board of Health would work for them, Homersham's London (Watford) Spring Water Company commissioned Edwin Lankester, a well-known sanitarian and naturalist, and Peter Redfern, lecturer in Physiology at King's College, to do a third microscopical examination of London's water in 1852. They too claimed that microscopy was superior to chemistry, and proclaimed theirs as the most complete microscopical survey to date. Their reports did have a sobriety lacking in Smith's and Hassall's; they made their point through charts of the numbers of species in different waters (the Watford spring had the fewest) and of the presence or absence of particular species in the different waters.[49]

As pioneering aquatic ecology or as a progressive step in the move from a chemical to a biological definition of the agents of disease, the work of these microscopists may seem important. As an approach to water analysis it was a failure, and the great problem for microscopists between 1850 and 1852 was how to respond to well-founded 'so what?' For their results to be useful in making water policy they had to make two assumptions. The first was that a constant relationship existed between an aquatic environment and the organisms that populated it. By the time Hassall testified before the Board of Health (probably in March or April 1850) he was able to suggest some characteristic organisms of ponds, streams, and lakes. Yet even supposing such a relationship existed and the details of it could be worked out, it still offered no new information: since everyone knew the Thames received sewage, to find in its water organisms characteristic of sewage would neither enlighten nor surprise policy-makers.

The second assumption was of a third, hidden element in the correlation: that the distinctions one made with the microscope somehow corresponded to the presence or absence of whatever it was that made water more or less unhealthy. One had either to suppose that some form of life was directly responsible for the harmfulness or that through microscopy one could make finer distinctions among waters than one could with chemistry, and that one of these fine classes would turn out to correspond to harmful water, even though the organisms in that class might not themselves be harmful. Smith, Hassall, and later microscopists who imitated them made both assumptions but they had great trouble justifying either.

The microscopists and their allies offered three resolutions to this problem of significance. These can be designated as the weak, the moderate, and the strong interpretation, according to the magnitude and immediacy of the danger each indicated in a biologically impure water supply. The interpretations were not mutually exclusive and many writers, including Chadwick, used all of them. In fact they formed a concentric series of rhetorical defenses. If the strong interpretation fell, one could still fall back on the moderate position or even the virtually invulnerable weak position.

The strong interpretation was that certain organisms were themselves the exciting causes of diseases, including epidemic diseases like cholera. The idea was not so much a portent of the germ theory of disease (whose origins would lie in pathology and fermentation theory rather than microscopy) as it was a return to a much older animalcular theory of disease. Neither Hassall nor any of the other reformers (with the exception of William Budd) was prepared to assign a particular bug to a particular disease; they were pointing out only that there were numerous diseases known to be caused by parasitic worms and fungi, and hence that it was plausible to think that various of the organisms in the London water supply might be capable of 'attacking the human frame from within,' as Hassall put it. Hassall himself believed that there was a germ of cholera, probably water-borne.[50]

There was, in fact, a likely candidate. William Budd had claimed to have identified a cholera fungus the previous fall. On 26 September 1849 he had written to the *Times* to call attention to the important discovery by the Bristol microscopists Frederick Brittan and J G Swayne of what Budd insisted was the cause of cholera, a fungus found in the excreta of cholera victims. Criticism came quickly. A B Granville, of mineral water fame, pointed out that Budd had offered no proof that the organism was the cause of cholera; it seemed more likely to be commonly but not causally associated with the disease. George Busk reported that the investigation of the London Microscopical Society showed Brittan's organisms to be neither exclusive to cholera nor in all cases even organisms.[51]

It may be true, as Margaret Pelling has argued, that the London medical establishment was overly quick to judge.[52] But along with its other failings, the animalcular/fungal theory in too many ways seemed not the kind of theory people were looking for. It offered a monocausal explanation when most medical men still understood disease multicausally, it failed (or appeared to fail) to come to grips with what to writers like Granville seemed the most important questions: why the disease appeared when and where it did, why it attacked some and not others. Those writing on the need for better water sometimes mentioned the possibility that some among the population of microscopic water organisms might cause disease, but they did not attach much weight to that possibility.[53]

The second interpretation of the significance of microscopic life, the moderate version, was that the creatures themselves were not harmful, but were reliable indicators of something that was. According to most authorities, this was decomposing matter. It was well accepted even before Pasteur that the function of at least some invisible animals and plants was to act as scavengers, purifying the world of matter that would otherwise become dangerous through its decay.[54] The idea was ancient, yet nicely adapted to pre-Darwinian biological thought, and especially to a natural theology of water purification in which organisms were perfectly adapted to their environments; each source of food had its consumer and every predator its prey. In the early '50s this view gained renewed visibility through the work of the chemist Robert Warington on the balanced aquarium, in which scavenging snails assimilated decomposing plant and animal matter and kept the water pure.[55] Because nature was so well balanced, discovery of the dominant species of microorganisms was assumed to indicate the presence of the particular food of this organism. Species that thrived on decaying organic matter signalled the presence of such matter. As Hassall put it, 'A knowledge of what constitutes the food of the infusoria has practical bearings upon the purity of the water of no inconsiderable importance.'[56]

Yet Hassall had trouble making this argument do any work. Almost everyone involved in the 1849–52 controversy, including scientists who testified in defence of the water companies in the 1851 and 1852 committee hearings, agreed that microorganisms were a sign of decaying matter (though a few noted that even the purest water contained some microscopic life). Not everyone saw the same implication, however. Where Hassall and the reformers saw organisms as proof of impurity, defenders of the companies saw them as proof of its absence, arguing that the microorganisms could be assumed to multiply to consume food as rapidly as that food appeared. Hence while the organisms might indicate the presence of impure matter, they also indicated the occurrence of purification. Had the impure matter in the water not been purified, the organisms would have been unable to survive, they pointed out: even Hassall's experiments showed that water saturated with hydrogen sulphide, which many regarded as the principal poison produced from rotting matter, killed microorganisms, just as a hydrogen sulphide atmosphere killed caged birds exposed to it.[57] There was a response to this objection: the organisms might not always be caught up in their purification operations, but this was not taken very seriously, since any really foul conditions would be at least as harmful to microorganisms as to humans.[58]

Vile Bodies

What with their strongest interpretation ignored for lack of evidence, their moderate interpretation collapsing under the weight of its own ambiguity, most of the reformers ended up relying on the third, 'weak,' interpretation: that the finding of microscopic life in water rendered that water unsuitable for a public supply on the grounds that, as Smith put it, microorganisms were 'disgusting.' It was also argued that if those most subject to the effects of urban squalor were to be persuaded to give up gin for water, the water had to be at least minimally appetizing.[59] The interpretation was 'weak' in two senses: first, it admitted that discovering microscopic life did not convey any additional analytical information—one knew nothing more about whether there really was anything dangerous in the water. Second, it could be (and was) objected to on the grounds that visceral reactions were too subjective a basis for changing water supplies. Yet however weak the argument, it was in practice the strongest of the interpretations.

Directly or indirectly, consciously or unconsciously, all the microscopists implied that there was something disgusting about living things in one's drinking water. (Lankester and Redfern, however, were a great deal more careful than the others in presenting their investigations as contributions to natural history.) The strongest imagery was Hassall's. By including graphic descriptions of the river banks from which he had taken samples, Hassall managed to convey the idea that any living thing in the waters that lapped these banks must also be impure. In his description of microscopic life of the inner London Thames, for example, he alluded to the 'carcasses of dead animals, rotting, festering, swarming with flies and maggots,' that lined its banks.[60] Even the water supplied by the New River Company from springs near Hertford was vitiated during its slow flow toward London; in one place it received water from an 'unclean and weedy ditch' in which algae were 'rising up into the water like clouds, and affording a nidus for the shelter, growth, and development of entomostracae, Infusoria, etc.'[61]

It worked. In parliament Sir Benjamin Hall used Hassall's drawings to assail opponents to water reform, saying that 'he never saw such odious, ugly things as his Hon Friend's constituents [in Lambeth] were continually eating and drinking.'[62] The disgust in Hassall's descriptions was taken up in reviews in medical and literary periodicals. Charles Kingsley noted that the supplies 'swarm with living animalcules.' Having presented his versions of the moderate and strong interpretations of these creatures, he went on to write of the 'animalcules which haunt the sewer mouths' and the 'filth-bred monsters'; of the River Lea, 'swarming with organic life'; of London's cisterns, each 'an alembic for further putrefaction, further multiplication of these wriggling monsters,' and finally of the deserved fate of Londoners for tolerating such a water supply:

> you are literally filled with the fruits of your own devices, with rats
> and mice and such small deer, paramecia and entomostraceae, and
> kicking things with horrid names, which you see in microscopes at the
> Polytechnic, and rush home and call for brandy—without the water—
> with stone, and gravel [i.e. bladder stones, attributed to hard water],
> and dyspepsia, and fragments of your own muscular tissue tinged
> with your own bile [another of Hassall's microscopical discoveries].[63]

The defenders of the water supply saw such arguments as the sacrifice of reason to emotion. Alfred Swaine Taylor, at the height of his career and London's leading forensic chemist, admitted that water with things growing in it 'looks offensive,' and even that 'the water acquires a taste and is injured,' yet attached little significance to these factors. There was no evidence that consuming 'microscopic animals' had ever done anyone any harm, he insisted, and added, 'I believe we should eat nothing and drink nothing if we used the microscope before hand to settle the point.' Swallowing animalculae was not more harmful than eating fish.[64] W T Brande (now an ally of the companies) and the engineer Thomas Hawksley (a former ally of Chadwick's who had left the fold) took much the same line. Brande maintained that it made no difference whether the organic matter in water was alive or dead.[65] Hawksley observed that if Londoners were going to be so perversely 'fastidious' they would have to pay for that luxury.[66] In their view, science, chemistry in particular, had progressed so far that scientists could now say exactly what was harmful in water. To cater to ignorance and prejudice would be to regress.[67]

At odds then were two opposing frameworks for assessing water quality. Those who found Hassall's exposés persuasive took the view that instincts were a reliable guide to safety. This was an ancient idea, yet a central tenet of the early public health movement. Numerous tracts argued that the Creator had endowed us with instinctive revulsions to guide us to dispose of wastes correctly.[68] Poised against this was the view of Brande and the chemists, who saw in science a way to test our instinctive judgments, to show through reason that what appeared unsafe might be safe, but also to reveal hazards we would not otherwise have recognized. As counsel for the New River Company put it in a question to Taylor in 1851, 'however *offensive to the imagination* it may be that privies should be emptied into the river from which water is taken to drink, *practically* the effect depends on the quantity of water, and the facility it has for decomposition?'[69]

A Role for Chemistry

The utility of chemistry and the credibility of chemists were under scrutiny during the 1849–52 controversy quite as much as were microscopy and microscopists. If the chemists fared better than the microscopists it was probably owing to their having a longer history of useful service in advising governments on technical matters and to their being the professionals who had traditionally been called on to assess the quality of water. During the 1849–52 controversy, chemistry—as a body of knowledge and techniques and as a group of practitioners—saw service mainly on the side of the water companies. They hired the most prominent of London's consulting chemists: Brande, A S Taylor, Arthur Aikin, and J T Cooper. To be sure, the reformers employed chemists—Chadwick relied on Robert Angus Smith and Lyon Playfair, and consulted A W Hofmann; the Watford Spring Company hired Thomas Clark, the Aberdeen chemistry professor, and John Stenhouse and Dugald Campbell of St Bartholomew's Hospital, while W A Miller of King's College, J E D Rodgers of St George's Hospital, and Harman Lewis of the Westminster Hospital advised a company desiring to supply parts of south London with Wandle water.[70] Yet none of the reformers' chemists had any strong claims to make; all appeared confused about how to document the impurities in water. The neutral investigation commissioned in 1851 by the Home Office also employed chemists: Miller and Hofmann as well as Thomas Graham of University College, another prominent academic chemist.

There was no deep-seated scientific reason why chemistry should come down so heavily on the side of the companies; the concepts and methods of that science would not automatically lead one to conclude that London's water was good. Instead, the splitting of science into rival camps of microscopists and chemists had more to do with the strategy of the reformers. Of their two main approaches to the question of impurity, one, the microscopical approach, denied the authority of chemistry, while the other, hardness, scarcely required it. Chadwick did have theoretical grounds for doubting the capacity of chemistry to monitor the changes he was most interested in, but the most important factor probably was the rules of evidence adopted by the select committees of 1851 and 1852.[71] The committees refused to grant the General Board of Health standing as an interested party, and it was unable therefore to present testimony to rebut the chemistry-based arguments of the companies' witnesses.

The companies' chemists put up a united front. While in 1828 there had been great uncertainty about how to measure and even how to conceive the insalubrity of water, by the early 1850s ideas of how water might be harmful were converging. Ironically, the impurity that seemed most compatible with the authority of chemistry was decaying matter, precisely the poison around which Chadwick had promoted the sanitary movement. In 1828 Pearson and Gardner had defended the companies' waters on the grounds that they contained no unusually large concentrations of medically active salts and it had been the companies' critics, especially William Lambe, who had seized on decaying organic matter as the harmful constituent and condemned water upon finding traces of organic matter in it.

By 1850 there was widespread acceptance not only of the primacy of decaying matter, but of Chadwick's particular conception of its action. Unlike several other sanitarian writers who took Liebig's view that decomposing matter was harmful because it generated decomposition in the victim's body, Chadwick and his medical theorist Thomas Southwood Smith were mainly concerned with the products of putrefactive decomposition.[72] These they regarded as poisons which worked either as predisposers in weakening a person's resistance to an exciting cause of disease, or as exciting causes themselves. They were mainly concerned with the malodorous products of anaerobic decay: hydrogen sulphide, phosphoretted hydrogen, some ammonia compounds. In Chadwick's view these compounds, derived either from decomposition taking place in the water, or absorbed from the foul urban atmosphere, constituted the harmful substances in water.

Using adaptations of mineral water chemistry, Brande, Taylor, and Cooper tested for these compounds, especially hydrogen sulphide, which they saw as an exclusive indicator of decayed animal matter and thus of sewage matter, which many sanitarians were beginning to regard as more dangerous than plant matter.[73] Hydrogen sulphide was easy to detect, qualitatively by its odour or by the black film that formed on silver exposed to it, quantitatively through the formation of a brown precipitate with lead acetate.[74] Yet Brande, Taylor and company went further, again founding their argument in orthodox Chadwickian medical theory: not only was hydrogen sulphide an indicator of sewage, it was what made sewage dangerous, and therefore they were directly measuring the harmful matter in sewage. Taylor even claimed to have isolated the chemical substance responsible for dysentery.[75]

It need hardly be said that the quantity of hydrogen sulphide that Taylor and his colleagues found in the water was not enough to condemn it.[76] Yet just as unsettled questions undermined the reformers' attempts to draw strong conclusions from microscopical evidence, there were also glaring ambiguities in interpreting the hydrogen sulphide results. The results had to be reconciled with the obvious fact that the rivers Thames and Lea received enormous quantities of organic matter in sewage. Much of this organic matter remained in the water. Water from near the Tower of London (well below any of the intakes) contained eight grains organic matter per gallon, but even the water the West Middlesex company took from near Barnes contained three to four grains.[77] All this organic matter was not dangerous, insisted Taylor and his associates, since it was either not decomposing or not derived from animal sources (or both). While these answers did forestall serious governmental meddling with the water supply, they did not really resolve the issue of how to measure water quality, even in the context of the sanitary science of the early 1850s.

The confidence and brashness of the companies' chemists contrasts sharply with the sober and tentative conclusions of the only neutral investigation during the controversy, the 1851 report to the Home Secretary by Graham, Miller, and Hofmann. In January 1851, Sir George Grey, the Tory Home Secretary, had consulted the three out of a mounting sense of bewilderment. Buffeted by claims of the dangers and benefits of hard water or animalculae and of the amounts of water that could be obtained from various sources at various costs, he sought the help of the three chemists.[78] The team followed the analytical protocol Hofmann was establishing at the Royal College of Chemistry with the significant exception of adding an elemental analysis of the organic residue to determine the nitrogen–carbon ratio and hence to estimate how much of the organic matter derived from potentially harmful animal matters, such as sewage, 'the existence of nitrogen ... being generally supposed to imply ... animal origin.'[79] One should not make too much of this: there was still no definitive epidemiological link between disease and human faecal contamination, only a slow shifting of sensibility to the view that among types of filth, some types were significantly more dangerous than others. The London waters contained between three hundredths and one tenth of a grain per gallon of such organic nitrogen—'a minute and probably unimportant portion of animal organic matter,' they concluded.[80]

While the three chemists found a number of 'serious evil[s]'— hardness, turbidity, contamination with vegetable organic matter— they found greater problems, mainly with respect to quantity, in each of the alternatives, and because of these doubts about the alternatives, their report was taken as an endorsement of the existing supply, though in fact it was highly ambivalent.[81] One gets the feeling that the chemists were concerned about the water but unable to find an analytical warrant for their concern. Their report nicely captures the state of water analysis in the early '50s. There was a deep-seated feeling that something was wrong with a great many water supplies. Some believed water contributed somehow to the spread of cholera, others decided the palpable impurities in their water were no longer tolerable, still others simply felt a long-standing anger at being at the mercy of the water monopoly. Yet it was agonizingly unclear which of these problems was most serious, why and how they were serious, and what kinds of remedies were workable. It was not yet clear just what an urban water supply should be, who should control it, how it should be paid for, how water should be distributed, or what standards of quality it ought to meet. Some of the companies' opponents treated water as a necessity of life, and hence a basic human right. The companies, on the other hand, and the government, viewed water supplies as if they were some sort of new category of railway (indeed the agency charged with administering the 1852 Metropolis Water Act was the Board of Trade's Railway Department), that is, as projects in which the public good guided and tempered but did not ultimately control the visions of capitalists.

There was an analogous lack of consensus as to what standards of quality a public water supply ought to meet. W H Wills insisted that 'all chemists agreed that a water containing from eight to ten grains of sulphate of magnesia or soda, to the imperial gallon, is best suited for ... domestic purposes.'[82] But even if chemists agreed on that (and many probably did not) they agreed on little else. Water was objected to both as being too impure and too pure: very soft water was objectionable because it might poison the public with lead dissolved from lead pipes. The 1849–52 water controversy, along with those that preceded it and followed it, was thus at its deepest level a problem of determining what a public water supply was to be. The companies' victory in part reflects the fact that they represented the *status quo* of water supply. Yet it was also due to the acceptance of analytical chemistry over analytical microscopy and the hardness standard. In failing to take seriously the standards and analyses of Chadwick, Hassall, and the other reformers, public and government were acquiescing to the dominance of chemistry— despite great uncertainty about what, if anything, in water might be harmful, despite the shallow, glib, and self-serving performance of the companies' chemists who appeared to have invented something to analyse for, which they could then show to be absent.

To understand why chemical standards and processes should be so compelling we need to reflect once more on the status of chemistry in the mid nineteenth century. Its stock was high: the body of chemical knowledge had been thoroughly reorganized; discoveries of new elements and compounds, and new explanations of processes—especially the biological processes of respiration, nutrition, and recycling—had made chemical knowledge a far more prestigious part of natural science than it had been a hundred years earlier. Through the efforts of men like Brande, chemistry was becoming increasingly visible in technology and in medicine. Most importantly chemistry was being looked to, especially in biology and medicine, and especially in Britain, as the science that offered the fundamental explanation of phenomena. In an 1851 review of Hassall's *Microscopical Examination*, Nathaniel Beardmore, the engineer to the River Lea Trust (an organization with a vested interest in the East London Water Company and hence in the existing water supply) wrote that Hassall 'would have done great benefit to science by his facts, if he had given us chemical analyses of the various waters experimented upon, so that the constituent elements of the Fungi, algae, Diatomaceae, etc. should be determined; and we should then have known the chemical nature of the inhabitants of our waters.'[83]

To most modern readers the argument will seem absurd. It calls for us to sacrifice a specific and detailed knowledge for a far more general knowledge. It embraces the idea that it is only as amounts of carbon, oxygen, hydrogen, nitrogen, and a few other elements, that organisms have significance in the world, and is reminiscent of childish arguments that human beings are worth the prices of the trace elements their bodies contain. Yet Beardmore's suggestion is nothing more than a brazen version of Brande's 1851 statement that it made absolutely no difference whether the organic matter in water was alive or not, and in gentler form it was the perspective of most of the chemists who testified. To them health and disease were states of chemistry, chiefly of the chemistry of decaying organic matter and the products it yielded.

1   In *HLRO, Minutes of Evidence, Commons, 1878*, v. 5 (Cheltenham Corp Water Bill), 12 March 1878, p 88.
2   S Finer, *The Life and Times of Sir Edwin Chadwick*; R A Lewis, *Edwin Chadwick and the Public Health Movement, 1832–1854*.

3   Finer, *Chadwick*, p 390. The investigation had been under way since the summer of 1848. Initially it was in the hands of the Metropolitan Sewers Commission; the GBH took over the project in the fall of 1849 though without legal mandate to do so.

4   Richard Lambert, *The Railway King, 1800–71: A Study of George Hudson and the Business Morals of his Times* (London: G Allen and Unwin, 1934).

5   *GBH MWS*, Appendix II, pp 3–15.

6   *Times*, 13 Dec 1848, 8d.

7   *Times*, 11 Oct 1849, 4b.

8   *Times*, 11 Oct 1849, 4b. It is probable that Chadwick orchestrated this editorial, with the aid of his associate, F O Ward (cf Lewis, *Chadwick*, p 260; Finer, *Chadwick*, pp 390, 393–4).

9   *Times*, 25 Dec 1849, 3a; 26 Dec 1849, 7a; 27 Dec 1849, 6d; 28 Dec 1849, 6c; 29 Dec 1849, 5c.

10   *Times*, 23 Oct 1849, 8e–f; Lewis, *Chadwick*, p 271.

11   *Times*, 11 Dec 1849, 4e; 12 Dec 1849, 5d; 1 Jan 1850, 5a; 5 Feb 1850, 5a; 12 Feb 1850, 8b; 13 Feb 1850, 8f; 18 Feb 1850, 4f; 19 Feb 1850, 6c; [W H Wills], 'The Troubled Water Question,' *Household Words* 1 (1850): 49.

12   Hassall noted in his autobiography that his approach had a decided effect in heating up the controversy (*The Narrative of a Busy Life. An Autobiography* [London: Longmans, Green and Co., 1893], p 68).

13   Hassall, *A Microscopical Examination*, pp 9–10; *S C Metropolis Water Bill, 1851*, QQ 3937–38.

14   But see [Charles Wall], 'Metropolis Water Supply,' *Fraser's Magazine* 10 (1834): 562, 566–7.

15   *Times*, 13 Nov 1850, 4e; 16 Nov 1850, 5a; 20 Nov 1850, 4d; Lewis, *Chadwick*, pp 263–5.

16   Thos Graham, W A Miller, and A W Hofmann, 'Chemical Report on the Supply of Water to the Metropolis,' pp 386–7. See the comments in the *Times*, 27 June 1851, 5b.

17   Lewis, *Chadwick*, pp 269–78; *Times*, 8 March 1851, 4e; 28 May 1851, 4d; 4 June 1851, 4d; 7 June 1851, 5a; 9 June 1851, 4f.

18   Lewis, *Chadwick*, pp 326–8.

19   *Times*, 23 Oct 1849, 8e; 13 June 1851, 8f; 13 Feb 1850, 8f; *Hansard's Parliamentary Debates*, 17 June 1852, c 862. Sir Benjamin Hall got double duty from the 'larger and fatter' quip, using it both in a Marylebone meeting in 1851 and in the debate on the 1852 bill.

20   *Times*, 23 Oct 1849, 8e.

21   *Times*, 11 Dec 1849, 4e.

22   *Times*, 14 Feb 1850, 8f.

23   *Times*, 21 April 1851, 4d; *GBH MWS*, pp 16–17.

24 Margaret Pelling, *Cholera, Fever, and English Medicine*, pp 146–202, 207–19; Budd in *Times*, 26 Sept 1849; A B Granville in *Times*, 28 Sept 1849, 5d; *Times*, 24 Oct 1849, 4a. See also [Wall], 'Metropolis water supply,' p 569.

25 A miasma properly so called was an exciting cause of disease emitted by decaying vegetable matter from a swamp. What concerned most writers in 1849 were predisposing causes emanating from an unhealthy environment (George B Wood, *A Treatise on the Practice of Medicine*, 2nd edn [Philadelphia: Grigg, Elliott & Co, 1849], pp 139–46; Charles Williams, *Principles of Medicine* ed with additions by M Clymer [Philadelphia: Lea and Blanchard, 1848], pp 61–2). According to orthodox Chadwickian theory, severe predisposing causes might become exciting causes (T Southwood Smith, *Treatise on Fever* [Philadelphia: Carey & Lea, 1830], pp 381–5).

26 The importance of predisposing causes as distinct from miasmatic explanations has been insufficiently appreciated. Even Pelling (*Cholera*, p 59) finds it hard to accept the primacy of predisposition as anything other than a ruse to persuade the poor to give up bad habits. Yet the concept of predisposition was central to orthodox medicine as well as to Chadwickian sanitarianism (cf Wood, *Treatise*, pp 126–39; Williams, *Principles*, pp 22–67; W B Carpenter, 'On Epidemic Diseases,' *Braithwaite's Retrospect* #27 [July 1853]: 17–21; R A Smith, 'On the Air and Water of Towns,' *Report of the 18th Meeting of the BAAS*, [Swansea] for 1848 [London: J Murray, 1849], pp 16–18; Wm A Guy, *Hooper's Physician's Vade Mecum: or Manual of the Principles and Practice of Physic*, 6th edn [London: Renshaw, 1858], pp 545–6; 'Water Supply and Disease,' *Builder* 12 [1854]: 57–8).

27 *Times*, 23 Oct 1849, 8e. Challis' 'decisive influence,' (*Times*, 11 Dec 1849, 4e) can be interpreted in the same way.

28 *Times*, 14 Sept 1848, 4e; 18 Sept 1849, 5f.

29 Ward, 'Metropolitan Water Supply,' p 482.

30 R D Grainger in *Report of the GBH on the Epidemic Cholera of 1848 and 1849*, App B, p 94.

31 *Report of the GBH on the Epidemic Cholera of 1848 and 1849*, pp 90–4; Ward, 'Metropolitan Water Supply,' p 482. The GBH report also argued, as did Snow, that oral rather than pulmonary intake was likely in light of the rapidity of action. See *GBH MWS*, pp 16–17, 34, 47. See also the testimony in *GBH MWS*, appendix III of L Playfair, p 78; H Garvin, p 62. Chadwick long continued to regard the role of water in setting up cholera as predisposing only (comments in E Byrne, 'Experiments on the Removal of Organic and Inorganic Substances in Water,' p 34).

32 Pelling, *Cholera*, pp 205, 218–9, 226–7; A B Granville to *Times*, 28 Sept 1849, 5d; 'Review of Snow, On the Mode of Communication of

Cholera,' *Lancet*, 1849, ii, p 318; E A Parkes, 'Mode of Communication of Cholera,' *British and Foreign Medico-Chirurgical Review* 15 (1855): 449–63; 'Review of Snow, Mode of Communication,' *Builder* 13 (1855): 49–50; B W Richardson on Snow in *JPH&SR* 1 (1855): 130–40.

33   John Sutherland in the *GBH MWS*, Appendix III, pp 7–8; Saunders, *A Treatise on Mineral Waters*, pp 390–3. Saunders' book was the major source of articles in J M Good's *Pantologica* (1813), s.v. water, *The British Cyclopedia* (1835), s.v. water; *Rees Cyclopedia*, s.v. water; 'Domestic Chemistry III (Domestic Waters)' *Knight's Penny Magazine* 7 (1838): 54.

34   Edwin Chadwick, *Report on the Sanitary Condition of the Labouring Population of Great Britain* ed with an introduction by M W Flinn (Edinburgh: Edinburgh University Press, 1965), p 79.

35   Chadwick, *Sanitary Report*, p 120.

36   *Ibid*, pp 148, 139, 422.

37   *GBH MWS*, p 312.

38   *GBH MWS*, pp 50–9, 64–5, 66–9. The government chemists, Graham, Miller, and Hofmann, who likewise put great stress on softness, agreed that soft water made more tea, but hard water made better tea (Graham, Miller, and Hofmann, 'Chemical Report,' pp 388–89).

39   *GBH MWS*, pp 70–80; *Times*, 30 Dec 1848, 4c–d.

40   *GBH MWS*, Appendix III, p 154; Playfair in *Builder* 9 (1851): 765; Ward, 'Metropolitan Water Supply,' pp 473–5; *Times*, 6 Jan 1851, 5a; Samuel Homersham, *Review of the Report by the GBH on the Supply of Water to the Metropolis*; contained in a report to the directors of the London (Watford) Spring Water Company (London: John Weale, 1850); Graham, Miller, and Hofmann, 'Chemical Report,' pp 381–5, 387–94; Edwin Lankester, 'Drinking Waters of the Metropolis,' p 467. See also Charles Kingsley, 'The Water Supply of London,' pp 243–4.

41   *GBH MWS*, p 82; *Times*, 4 June 1850, 5e; F Mowatt in *Hansard's Parliamentary Debates*, 17 June 1852, c 845. On softening see Lankester, 'Drinking Waters,' p 467; Graham, Miller, and Hofmann, 'Chemical Report,' pp 393–5.

42   *RCMWS*, 1828, pp 36–58. See also [Wall], 'Metropolis Water Supply,' pp 566–7.

43   Smith, 'On the Air and Water of Towns,'* pp 16–31.

44   *GBH MWS*, Appendix III, pp 94–5.

45   *GBH MWS*, p 41. On Smith see A Gibson and W V Farrar, 'Robert Angus Smith, FRS, and Sanitary Science,' *Notes and Records of the Royal Society of London* 28 (1974): 241–62.

46   On Hassall see his *Narrative of a Busy Life**; E G Clayton, *Arthur Hill Hassall: Physician and Sanitary Reformer* (London: Balliere, Tindall, and Cox, 1908); and Ernest A Gray, *By Candlelight: The Life of Dr Arthur Hill Hassall, 1817–1894* (London: Robert Hale, 1983).

47 Hassall, *Microscopical Examination*, p 1.

48 Hassall, *Microscopical Examination*, pp 10-16.

49 E Lankester and P Redfern, *Reports to the Watford Company*.

50 Hassall, *Microscopical Examination*, pp 28, 31, 57. Lancet Analytical Sanitary Commission, 'Results of analyses of the solids and fluids consumed by the public,' pp 222–3; Hassall, 'Memoir on the microscopical examination of the water,' p 235; Hassall in *GBH MWS*, Appendix III, pp 57–8; Redfern in *SC Metropolis Water Bill, 1852*, QQ 11368–11393; Pelling, *Cholera*, p 220.

51 *Times*, 26 Sept 1849, 4f–5a; 27 Sept 1849, 6d; 28 Sept 1849, 5d; 24 Oct 1849, 4a; Pelling, *Cholera*, pp 163–88.

52 Pelling, *Cholera*, pp 175–78.

53 Pelling, *Cholera*, pp 173–74; Granville in *Times*, 28 Sept 1849; Kingsley, 'Metropolis Water,' p 242. Even J G Swayne ('An Account of Certain Organic Cells peculiar to the Evacuations of Cholera,' *Lancet*, ii, 1849, pp 368–71) regarded predisposition as central. Without debilitation due to general filth these cells would have no effect: in a normal stomach they would simply be digested. Hence their discovery shed little light on the cause of cholera. See also comments of Budd, Snow, and Lankester, pp 371–2, 413, 460.

54 C Hamlin, *What Becomes of Pollution*, chapter 4.

55 Hamlin, 'Robert Warington and the Balanced Aquarium' pp 131–53.

56 *GBH MWS*, Appendix III, p 40.

57 Hassall, *Microscopical Examination*, p 30; *GBH MWS* Appendix III, p 41.

58 Lankester, 'Report to the Watford Spring Company,' p 3. See also Gideon Mantell, *Thoughts on Animalcules; or a Glimpse of the Invisible World revealed by the Microscope* (London: John Murray, 1846), p 85; *S C Metropolis Water, 1852*, Q 552.

59 R A Smith and Thomas Clark in *GBH MWS*, Appendix III, p 97, 172.

60 Hassall, *Microscopical Examination*, p 6.

61 *Ibid*, pp 19 24, esp. p 20.

62 *Hansard's*, 3rd series, 122, c 861.

63 Kingsley, 'Metropolis Water,' pp 241–6. See also *Lancet*, 23 Feb 1850, p 246; 'Metropolitan Water Supply,' The *Builder* 9 (1851): 494; Ward, 'Metropolitan Water Supply,' pp 493–4; W O'Brien, 'The Supply of Water to the Metropolis,' 91 (1849–50): 382; *Times*, 22 May 1850, 4d; 27 June 1851, 5b. According to Wall, the same tactic had been used in the controversy of the 1820s: he wrote of meetings 'where microscopic entomologists attended and exhibited specimens, either from nature or in large drawings, of the many frightful hydra-headed and millipede insects taken out of the water-cisterns of the metropolis. After which they were placed in the windows of the picture-shops throughout the

126 A Science of Impurity

towns, in order to drive the few remaining water-drinkers to the public-house' ([Wall], 'Metropolis water supply,' p 562).

64 *S C Metropolis Water, 1852*, Q 539. See also Cooper, QQ 690–2; *S C Metropolis Water, 1851*, QQ 12200, 12209.

65 Taylor, *S C Metropolis Water, 1851*, Q 3842.

66 *S C Metropolis Water, 1852*, QQ 1248–50. On Hawksley's relations with Chadwick see Lewis, *Chadwick*, pp 120, 132–3.

67 W T Brande, 'Analysis of the Well-Water at the Royal Mint,' pp 350–1.

68 C Hamlin, 'Providence and Putrefaction,' pp 385–411.

69 *S C Metropolis Water, 1851*, Q 12359, italics mine.

70 For Brande, Taylor, Aikin, Cooper, Clark, Stenhouse, Campbell, Miller, Rogers, and Lewis, see *S C Metropolis Water, 1852*; for Smith, Playfair, and Hofmann, see *GBH MWS*.

71 *GBH MWS*, pp 31–4.

72 Pelling, *Cholera*, pp 1–80, 101–7; Finer, *Chadwick*, pp 297–8; John M Eyler, *Victorian Social Medicine*, pp 97–107.

73 *S C Metropolis Water, 1852*, QQ 509, 57, 625–6, 645, 789–91; *S C Metropolis Water, 1851*, Q 693.

74 *S C Metropolis Water, 1852*, QQ 417, 435, 790. The test was commonly used by mineral water chemists for sulphurets (Nicholson, *British Encyclopedia*, 1809, s.v. water).

75 *S C Metropolis Water, 1851*, Q 10268.

76 *Ibid*, Q 428.

77 *Ibid*, 1852, QQ 576, 790.

78 See Graham, Miller, and Hofmann, 'Chemical Report,' pp 375–6, for the text of Grey's instructions.

79 *Ibid*, p 380.

80 *Ibid*.

81 *Ibid*, p 386. See the *Times*' reaction (27 June 1851, 5b).

82 W H Wills, 'The Troubled Water Question,' *Household Words* 1 (1850): 52.

83 [N Beardmore], 'Water-Supply,' *Westminster Review* 54 (1851): 190.

5   Nitrogen and Nihilism, 1852–68

Sanitary pursuits produce a kind of intoxication which raises the intellect of a genuine theorist above the vulgar rules of induction.[1]

Saturday Review

It is hard to escape the conclusion that the arguments of the water reformers in the 1849–52 controversy backfired. The attractions of soft water had turned out not to be compelling for very many people, especially in light of opposing arguments that soft water dissolved lead in pipes and thus could be poisonous and that people who drank soft water got soft bones.[2] As for the microscopic inhabitants of the water, however disgusting they might be, there was no rational reason to object to them. Hassall himself accepted two key arguments of the companies' chemists: that the activity of microscopic organisms was one of the main ways dirty water became pure, and that the presence of such life was incompatible with the presence of anaerobic decomposition products, the sulphides and the ammonia that most people agreed were the most harmful materials in polluted water. As far as theories of water quality were concerned then, the companies were vindicated.

Yet there was never a time during the early and mid '50s when Londoners could sigh with relief that their water had passed some crucial test of purity and was henceforth safe. Whatever the theories might say, water consumers and medical men who had studied the habits of cholera knew better: the water was not good. In 1853–4 their fears were confirmed. Cholera returned to Britain and London before the companies had fully implemented the Metropolis Water Act of 1852. It was during the 1854 epidemic that John Snow carried out his famous epidemiological investigations and showed the disease to be transmitted by water specifically polluted by the excreta of a cholera victim. In one of these investigations Snow studied the prevalence of cholera in the area around London's Broad Street pump. He discovered that cholera developed only in households which used water from the pump. He was also able to show that the pump water had probably been contaminated with excreta from an employee of a nearby brewery who had the disease. Snow carried out a more extensive investigation of cholera in south London, where he discovered a high positive correlation between cholera and the service area of the Lambeth water company, which had not yet moved its intake upstream as required by the 1852 act. The correlation was especially striking in areas at the edge of the company's district where one side of a street was served by the Lambeth and the other by the Southwark and Vauxhall Company, which had improved its intake.[3]

These results appeared in the second edition of Snow's essay *On the Mode of Communication of Cholera* and historians have commonly regarded them as the proof that cholera was a water-borne disease. Some have also credited him with the recognition that the agents of the disease were specific living germs.[4] In fact Snow's work did not lead to the revolution in water science usually attributed to it. His epidemiological results were certainly striking, but others, including Chadwick, had found equally striking evidence linking outbreaks of cholera to contaminated water. Snow simply drew a stronger conclusion from his research in asserting that something in bad water was *the cause* of cholera, rather than an important contributory factor. Moreover, his 'germ theory' was so ambiguously stated and so heavily embedded in Liebigian pathology that it could not be distinguished as a new and significant hypothesis.

Nevertheless there was an enormously important change taking place in concepts of water quality and disease causation during the mid 1850s. John Snow was not the main instigator of these changes; his work both reflected and contributed to them. The changes were not mainly empirical—chemists and microscopists did not finally discover the harmful matters they were hunting. Instead, it was during this period that the modern concept of disease specificity and the corresponding emphasis on a single exciting cause of each disease were coming to be applied to the kinds of diseases bad water was thought to cause. The two ideas reinforced one another. If there were many species of morbid agents, each capable of acting as an exciting cause of disease, it seemed more likely (though by no

means necessary) that there was an equal number of specific diseases. Conversely, if the various filth fevers that sanitarians worried
about really were specific diseases it seemed likely that they were
specific because they had specific causes. Hence where Chadwick
and the early sanitarians had assumed that each outbreak of disease
was a version of a common filth disease and had felt little obligation
to restrict the number of operating causes in any particular case,
sanitarians in the 1860s and 1870s were beginning to look for single causes and tending to regard multi-factoral explanations as the
lumping together of an important, exciting cause with a collection
of less important predisposing causes.[5]

This transformation took place gradually among British sanitarians, and it never appeared as an 'all-or-nothing' dichotomy. On
the contrary, there was a great deal of middle ground occupied by
a great many sanitarians during the middle decades of the century.
This transformation is sometimes seen as the replacement of anti-
contagionism by the new contagionism of the germ theory, but that
was only one aspect of it. More significant than the change in ideas
of how diseases spread was the change in the concept of what a
disease was.

With this new style of explanation came a new conception of
the utility of water analysis (and new processes of analysis based
on that conception), new sets of arguments about how one ought
to infer the salubrity or harmfulness of the water from the results
of analysis, and new sets of standards public water supplies were
to meet. In regard to water quality, this transformation began in
the mid 1850s, though nearly twenty years would elapse before it
was relatively complete. No single individual was responsible for it,
but as Margaret Pelling has demonstrated, the central figure was
the German chemical theorist Justus von Liebig.[6] We can see this
transformation taking place in the ideas of John Snow, in the great
body of sanitary literature of the '50s and early '60s, and becoming
embodied in water analysis in the work of Liebig's pupil August
Wilhelm Hofmann, the professor of chemistry at the Royal College
of Chemistry.

The Threat of Zymosis

What Liebig proposed in his 1840 work on *Chemistry in its Application to Agriculture and Physiology* may seem a subtle and minor
variation on the filth theory of Chadwick and Southwood Smith,

but it had far-reaching implications. Liebig too believed that filth—decaying organic matter—caused disease. But he believed that its morbid action was due not to the poisonous substances it produced, but to the actual process of decay that was occurring and which could spread to the tissues of a human being and there reproduce itself. Liebig's theory of pathology was a special case of his explanation of organic decomposition. In his view organic matter was susceptible to decomposition when it was dead or weakened, when a vital force no longer kept its elements strongly bound into large organic molecules. If such susceptible matter—the tissues of a person's body, for example—came into contact with nitrogenous organic matter already undergoing decay (a 'ferment' in Liebig's terminology), it too would take up that particular process of decay.

In Liebig's view a great many diseases were properly regarded as peculiar forms of fermentation of parts of the body. One caught these diseases upon being exposed to matter undergoing the particular form of decomposition that characterized the disease. The theory accounted for the fact that the so-called filth diseases were not always present in proportion to the amount of filth, for the fact that epidemics represented particular species of disease, for the seemingly random generation of epidemics (a result of the chance appearance of a virulent form of putrefaction), for the production of contagious matter in the sick organism and for the transmission of the disease to others, either directly or through filth in the environment. In short, it explained most of the features that would be put forward as evidence for the germ theory.[7]

Through most of the 1840s this zymotic theory did not attract much attention. Most sanitarians took the view that the products of decomposition were themselves directly poisonous and that some kind of filth fever would occur sooner or later if one were continually exposed to an environment of decay. Among the earliest of the British converts to the zymotic idea (and the coiner of the term 'zymotic') was William Farr, who as the statistician in the Registrar General's Office was continually confronted with the puzzling data of differential mortality. Farr recognized that the laws of gaseous diffusion made it improbable that putrefactive products could exist in concentrations high enough to cause diseases, and he suggested that particulate agents must be responsible.[8]

A survey of British sanitarians in the late '40s or the '50s would have found a few confirmed Chadwickians, a few equally staunch Liebigians, and a great many either holding some combination of

the two views or ignorant of (or wholly uninterested in) such theoretical niceties. There was a great deal of middle ground. In 1848
for example P B Ayres observed that the sulphides of ammonia and
hydrogen did not cause fevers, though they clearly were harmful
to health. Ayres believed some sort of ferment must be responsible for fevers.[9] This view was contradicted by B W Richardson and
others who claimed on the basis of experiments on guinea pigs and
dogs that such poisons produced exactly the symptoms of typhoid
fever. Richardson tried his hand at developing a pathological explanation of how this happened, one based on the effects of excess
ammonia in the blood, but distinguished his explanation from the
Chadwick/Smith view. T Herbert Barker confirmed Richardson's
claim that the gases of decay caused typhoid symptoms, yet insisted
that a quite different zymotic process was responsible for typhoid
fever itself.[10]

By the late '50s the zymotic theory was becoming the dominant
explanation of the mechanism by which decomposing matter caused
such diseases as cholera and typhoid fever. Its ascendancy was likely
the result of a number of factors—Chadwick's fall from power, experience with the unpredictable waxing and waning of cholera, and the
failure of the Great Stink of the Thames of June and July 1858, the
most horrible stench in memory, to produce any unusual outbreak
of disease. On that occasion some medical commentators had come
to the conclusion that however offensive the Thames might become,
and no matter what might be the temporary consequences of exposure to the air near the river, there would not be an epidemic disease
unless a special form of decaying matter were present.[11]

John Snow's writings were another factor. In 1853 Snow had published a long essay 'On Continuous Molecular Changes, more particularly in their relation to Epidemic Disease.' This essay provides the
background for what has been seen as Snow's articulation of a germ
theory of cholera in the second edition of his better known work *On
the Mode of Communication of Cholera*, which appeared in 1855. In
the former work he followed Liebig in arguing that there existed a
class of processes of decomposition which included the processes that
were the essences of zymotic diseases. Like Liebig he pointed out
that a particular zymotic process could reproduce itself indefinitely,
unlike a normal process of chemical change. For this reason, wrote
Snow, 'the material cause of every communicable disease resembles
a species of living being.'

both ... depend on, and in fact consist of, a series of continuous molecular changes, occurring in suitable materials. The organised matter, as we must presume it to be, which induces the symptoms of a communicated disease, ... can hardly ever be separately distinguished, like the individuals of a species of plant or animal; but we know that this organised matter possesses one great characteristic of plants and animals—that of increasing and multiplying its own kind.[12]

While Snow admitted that the disease agent 'resemble[d]' a form of life or was probably similar to a cell, he would not say it was life.[13] From the view of the Liebigian pathology it was of relatively little importance whether the poison was living or non-living organic matter for the same kinds of processes of decomposition could go on in each. More important was the new view of the pathology and etiology of cholera: the idea that there was a unique, particulate poison that was swallowed and set up a process of decomposition within the victim's body which effectively reproduced the poisonous process of decomposition, both in the body and in its evacuations, thus permitting the disease to be transferred when another person ingested food or drink contaminated with those evacuations.

Besides making it clear that water contamination needed to be taken far more seriously, Snow's theory and the zymotic philosophy in which it was embedded had an enormous impact on the philosophical aspects of water analysis. Consider some of the implications of the switch from the Chadwick/Smith view of water contaminants as products of putrefaction to the Liebig/Snow view of contaminants as obscure putrefactions. From the viewpoint of water analysts, both those who used chemical means and those who hoped to develop microscopical analysis, a great deal was being given up. With the change from substance to process went much of the utility of quantitative measures of contamination. Those who held the old view could justly say that water was safer or more dangerous according to the concentration of putrefactive products in it, but what made the difference according to the zymotic view was some wholly uncharacterizable qualitative property. In some cases water heavily contaminated with decomposing material might prove relatively harmless while that with the tiniest trace might be lethal. The switch from matter to process could even be seen as denying utterly the relevance of chemical or microscopical analysis. The 'scientific inquirers' appointed by the General Board of Health to look into the 1854 cholera took little serious interest either in the usual chemical species or in microscopical entities. They were much more interested in 'the chemistry of organic decomposition during the epidemic prevalence of cholera—especially ... the successive transformations of animal refuse at such times.' Over a decade later, when John Burdon Sanderson began to study 'microzymes' (bacteria and similar organisms), he was interested in them as possible indices of dangerous putrefactive change: the germs themselves were not the entities of primary interest.[14]

By defining the poison as a process, Liebig had with a single stroke of definition cut it off from anything his own science of chemistry was capable of detecting. The prospect of fruitful experimentation was curtailed. It was of course possible to discover the properties of chemical poisons by their effects on animals. There was plenty of evidence of the poisonousness of putrefying material when injected into the blood but without an independent means of distinguishing different modes of putrefaction it was not clear how research was to be extended along those lines.[15] We can get some impression of the enormity of the dilemma sanitary scientists faced if we recall the responses of more recent times to hypotheses which claim that the diversity of nature can only be explained by hitherto unknown forces which just happen to be completely undetectable.

Zymosis and Analysis

Chemists and sanitarians adopted Liebig's zymotic theory nonetheless and chemists quickly adapted analysis to its constraints. Doing so, however, meant adopting a very different view of what analytical results meant. In 1856 Liebig's student August Hofmann pointed out these implications. Hofmann had been one of Liebig's prize pupils in the early 1840s. In 1845 he had accepted an invitation to head the newly established Royal College of Chemistry in London, the creation of a group of agriculturalists and industrialists enormously impressed by the economic potential of chemistry.[16] Hofmann brought to London Liebig's mix of rigorous pursuit of pure research and concern for the manifold applications of chemistry.

As local representative of the most advanced analytical organic chemistry, Hofmann did well as a consultant on aspects of sanitary chemistry among a great many other areas. He had presented evidence to Chadwick's water supply investigation and also served with Graham and Miller as one of the 'government chemists' in 1851. In 1856 Hofmann and Lyndsay Blyth, one of his students, were commissioned by the General Board of Health (now without Chadwick) to discover how much the quality of London's water had changed as a result of the removal upstream of the companies' intakes. In many respects, the Hofmann–Blyth report was a follow-up to the Graham, Miller, Hofmann report five years earlier. The expectations of William Cowper, president of the General Board of Health, were little different from those of Sir George Grey, who had commissioned the earlier report. Like Grey, Cowper focused on hardness; he wanted also to know 'the total admixture of matters foreign to ... [the water's] chemical composition, distinguishing the suspended from the dissolved, the mineral from the organic, and among the latter to specify, as far as may be practicable, those which are putrefiable.'[17] None of these concerns—hardness, minerals, or organic matter—stood out as most important; together they were the ensemble of chemical criteria for determining the appropriateness of a public water supply.

Hofmann and Blyth began by answering a different question than Cowper had asked. They responded to his request to 'specify' whether the organic matter in London's water was 'putrefiable' by noting the difficulty of distinguishing among types of organic matter in water. 'Very little is known of the nature of the ill-defined substances which constitute the organic matter generally found in water,' they wrote.[18]

In fact, simply measuring the organic matter in water was problematic. Mineral water analysts sometimes included organic matter measurements in their reports. Their usual method was to list as organic matter the weight lost after an evaporative residue had been ignited. Yet even the mineral water chemists had recognized that the figure they obtained 'cannot be taken to represent the amount of [organic matter] present; nor can it be ascertained with positive accuracy.'[19] The problems were twofold: some volatile organic matters would be lost during evaporation, leading to too low a figure; or some of the weight lost upon ignition might be water of hydration or carbon dioxide driven off by the heat from its combination with mineral matters, in which case the result would be too large.[20]

With regard to discriminating different kinds of organic matter in water the situation was even worse. Hofmann and Blyth mentioned the hypothesis of the Dutch chemist Mulder that water extracted from decaying matter a series of 'humic' acids, but noted that not only were there no sure methods for distinguishing these acids, they were not even proved to exist.[21] Yet it was the quality of putrefaction that was the crucial issue. In a statement that would be quoted fre-

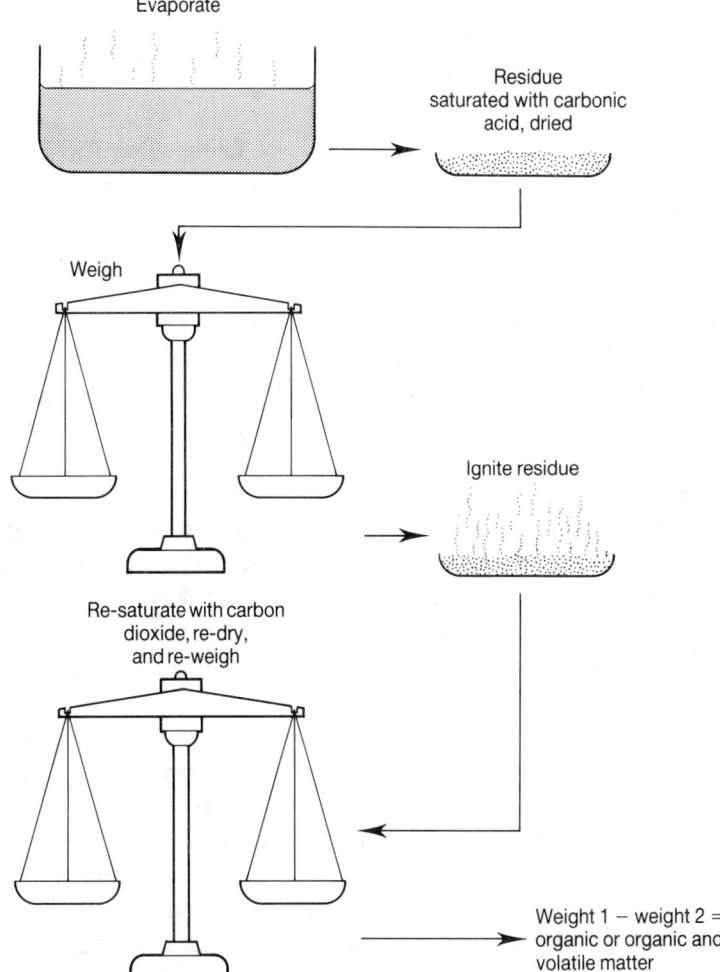

Figure 5.1 One version of the incineration or ignition process. Saturation with carbonic acid was a means of correcting for one source of error, a way of helping to ensure that the weight lost was due to organic matter (from Miller, 'Analysis of Potable Waters,' 1865).

quently in the next two decades Hofmann and Blyth made clear the implications for water analysis of Liebig's zymotic theory of disease:

> it is now generally admitted, that the substances which constitute the organic matter of water act injuriously by no means in conse-quence of being poisonous themselves, but by undergoing those great processes of transformation, called decay and putrefaction, to which all vegetable and animal matter is subject, when no longer under the

control of vitality .... These putrefactive processes either give rise
to the formation of poisonous bodies, or they act simply as ferments,
generating similar processes of decomposition in the substances com-
posing the animal organism.[22]

Hofmann and Blyth abandoned hope of using chemical analysis
to distinguish among types of putrefaction and consequently among
types of disease. Instead they withdrew to the less ambitious position of simply trying to find some index of the potential of a water
to enter into dangerous putrefaction. This index was to be the concentration of organic nitrogen and again the idea came from Liebig's
zymotic theory. In Liebig's view ferments were always nitrogenous
organic matters (yeast for example). These incited decomposition
in a fermentable material. Since specific diseases were nothing other
than specific decompositions, it followed that the amount of nitrogenous organic matter a water contained was equally a measure of the
amount of potentially disease-causing ferment. Hofmann and Blyth
concluded: 'could this nitrogen be estimated with any degree of accuracy, such an estimate would certainly afford the most satisfactory
element in the examination of the organic matter.'[23]

But there was a hitch: it was very hard to distinguish organic nitrogen, which had yet to putrefy, from inorganic forms of nitrogen,
such as ammonia and nitrates, which had completed their decomposition and were therefore safe. Knowing no way to measure the
organic nitrogen directly, Hofmann and Blyth recommended determining the total nitrogen, and then the nitrogen in inorganic forms,
and then subtracting the inorganic forms from the whole to obtain
the quantity of organic nitrogen. Yet even with this simplification
the procedure would require 'the greatest circumspection' and 'all
the appliances which the modern progress of analytical chemistry
can suggest' since the concentrations of these materials in potable
waters were often very small.[24]

Owing to Blyth's illness the two were unable to act on their proposal, and no procedure was available for determining the organic
nitrogen in potable water until Edward Frankland took up the problem in 1867. Yet their work was significant nevertheless, not so much
in terms of the analytical techniques they proposed, but in their conception of the meaning analytical results were to have. For in two
significant ways Hofmann and Blyth were departing from patterns of
interpretation of analytical results that had not been seriously questioned during the Chadwickian hegemony. First they had abandoned
the assumption that one was directly measuring harmful substances.

Hofmann and Blyth assumed that one could only measure an indicator, some entity that showed a high positive correlation with a harmful substance and could be conceived as a necessary condition for a water-generated disease outbreak. The chemists who had testified before the select committees in 1851 and 1852 had been looking for hydrogen sulphide and other products of putrefactive decay, believing that these were directly 'injurious to the health' (the phrase is a significant contrast to conceptions of an 'agent of disease' or 'morbid poison') since they contributed to the environment of decay which was responsible for illness.

Yet it was not the use of indicators that was so radical: after all, Hassall had used an indicator argument to explain the significance of microscopic life. Hofmann, however, was going even further in proposing organic nitrogen as an indicator not of harmfulness but of potential for harm. In measuring organic nitrogen one would be measuring the potential of the water to undergo a process of putrefaction which had the potential to occur in a pathogenic mode. Hofmann's results would thus be two steps removed from a direct measurement of harmfulness.

The tradition that Hofmann was beginning was one of recognizing a broad gulf between what scientists could discover through analysis and what needed to be known to make sound policy decisions about water supplies. In this tradition water could no longer be shown to be safe or harmful. It could only be shown to be more or less dangerous and it would then be up to the wise men of government to decide whether it would be used or not. It is noteworthy that in the Hofmann–Blyth report there were analytical results, but no recommendations. The results—of tests for hardness, total solids, and organic matter—showed significant improvement (a drop of about 50 per cent in organic contamination) since 1851, but Hofmann and Blyth would say only that the water was improved, not that it was safe.

A third result of the rise of the zymotic theory and of the Hofmann/Blyth approach to water analysis was an increase in the distance between an evaluation based on the traditional sensory standards and one founded on the new zymotic sensibility. Already during the 'great stink,' letters had appeared in the *Times* of outrage at 'those learned doctors' who maintained that the stink was less serious than the writers' own noses indicated.[25] In a long and angry review of the Hofmann/Blyth report the *Lancet* complained that the new emphasis on obscure poisons had led the authors to ignore the obvious:

> This, then, is ... what chemistry can do towards the solution of
> the most important question connected with the impurities of water
> .... Messrs Hofmann and Blyth can give us nothing more than a
> *doubtful* determination of the *quantity* of organic matter; concern-
> ing its nature, they can only amuse us with profitless jargon about
> those imaginary entities, crenic and apocrenic acids [two of the humic
> acids].[26]

By contrast with chemistry, microscopic investigations could extend
our senses rather than betraying them.

> And yet there are persons who can tell us something of the nature of
> this organic matter .... We possess an instrument—albeit despised
> ... by chemical dogmatists—which ... enables us to determine that a
> portion of the organic matter—probably the most important—exists
> in the form of living and dead animal and vegetable structures. We
> may see, if we only look, in Thames water, shreds of muscular fibre,
> the various tissues of vegetable productions that have served for hu-
> man food and clothing, a whole world of aquatic Flora and Fauna in
> a living state; we may, by further study of the conditions of life of the
> several varieties of these animal and vegetable forms, arrive at im-
> portant conclusions as to the quantity and nature of the unorganised
> organic matter held in solution.[27]

Yet when it came to saying exactly what this microscopical evidence
actually demonstrated, the *Lancet* could only speculate in much the
same way as Hassall had:

> we know ... that some ... of the entozoa which infest man ... find
> their way into the human body ... in the water we drink; we also
> have good reason to infer that some diseases associated with the
> development of fungi owe their origin to the imbibition of sporules
> with water; and the argument is thus far made good,—that *one* mode
> most influential in the propagation of cholera is water contaminated
> with sewage.[28]

To the *Lancet* the effect of Hofmann's and Blyth's equivocation
about undetectable zymotic processes was to redeem the Thames
as a source of water and uphold the companies' monopoly. In fact,
such a conclusion was contrary to the spirit of the report. Far from
vindicating the Thames the Hofmann–Blyth conception of the zymotic poison would provide a basis for condemning contaminated
water that was virtually unassailable by any means of analysis that
either chemists or microscopists could offer.

The views of Hofmann and Blyth of the undetectability of poisons became increasingly prominent in the late 1850s. In 1856 William Ranger, Henry Austin, and Alfred Dickens, three civil engineers commissioned by the General Board of Health to examine the condition of the upstream Thames, took a much more pessimistic view than had Graham, Miller, and Hofmann five years earlier. They flatly stated that 'chemical analysis does not at present convey an exact understanding of the danger to health which a particular water may occasion.' Water supplied to those areas of south London hardest hit with cholera had been analytically the purest of all the London waters.[29] What harm London might suffer from the sewage of its upstream neighbours remained an open question: 'To what extent the danger of such pollution may be obviated by the atmosphere, or by other causes, in so great a length of flow [50 miles] ... we do not suppose that anyone is capable of determining.'[30]

Frederick Crace-Calvert, a Manchester chemist, took a similar position, writing that the filthy River Medlock could be so far purified with lime that its water would appear superior on analysis to many domestic supplies. Yet Crace-Calvert would still not advise drinking it since all of the dangerous 'nitrogenous residue' might not have been removed.[31] Robert Dundas Thomson, nephew of the Glasgow chemist Thomas Thomson and student of Liebig, took a similar view. R D Thomson had served as chemist to the investigation of the 1854 cholera by the Medical Council of the General Board of Health. He too was critical of the complacency of the 1851 chemists' investigation. He regarded organic nitrogen and its decomposition products as the best measure of impurity yet was unwilling to trust analysis to show whether or not a water was safe.[32]

As these examples suggest, doubts about the adequacy of analysis were usually raised to support the claim that a water was dangerous. In 1850–52 chemistry had been used mainly to counter traditional sensible standards. Then the chemists had argued that the foulness of a water was illusory, and that only they as chemists had the ability to examine water at the invisible and insensible level at which it affected health. Chadwick and many of his followers had refused to accept this argument and, as we have seen, Hassall's microscopic analysis had been in part an attempt to find a means of analysis more nearly consistent with sensible standards. Hofmann and like-minded chemists achieved what Chadwick had hoped to achieve by a quite different method. The new version of the water-borne poison was so obscure, abstract, and unmeasurable as to make avoidance of once-polluted water the only sound policy.

The Burden of the Burden of Proof

Before the mid 1860s such criticisms were occasional and passing. They were reservations or qualifications, not dogmas forcefully and systematically put forward and incorporated into policy. During the '60s, however, matters of water and sewage were almost constantly under consideration by royal commissions of inquiry. A Royal Commission on the Sewage of Towns consisting of the Earl of Essex, H Ker Seymour, sanitary engineers Robert Rawlinson and Henry Austin, agricultural chemists John Thomas Way and J B Lawes, and medical men Thomas Southwood Smith and John Simon sat from 1857 to 1864 and produced three reports. A Rivers Pollution Prevention Commission was established in 1865—some implied so that the government could postpone taking any action on the recommendations of the Sewage of Towns Commission.[33] The commissioners, Rawlinson, Way, and an agriculturalist John Thornhill Harrison, produced three reports in three years and then resigned, their disagreements about the merits of dry earth privies having become so great that they could no longer work together.[34] They were quickly replaced by another Rivers Commission—again despite widespread sentiment that it was time to stop studying and start acting—which issued six reports between 1868 and 1874.[35] For most of its life it had only two members, the chemist Edward Frankland, focus of the next two chapters, and the agriculturalist John Chalmers Morton.

What is remarkable about these nearly 20 years of investigation is their unanimity: governments came and went and the names of the commissioners changed but they continued to spout the same water-and-sewage doctrine. Sanitary reform had to go forth by bold actions. River pollution was to be prevented by making sure that sewage and industrial waste products were recycled on land. Cities and towns were not to use water supplies known to have received sewage, no matter how pure these might seem upon chemical analysis.

In articulating and defending these positions, members of the commissions—especially Way, Rawlinson, and Frankland— developed arguments that held contaminated water to be harmful no matter what analysis or even epidemiological evidence seemed to indicate. With respect to water analysis, I have labelled their position 'analytical nihilism,' as it involves the steadfast refusal to acknowledge as adequate any diagnostic techniques that were touted as being capable of distinguishing safe from harmful water. Usually those who were analytical nihilists were also 'purificatory nihilists,' denying that any method of water purification—e.g. boiling, filtration, or long flow in a river—could be relied upon to have removed whatever noxious materials had entered the water with sewage. These positions, simply by virtue of the way they were stated, could never be disproved by evidence. Those holding them did not deny that sometimes water really became pure and really was pure when analysts said it was, they simply refused to accept actual purification as a necessary consequence of water having undergone a process of purification or to accept negative analytical results as meaningful. What is remarkable is that rather than retiring behind the invulnerable shield that such a position provided, the analytical nihilists, and especially Frankland, tried to develop a science of assessing water quality which incorporated these limitations and yet nevertheless provided the public with useful information.

For Frankland, and probably for Way and Rawlinson, the most eloquent and forceful exponent of these views was Sir Benjamin Brodie bt, the younger, the Oxford professor of chemistry and son of one of the most prominent surgeons in early nineteenth century Britain. Unlike most British chemists, who combined pure science with practical work and consulting, Brodie didn't need to make a living and worked almost exclusively on chemical theory, and in particular on austere and mathematical aspects of atomic theory.[36]

Brodie first made clear his views about water quality on 6 November 1865, as a witness at the Oxford hearings of the first Rivers Pollution Commission. He was not testifying as an expert in sanitary matters; the Commission was unselective about witnesses, interviewing anyone who might have anything remotely relevant to contribute, including Oxford professors in fields like chemistry which bordered on sanitary matters. Under questioning from Way, Brodie denied the generally accepted view that sewage poured into the Thames by Oxford and other towns became oxidized, and therefore harmless, long before reaching the intakes of the London water companies. In the laboratory, Brodie pointed out, organic matter could only be oxidized by powerful oxidizing agents, hence 'to think to get rid of the organic matter by exposure to the air for a short time is absurd.' Brodie admitted that rivers seemed to grow purer as one travelled downstream from a sewer outlet, but thought it highly imprudent to mistake appearance for reality. Since the most definite thing one could say about the 'poisonous qualities' of the organic matter in water was that they were due to the 'quality and nature' of the matter, it seemed absurd to place any faith in chemical analyses, since no one knew what the harmful qualities and natures were.[37]

Slightly more than two years later, in March 1868, Brodie enlarged on these views before the 1867–68 Royal Commission on Water Supply, a large and sedentary commission appointed to consider a new set of alternative water supplies for London. Again Brodie startled his hearers with his peculiarly blunt way of stating truths over which others were successfully equivocating. He argued that the question of when sewage-polluted water became safe to drink had to be left to common sense: 'if you ask whether it is wise to drink water into which you have put sewage, knowing that you have no positive means of getting that sewage out of it, that is a question which anyone can answer for himself.'[38] As an analogy to the purification that was supposed to be occurring in sewage-contaminated rivers Brodie asked the Commissioners to imagine a glass of water to which one per cent of sewage had been added and the mixture then aerated to simulate what took place in a stream. 'The question is at what time would that tumbler of water become in such a state that any one of us would be willing to drink it off, how many days, weeks, months, or years would elapse? That I understand to be the problem, and I am sure I cannot solve it; but I can only say that when you have once put sewage into the water I should be rather reluctant to drink it.'[39]

Brodie had equally devastating things to say about water analysis. Beyond the fact that analysts were in the position of trying to detect something which they did not know how to identify and which was both continually undergoing dilution and presumably being destroyed, there were limitations inherent in chemical analysis. 'Chemical analysis must be limited by our power of weighing and measuring; we can only do those two things. We can weigh and we can measure, and we can do that with a certain accuracy, and there we stop; but that accuracy is not capable of being multiplied *ad infinitum*. ... I think that it is impossible absolutely to answer those questions.'[40]

Thus Brodie had stepped into an area of scientific discourse that had been characterised by confident yet contradictory assertions for more than forty years. Having stepped in he had immediately put forth basic and simple principles of water policy: that in the face of ignorance about the nature of impurity (and consequently ignorance of the nature of purification and of appropriate analytical methods) it was unwise to trust either analysis or purification. The only sure way of discovering the effects of waters on human health was through epidemiological studies of populations using different waters.

It is hard to escape the force of Brodie's statements. His answers were forthright and sober. He had the knack of making his conclusions appear so obvious that only a fool would take issue with them. And Brodie's arguments do seem unanswerable, especially in the context of the 1860s, when almost every chemist or medical man who testified before the Water Supply Commission admitted to not knowing the specific nature of water-borne morbid poisons. And yet Brodie's testimony was profoundly unsatisfying. The trend of questioning itself shows the Commissioners' growing frustration with Brodie; whatever the validity of his arguments, the Commissioners did not find them useful.

There are several reasons for the rejection of Brodie's perspective, both by the Commissioners and by his colleagues. First, the intransigent skepticism in which Brodie indulged himself seemed an insubstantial foundation for so momentous a social action as establishing a new water supply. Brodie might be able to knock down with ease any number of arguments supporting various philosophies of water supply, but he offered the Commissioners little help in choosing the best among a number of imperfect alternatives. Second, Brodie's arguments were clearly extrapolations of conclusions from the laboratory, where substances were pure and variables could be controlled. He avoided the empirical aspects of the water issue altogether. He dismissed purification as only an apparent phenomenon or at best an incomplete one, and likewise rejected water analysis because it was not yet a perfected science. For Brodie one either knew something or one didn't, and he objected to the making of important public health policy decisions on what seemed to him little more than a mixture of plausible hypotheses, misleading appearances, and wishful thinking. The Commission and the great mass of work-a-day sanitary chemists took quite a different view. They accepted the partial and imperfect knowledge that could be derived from empirical investigation as the best available guide for making sound decisions, and one far more satisfactory than the admission of ignorance.

Third, the position that Brodie took had implications that were fundamentally at odds with the social aspirations of British chemists. Men like Brande and Taylor had made their careers arguing the applicability of chemistry to industry, commerce, engineering, and public service. According to Brodie one could do no such thing; in his hands the application of the science showed only its inapplicability. Unlike Brodie, who inherited his father's fortune, these chemists had livings to make.

Yet while Brodie's position might be unacceptably radical with respect to a philosophy of government based on the balancing of opposing views, it was taken up by the successors of Lambe and Wright, Chadwick and Hassall, those who believed the time had come for the massive building programme of the sewage farms and water supplies that would safeguard the nation's health. The great advantage that these reformers found in Brodie's position was that it so clearly placed the burden of proof on those who held polluted water to be safe. The defenders of river water would henceforth have to prove a negative, to show that something unknown was absent, and this was an impossible task. The burden of having to demonstrate the harmful constituents in a water under which Lambe, Chadwick, and Hassall had laboured was being lifted; and as the burden of proof shifted to the other side the reformers could sit back and snipe away at their opponents with philosophy.

It is in the hearings of select committees considering reform of the conservancy boards that administered the rivers Thames and Lea, in June 1866 and April 1868 respectively, that Brodie's nihilism first appears in the hands of working sanitarians as part of a coherent alternative philosophy of water supply. In both hearings Brodie's arguments were articulated by Robert Rawlinson and John Thomas Way, two of the three members of the first Rivers Pollution Commission.

Rawlinson was a civil engineer who had learned the trade from Robert Stephenson during the railway mania of the '40s. Active in the campaign for a pure water supply for Liverpool, he had attracted Chadwick's attention and became one of the first of Chadwick's sanitary engineers. From 1858 until 1888 he was chief engineering inspector for the Local Government Act Office and later the Local Government Board, responsible for approving towns' requests for improvement loans.[41] Way, a student of Thomas Graham, was mainly a consulting agricultural chemist and had spent ten years as consulting chemist to the Royal Agricultural Society.[42] He had become involved in sanitary chemistry through his studies of what happened to manure or sewage when it was spread on arable soil. Neither man had medical training, nor was either expert in pathology or epidemiology, yet both were to take strong positions on the medical implications of water analysis.

One of the concerns of both select committees was to determine whether the sewage of upstream towns adversely affected the London water supply. During both hearings counsel and witnesses representing towns and industrialists on upstream stretches of the rivers were present to make the case that sewage quickly and completely disappeared during a short period of flow in a river. Rawlinson and Way argued that this was not necessarily the case. As members of the Rivers Pollution Commission they had thoroughly investigated the Lea and the Thames. They insisted that both rivers would remain unsuitable as sources of water for London as long as they continued to receive the sewage of upstream towns. Asked to back up his contention that some trace of sewage remained in London's water, Rawlinson would say only that 'I should decline to take any analysis, or to take any evidence that it was not impure.'[43] As chemistry was inadequate, the only way to make conclusions about the fates of morbid poisons was through rational inference: 'it is a thing that you can only follow in your imagination,' Rawlinson insisted, 'if all the chemists in Europe told me that after the sewage had gone into the water that there would be no injurious effects from it, I simply would not believe it.' He admitted 'I am now daily drinking Thames water, knowing ... all the abominations that go in ... under certain conditions this effete matter may go on to corruption, and if it pass into my system it will do me serious injury.'[44] Without more knowledge of what it was in water that could be harmful there was no reason to trust any form of purification and no reason to think that contamination even in the highest tributaries might not harm Londoners.

Way took the same tone with respect to the Lea. The Lea contained matters 'known under certain circumstances [i.e. special modes of putrefaction] to be injurious to health.' Though he admitted he could find 'nothing positively injurious to health,' Way condemned its water anyway. He happily admitted the 'defective condition of water analysis' and noted that the most precise characterization that medical men could give of the morbid poisons of zymotic diseases was that they were comprised of 'some indefinable matters, most probably of an animal character, which are capable of setting up a kind of fermentation in the blood, and producing disease.' They might be 'infinitesimally small.' When counsel representing Luton presented him with a copy of A S Taylor's analysis of Lea water which purported to demonstrate that the water was safe, Way expressed his admiration for Taylor but would not alter his opinion. When pressed to explain why there was no epidemiological indication of the contaminated Lea water, Way replied that in most cases bad water was only a predisposing cause: it bore some responsibility for the fact that London's mortality rate was higher than it need be, but one didn't need to isolate its effects to be confident that they existed.[45]

Way acknowledged that the case he was presenting was not based on his scientific expertise: as one of his examiners put it, on the issue of when contaminated water became safe there was no difference between science and 'general inference.'[46] Indeed, the position he took was a deduction from a set of principles for making practical decisions in the face of uncertainty. He admitted that the purification of the water appeared to take place, and that in theory—in 'abstract'—it ought to take place. But he went on: 'I think if it [the spontaneous purification of contaminated river water] is urged as a reason why that which is known to be in the water has been taken out, we should call upon other people to show that it has been taken out.'[47]

Here Way, like Rawlinson and Brodie, was transferring the burden of proof, putting the defenders of river supplies in the impossible position of having to prove that the unknown harmful materials or conditions in sewage never survived long enough to cause harm to London water drinkers. Not surprisingly, one of the issues which came up towards the end of Way's testimony was the great problem of just what scientists were to contribute in such cases where the phenomena—the causal agents of cholera, typhus and so forth—could not be 'put ... upon [the] table.' Way's reply was much like Brodie's. The only valid knowledge about the effects of waters was that obtained by 'induction,' by epidemiological investigations of the populations that drank different waters. Just as the successors of Newton had gotten used to thinking of gravity as a lawful phenomenon even without knowing how it worked, so too sanitarians would have to learn to give up their dependency on analysis to validate phenomena.[48]

The 1868 hearings of the select committee on the Lea Conservancy Bill are significant not only for the emergence of nihilism as a systematic approach to matters of water purity, but also as a last stand in the campaign to adapt mineral water analysis, mineral water standards, and the certainty of mineral water chemistry to potable waters. Alfred Swaine Taylor, still the greatest forensic chemist of the day, testified as an expert witness for the town of Luton. Taylor had analysed Luton's sewage effluent and the water of the Lea and found them unobjectionable. Yet his testimony was based on the chemistry of an earlier age before Liebig's chemistry and Snow's epidemiology had made sanitarians so concerned about minute quantities of organic matter undergoing peculiar processes of putrefaction. The contrast between Taylor's brazen assurance that well-established processes of inorganic solution analysis could easily show whether water was safe and Way's insistence that no analytical processes could ever warrant the conclusion that a contaminated water had become safe is striking indeed.

Taylor was candid with respect both to what ingredients distinguished safe from unsafe water and to the ways that chemistry could serve public health policy making. His conceptions of purity and impurity had changed little over the years. Water was more or less impure according to the quantity of 'solid contents': 'If I find the water containing upwards of 40 or 50 grains in an imperial gallon, of solid contents, and four or five grains of organic matter, and especially if there are nitrates with it, I do not care to inquire what the influence is upon the population; I say at once, it is not wholesome water.'[49] Taylor objected to nitrates, not because he viewed them as the final products of organic decomposition and hence as indicators of the presence of putrefying matter, but because nitrates directly irritated the bowels and were thus the exciting cause of dysentery and a predisposing cause of cholera.[50]

Hence Taylor was thinking about the health problems brought on by bad water in a very different sense than were Snow or Way. In his view a bad water was always exerting its pathological influence though this might be manifested only when reinforced by other causes. Way, by contrast, while acknowledging that bad water might act continually as a predisposing cause, accepted with Snow, Hofmann, and Brodie that water might also occasionally carry the specific morbid poison or 'seed' of a zymotic disease. A corollary of Taylor's view that the harmfulness of a water was a direct consequence of the physiological activity of the dissolved salts it contained was that chemistry was entirely competent to determine the harmful constituents in water. Taylor put it bluntly: 'I deny the existence of what cannot be discovered.'[51] When asked if hydrogen sulphide and ammonium sulphide were 'the only noxious things in sewage,' Taylor agreed that 'so far as science will enable us to say,' these were 'the only very poisonous ingredients.'[52]

For Taylor what science could say and what was in fact the case were the same; he would not concede that there might be unknown ways that water could be harmful. Whenever he had investigated a case where epidemiological inference indicated that drinking water was responsible for an outbreak of disease, he had discovered the ingredient responsible, Taylor claimed. Yet on closer examination this turned out not to be quite true. Under aggressive questioning he admitted that his claim for the adequacy of analysis was not an inductive inference, the result of discovering the presence of a certain chemical in every occurrence of a certain set of symptoms, but an axiom: 'In other cases they [waters] have been said to produce illness, but on analysing I have found them perfectly wholesome, and I have told the parties that the illness must be referred to something else.'[53]

This statement alone reveals how greatly Taylor's conception of the utility of chemistry differed from Way's. For Taylor chemistry was primary and epidemiological hypotheses had to be modified in light of the findings of analysis. For Way epidemiological arguments, like those made by Snow, were irrefutable, truly empirical. The failure of chemistry to confirm them only showed its inadequacy and the necessity of conceiving a model of a morbid poison that transcended the limited abilities of chemical (and microscopical) analysis. During the 1870s and 1880s most sanitarians adopted Way's position, one which would seem to lessen the importance, or even wholly negate the utility, of any analytical confirmation. Yet this did not happen. Even as increasing numbers of analysts admitted their inability to detect the really harmful substances in water, water analysis remained important, and even became more important. It did so on the basis of a new set of conventions about what analysts should be seeking and what kinds of interpretations their measurements could legitimately support.

1   *Saturday Review*, reprinted in *CN* 18 (1868): 214.
2   H Letheby in *S C East London Water Bills*, QQ 2253–64; in *R C Water Supply*, Q 3934; 'Analysis of London waters,' *CN* 12 (1865): 302.
3   On Snow see M Pelling, *Cholera*, pp 204–29; Frazer, *A History of English Public Health*, pp 64–9; P E Brown, 'Another Look at John Snow,' pp 646–54; *idem*, 'John Snow—the Autumn Loiterer,' pp 519–28; H

Whitehead, 'The Broad Street Pump: An Episode in the Cholera Epidemic of 1854,' *MacMillan's Magazine* 13 (1865–6): 113–22; *idem* 'The Influence of Impure Water on the Spread of Cholera,' *MacMillan's Magazine* 14 (1866): 182–90; John Snow, 'On the Mode of Communication of cholera'; J Snow, 'Cholera and Water Supply in the South Districts of London, in 1854,' *JPH&SR* 2 (1856): 239–57.

4   Frazer, *English Public Health*, p 65; W H Frost in *Snow on Cholera*, p ix. But see Pelling, *Cholera*, pp 207–11.

5   A clear example of these gradually changing ideas can be found in the writings of John Simon, *Public Health Reports*, ed E Seaton, 2 vols (London: The Sanitary Institute, 1887), II, pp 151, 157–8, 294, 413–5, 534–7, 563–85; C-E A Winslow, *The Conquest of Epidemic Disease*, pp 259–66. Simon did not begin to regard bowel diseases as primarily water borne until the end of the '60s (*12th Report of the MOPC*, pp 21–32). See also Pelling, *Cholera*, p 206.

6   Pelling, *Cholera*, chapters 3 and 4.

7   Justus Liebig, *Chemistry in its Application to Agriculture and Physiology*, ed from the mss of the author by Lyon Playfair, 2nd edn (London: Taylor and Walton, 1842), pp 364–70.

8   Eyler, *Victorian Social Medicine*, p 104; Pelling, *Cholera*, pp 106–7, 130; J K Crellin, 'The Dawn of the Germ Theory,' p 61.

9   P B Ayres, 'On the Nature and Results of the Putrefactive Fermentation of Animal and Vegetable Matters,' *Lancet*, ii, 1848, pp 445–7.

10 B W Richardson, *The Cause of the Coagulation of the Blood, being the Astley Cooper Prize Essay for 1856* ... (London: J Churchill, 1858), pp 118–21, 345–70; T Herbert Barker, 'The Influence of Sewer Emanations,' *JPH & SR* 4 (1858): 70–82; Barker, *On Malaria and Miasmata*, pp 176, 213–26; Charles Murchison, *A Treatise on the Continued Fevers*, pp 447–8; Henry Letheby, 'Report to the City of London Commissioners of Sewers, Sept 9 1858: Sewage and Sewer Gases,' *JPH & SR* 4 (1858): 279–93.

11 William Odling, *Report on the effects of Sewage Contamination upon the River Thames* (Lambeth: The Vestry/G Hill, 1858), pp 15–7; William Ord, 'Report on the Thames Nuisances of 1858–9,' in *Second Annual Report of the MOPC*, pp 54–6; Wm Budd, 'Observations on Typhoid or Intestinal Fever,' *BMJ*, ii, 1861, pp 485–7; B W Richardson, 'The Thames,' *JPH&SR 4* (1858): 142.

12 Snow, 'On Continuous Molecular Changes,' p 156; cf Pelling, *Cholera*, pp 207–12, 247–8.

13 Snow, *Mode of Communication*, p 15.

14 GBH Medical Council, *Report of the Committee for Scientific Inquiries on the 1854 Cholera*, p 37; *12th Report of the MOPC*, pp 20, 79, 162–3.

15 Murchison, *Treatise on the Continued Fevers*, pp 447–8; J Liebig, *Animal Chemistry, or Chemistry in its Application to Physiology and*

Pathology, ed from the author's mss by William Gregory, fr. the 3rd London edn, revised and greatly enlarged (New York: John Wiley, 1852), pp 137–53; *idem*, *Familiar Letters on Chemistry, in its Relations to Pathology, Dietetics, Agriculture, Commerce, and Political Economy*, 3rd edn, revised and much enlarged (London: Taylor, Walton and Maberly, 1851), #18, pp 228–30.

16 Roberts and Bud, *Science versus Practice*, pp 51–63.

17 A W Hofmann and Lyndsay Blyth, 'Chemical Report,' p 3.

18 Hofmann and Blyth, 'Chemical Report,' p 4.

19 J W Kynaston, 'Analysis of a Spring Water at Billingborough, Lincolnshire,' *Q. J. Chem. Soc.* 12 (1859): 60–2.

20 R Phillips in *S C (House of Lords) on the Supply of Water to the Metropolis*, Q 1095; B H Paul, *Manual of Technical Analysis: A Guide for the Testing and Valuation* (London: H G Bohn, 1857), p 223.

21 Hofmann and Blyth, 'Chemical Report,' p 4. On these acids see K R Fresenius, *Instruction in Chemical Analysis: Quantitative* ed by J Lloyd Bullock (London: Churchill, 1846), pp 466ff.

22 Hofmann and Blyth, 'Chemical Report,' p 5.

23 *Ibid*, p 5.

24 *Ibid*, p 6.

25 *Times*, 16 June 1858. See also 'A Sonnet upon a Scent,' *Punch* 37 (1859): 45.

26 *Lancet*, ii, 1856, p 576.

27 *Ibid*, p 576.

28 *Ibid*, p 576.

29 William Ranger, Henry Austin, and Alfred Dickens, 'Report on the Examination of the Thames,' pp 90–5. The reference is probably to the 1849 epidemic. The water companies used a similar observation to argue that the water could not have been the cause of the outbreak (*GBH MWS*, Appendix I, pp 33–9).

30 *Ibid*, p 91.

31 F Crace-Calvert, 'On the Purification of Polluted Streams,' *J. Royal Soc. Arts* 4 (1856): 506.

32 R D Thomson, 'Report on the Chemical Composition of Metropolitan Waters during the Year 1854,' in GBH Medical Council, *Report of the Committee for Scientific Inquiries on the 1854 Cholera*, Appendix 7, p 201. Cf Playfair in *GBH MWS*, Appendix III, p 78.

33 Lord Robert Montagu in *Hansard's Parliamentary Debates*, 8 March 1865, c 1358.

34 The *Builder*, 25 January 1868, p 59; *PRO* HO/74/3 pp 409, 412, 417, 426.

35 Fisheries Preservation Association, *On the Pollution of the Rivers of the Kingdom* (London: The Association, 1868), pp 50–2; *The Field*, ii, 1867, p 120.

36 On Brodie (1817–80) see *DNB* v 2, pp 1288–89.

37 *RPPC*, 1865, 1st Rept, QQ 1493–4, 1497–1502.

38 *R C Water Supply*, Q 6991.

39 *Ibid*, Q 6989.

40 *Ibid*, QQ 7011, 7014.

41 'Robert Rawlinson,' *Minutes of Proceedings, Institution of Civil Engineers* 134 (1897–8): 386–91.

42 E A Russell, *A History of Agricultural Science in Great Britain, 1620–1954* (London: George Allen and Unwin, 1966), p 116.

43 *S C Thames Navigation Bill*, Q 2713.

44 *Ibid*, Q 2725.

45 *S C Lea Conservancy Bill*, QQ 1880, 1900, 1939–43, 1919–23, 1927–8.

46 *Ibid*, QQ 1952–3.

47 *Ibid*, Q 2036.

48 *Ibid*, QQ 2056–61.

49 *Ibid*, Q 2405.

50 *Ibid*, Q 2405. See also W T Brande and A S Taylor, *Chemistry* (London: John Davies, 1863), pp 132–4.

51 *S C Lea Conservancy*, Q 2400. Taylor had made much the same argument in 1850, 'Return of the Water Companies,' in *GBH MWS*, Appendix I, pp 33–5.

52 *S C Lea Conservancy*, Q 2495.

53 *Ibid*, Q 2518.

6   Edward Frankland: The Analyst as Activist

'[the public] *almost invariably display an apathy as marvellous as it is culpable.* '[1]

A H Hassall

As more and more chemists came to admit that whatever it was that poisoned water was beyond their ability to measure, the issues of interpretation and advising became increasingly important. To those like Brande who claimed to know the characteristics of bad water, the recommendations one made were straightforward. The sort of transitory and inaccessible morbid poison Hofmann, Way, and Rawlinson envisioned raised more troublesome ethical questions. One might well, as had Hofmann and Blyth, recognize in a water nothing actively dangerous yet judge it unsafe. Such a judgement might be made independently of chemistry, simply from knowledge of the sorts of pollutants upstream towns dumped into a river. The problem was what one should say about that danger and how one should say it.

One response was that taken by Hofmann and Blyth in 1856, to report analytical findings yet also to point to the inadequacy of analysis. This didn't work well. During the '50s and early '60s a great many people were hectoring the public about dreadful and deadly impurities of one sort or another; simply to add one's voice to the crowd without demonstrating the physical existence of the impurity in question was not enough to surmount the din. Because it did not strongly condemn Thames water, the Hofmann–Blyth report, which was actually quite critical of that supply, was taken by the *Lancet* as an affirmation of the *status quo*.[2] An alternative, the sort of outrage-and-disgust approach taken by Hassall, was equally unworkable. However worked up one might manage to get at being sold sewage-polluted water, it was impossible to escape the ambivalence and ambiguity of the verdict supplied by the best contemporary science, which was that impurities were usually purified, and that the water apparently caused little serious harm.

In the cases of both of these alternatives, analysts were adopting a stance toward policy-making in which expert and layman were essentially on an equal footing. Hofmann and Blyth were willing to let the facts speak for themselves, making only the one major qualification that the particular facts they could offer failed to address the most important questions. Hassall, too, was calling attention to the facts of water supply, albeit by means of the most vivid language he could find. With the facts, the Members of Parliament and of the London vestries would presumably make wise choices. Yet from the reformers' point of view, they were not doing so.

By the late '60s a new approach to transmitting information on water quality to the public had become common, one which rejected the earlier egalitarianism for deliberate and sophisticated mystification. It was not a policy of deceit; it was instead the development of ways of colouring pieces of information, particularly quantitative information, to give the impression that the water was either remarkably safe or dangerously contaminated. Such tactics were not new, yet during these years they began to be employed with unprecedented subtlety and insidiousness. They were incorporated into new sets of analytical procedures, first for chemical water analysis and later for bacteriological. Implicit in such tactics was the view that the public could not be trusted to make wise decisions; it was necessary for experts to lead the public to make the right choices without its recognizing that it was being led. The person most responsible for this policy was the chemist Edward Frankland, who from 1866 until his death in 1899 served as the quasi-official 'government analyst' of the London water supply. It may seem striking that such a campaign should be undertaken by the official analyst, for most representations of the coming of science into public administration treat it as a movement from obscurity and arbitrariness to clarity and rationality. Yet other government scientists, like William Farr and John Simon, were equally activist. All were gadflies in government, occupying peculiar niches in bureaucracy that stood outside any rationalized administrative structure, and each shamelessly used his position as a forum for social change. And if their proposals were politically untimely and maladroit, they had an excuse not available to other public servants: they spoke, they claimed, the truth of science, and science would not be held hostage to political convenience. The ideal of neutrality and the claims of expertise carried sufficient political power that each had a long run. Of the three Frankland was probably the most successful in having a direct effect on change, if only because he concentrated on the narrower area of water policy.

The London Water Controversy, 1865–68

So long as water quality was not at the centre of public attention and so long as statements like those made by Brodie, Rawlinson, and Way were expressed as abstract ideals or arose only in peripheral policy issues, they could be disregarded. But as nihilism became more systematic, coherent, consistent, and institutionalized, it became impossible to ignore, and that began to happen in 1866. Beginning in the early '60s there had been renewed stirrings about the possibility of a new water supply for London. Along with the old plans for obtaining purer water from nearby sources were several more ambitious schemes for systems of conduits linking London with some distant watershed, even one as far away as the Lake District or the Welsh hills. Liverpool, Glasgow, Manchester, and Dublin had each utilized the technology of long distance water transport to obtain copious soft water, and there seemed no reason why the approach would not work for London, even though the conduits would have to be substantially longer. These schemes seemed unlikely to be objectionable on grounds of insufficient quantity of water, the principal objection to the alternatives considered in 1850–52.[3]

The two main contenders, J F Bateman's proposal to get water from the Severn watershed in Wales, and the Lake District scheme of G W Hemans and Richard Hassard, were announced in November 1865 and July 1866 respectively. It is not clear that either would have received a serious hearing had not epidemic cholera returned to the metropolis in July 1866. But the cholera—acknowledged to be water-borne except by diehard supporters of the water companies— along with the alternative schemes, and concerns raised in the first two reports (on the Thames and on the Lea) of the first Royal Commission on Rivers Pollution combined to keep debate on water quality (and water analysis) lively for the next three years.[4] During that period there were numerous inquiries, official and unofficial, into various combinations of these issues, with the most massive being that of the Royal Commission on Water Supply which sat from early 1867 to early 1869.[5]

The Registrar General's Water Analyst: Edward Frankland

Edward Frankland had only just become involved with the London water supply when the commotion began. In June 1865 he had been appointed official analyst of the London water supply by the Registrar General of Births, Deaths, and Marriages. Since 1857 the holder of the post had provided a monthly report to the Registrar General on the composition of the water supplied by the London companies, and frequently also on the composition of the water supplies in a few other large British cities. R Dundas Thomson, chemist and medical officer of St Marylebone, had been the first occupant of the position, holding it from October 1857 until shortly before his death in August 1864. In February 1865 Hofmann had taken over the analyses, but had soon quit to return to Germany. When Frankland succeeded Hofmann as professor of chemistry at the Royal School of Mines (which had absorbed the College of Chemistry), he also took on the post of water analyst.[6]

As a youth in Lancashire, Frankland had practiced chemistry as a hobby and been apprenticed to a pharmacist. Through local connections he had been sent in 1845 to Putney to study more advanced chemistry with Lyon Playfair and had gone from there to work in the laboratory of Robert Bunsen in Marburg and briefly with Liebig at Giessen. In 1851 Frankland became the first chemistry professor at Owens College, Manchester, having already established a solid research reputation through investigations of organometallic compounds. In 1857 he left Manchester to become chemistry professor at St Bartholomew's Hospital. He also held one of the chemistry professorships at the Royal Institution from 1863 to 1868, but spent most of his career (1865–85) as the professor of chemistry at the School of Mines, the most prestigious of London's chemistry professorships.[7]

The post that Frankland took up in 1865 was not one to which any great significance was attached. Its origins are obscure, but it is likely that William Farr, the statistician at the Registrar General's Office, wished to discover whether there was any significant relation between mortality and water quality, just as there might be between mortality and elevation, air quality, population density, or any other variable suspected of leading to illness. The post had no statutory sanction (as Frankland's critics would sometimes point out), and hence it was not quite correct to think of its occupant as the 'official' analyst, but the government did pay the analyst for his services.[8]

Even before Frankland took up the post the question had arisen of the propriety of publishing comparative analyses of different waters. In 1861 the engineer George Burnell had complained that Thomson's reports were purposefully misleading. For purposes of comparison, Thomson listed the constituents of distilled water. Next he gave standard figures for the composition of the soft and pure waters Glasgow and Manchester had obtained. Finally he gave the constituents of the companies' waters, and of water from one of London's wells. Burnell argued that the intent of this format was to 'lead to the conclusion that the London water companies supply a fluid of very objectionable quality' simply because their waters contained more dissolved matter than those of Glasgow and Manchester. Since both those cities had higher mortality rates than London, Burnell felt that there was 'both great injustice and a great want of the true spirit of philosophy, in the insinuations which are now constantly urged by the Registrar General.'[9]

It is hard not to sympathize with Burnell. By listing the constituents of distilled water, Thomson appeared to be suggesting that the London supplies were impure to the degree they differed from that standard. It is likely that Thomson knew exactly what effect he was trying to achieve, that his format was meant to remind Londoners that what they drank was less pure than it might be. Certainly Farr, Thomson's superior, shared with many sanitarians a sophisticated appreciation of the rhetorical utility of statistics. For them statistical presentations were not neutral; they were an opportunity to juxtapose facts in such a way as to highlight problems or reveal solutions.[10]

Initially, Frankland's analyses had no such overt political implications. During his first year in the post he found the water generally satisfactory. He was determining organic matter by Hofmann's improved version of the ignition process and also used a version of what was called the oxygen absorbed or potassium permanganate test. Potassium permanganate was an oxidizing agent widely believed to have a peculiar attraction to putrefying or putrescible matter. The Danish chemist G B Forchhammer had developed the test in 1850 and by 1865 several versions were in use. In all of them a known quantity of permanganate was added to the water sample. The amount of oxygen it lost (and in some versions the rate of oxygen

		Total impurity per gallon.	Organic impurity per gallon.
		grs. or°.	grs. or°.
Distilled Water		0·0	0·0
Loch Katrine Water, new supply to Glasgow		2·35	·605
Manchester Water Supply		3·33	·680
THAMES COMPANIES.			
Chelsea		17·92	1·12
West Middlesex		16·16	1·24
Southwark		16·48	1·12
Grand Junction		18·80	1·36
Lambeth		16·00	·96
OTHER COMPANIES.			
Kent		25·36	2·32
New River		15·28	1·04
East London		15·36	1·36

"The Table is to be read thus :—Loch Katrine water contains in the gallon 1·35 degrees or grain of foreign matter in solution, of which ·605 degrees or grains are of vegetable or animal origin."

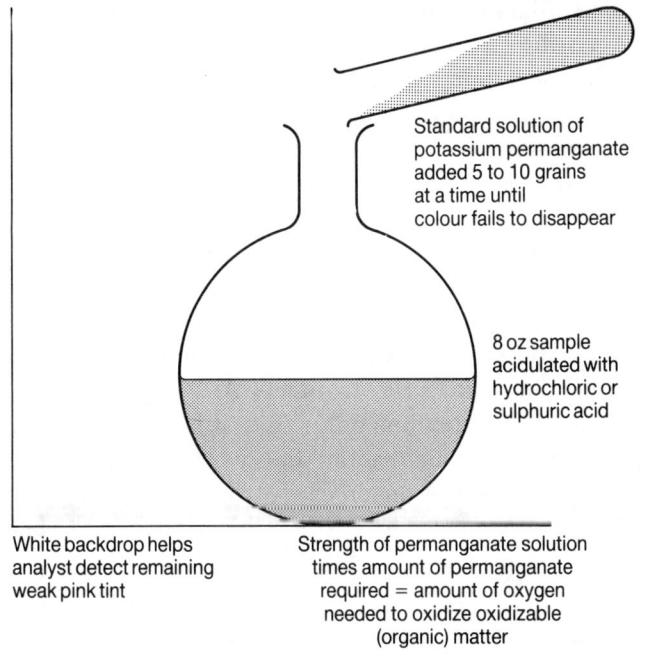

Standard solution of potassium permanganate added 5 to 10 grains at a time until colour fails to disappear

8 oz sample acidulated with hydrochloric or sulphuric acid

White backdrop helps analyst detect remaining weak pink tint

Strength of permanganate solution times amount of permanganate required = amount of oxygen needed to oxidize oxidizable (organic) matter

Figure 6.1 (*a*) R D Thomson's concept of degrees of impurity and his practice of comparing the soft Glasgow and Manchester waters with the hard London supplies drew criticism from the water companies' defenders similar to those that would soon be levelled at Edward Frankland. No matter what the results showed, the format itself was biased in the eyes of the critics. (RPPC, *6th Report*, p 250). (*b*) An early version of the potassium permanganate or oxygen absorbed process. Later improvements made the process more precise. Some also tried to use the process to distinguish animal from vegetable contamination (from Miller, 'Analysis of Potable Waters,' 1865).

loss as well) was understood to correspond to the putrescibility and hence the insalubrity of the water.

Although great hopes had been held out for it at the outset, the permanganate process was objected to for a number of reasons: most importantly, there was some doubt whether it allowed a sufficiently precise distinction between harmful and innocuous organic matter; some held that the test was popular only owing to the vivid colour reaction (when the bright pink fluid ceased to lose its colour, all the readily oxidizable material was gone). There was also recognition that the oxygen in the permanganate might be lost to inorganic compounds, such as nitrites.[11] Moreover the variety of versions of the test led to problems. As one chemist complained, 'solutions so different in strength are used, and there are such diverse ways of employing them, that it is difficult ... to institute any comparisons between the results arrived at.[12] Frankland's use of the process shows that he, like most chemists, was mainly concerned with putrefaction. According to this standard, London's water was normally safe since the most putrescible materials were also those quickest to decompose, leaving the water pure.[13]

When Frankland issued his first annual report in March 1866 he was already becoming suspicious of the process. The fact that the unquestionably superior water Glasgow obtained from Loch Katrine sometimes had higher oxygen-absorbed levels than did Thames water suggested the test might be misleading. Frankland's doubts were confirmed when cholera struck in July 1866. It appeared to be spreading in water supplied by the East London company, yet analyses showed nothing out of the ordinary. At the peak of the epidemic in early August the water was purer than it had been a month earlier and contained less organic matter than usual. Frankland recognized that analysis was useless on precisely those occasions when it was most needed. On August 4 he wrote to his superiors at the Registry:

> Chemical analysis ... does not reveal any exceptional degree of pollution in this water. It must be borne in mind, however, that chemical investigation is utterly unable to detect the presence of choleric poison amongst the organic impurities in water, and there can be no doubt that this poison may be present in quantity fatal to the consumer, though far too minute to be detected by the most delicate chemical research.[14]

Thus Frankland responded to cholera much as Hofmann and Snow had. When confronted with a contradiction between analytical and epidemiological evidence, one trusted epidemiology. Like Hofmann and Snow he was also thinking in terms of a specific morbid poison transmitted in a particular medium, and while it might be true that catching the disease was a consequence of many causes, that did not detract from recognizing water as the main means of its spread.[15] His call for treating water with potassium permanganate, a disinfectant as well as a reagent, shows that like Hofmann, he believed the cholera poison to be a unique mode of putrefaction.

But Frankland was changing his views. By the end of August he had lost confidence in permanganate disinfection and even in filtration through animal charcoal. These helped, yet could not be counted on to purify water, and even boiling might not work.[16] In November he presented evidence that the cholera poison was likely never to be detectable through chemical means: he had obtained a sample of the 'rice-water' evacuations of a cholera victim, diluted it with 500 parts distilled water, and filtered the mixture. Upon analysis, the still turbid mixture absorbed 0.04 parts/100,000 oxygen from permanganate, while normal Thames water took 0.07 parts. At a ratio of only 1 part cholera evacuations to 1000 parts water—far more concentrated than would usually be the case—cholera evacuations would be analytically undetectable, yet nevertheless deadly.[17]

The Context of 'Previous Sewage Contamination'

Though Frankland eventually demonstrated that human error had been responsible for setting off the east London epidemic—company technicians had improperly put into service a reservoir contaminated with water from the polluted lower Lea—the epidemic had been fundamentally a consequence of reliance on polluted water, and in the next few months Frankland took up the problem of water analysis. In a Royal Institution lecture in late March 1867 he reviewed the two leading processes for measuring organic matter in water, the ignition and permanganate processes, and found them unacceptable. His criticisms were not new, but Frankland put them more forcefully than others had. He was in a position to do so for he was perfecting a new method along the lines suggested by Hofmann and Blyth in 1856. He faced the same problem they had faced, of focusing on organic nitrogen but of having no way to measure it directly, and he took up Hofmann's solution of determining organic nitrogen as the difference between the 'total combined nitrogen' and the sum of the inorganic forms—nitrates, nitrites, and ammonia.

Frankland had little to say about this organic nitrogen, however. He devoted more attention to the inorganic nitrogen compounds, from which he calculated the 'previous sewage contamination' (PSC) of the water. One made the calculation by multiplying the total inorganic nitrogen in parts per 100,000 less 0.032 (a correction for rain-water nitrogen) by 10,000 (the dilution factor of average London sewage). Ostensibly the result indicated the amount of sewage which would have had to have been in the water to produce upon decomposition the inorganic nitrogen actually found in the water. This calculation would be the centrepiece of Frankland's format for presenting analytical results until 1876 when it was quietly dropped. It was the source of confusion and occasioned much criticism, and it is well to ask why Frankland developed it.

Even as he presented the term Frankland recognized that it was a misnomer. He admitted that all nitrogen compounds in water did not come from sewage or even from animal wastes, but maintained (at least in 1867) that 'animal or vegetable, no distinction founded upon this can be drawn between their respective noxious qualities.'[18] In the next few years he went to great lengths trying to establish that whenever nitrates were found in a water they could be traced to an animal source.

The term was also objectionable on the grounds that it implied that something actively harmful was in the water, some constituent of sewage. In fact, as was frequently pointed out, the constituents Frankland was measuring were precisely those which indicated that sewage had been purified. The expression was thus a double entendre. On one level it referred to a certain quantity of material that might once have been harmful but was now innocuous. At the same time, because Frankland expressed 'previous sewage contamination' in pounds per hundred thousand pounds of water, it evoked an image of large quantities of sewage in the water.[19] Month after month Londoners faced the startling image of several hundred or even a couple of thousand pounds of sewage (a figure they might reasonably be expected to be able to conceptualize) in a hundred thousand pounds of their water, a quantity far more difficult to imagine. Frankland surely had fun with this irony: it would be 'consolatory to the drinker of Thames water to know,' he maintained, 'that the whole of the fecal matter is so completely oxidised before it reaches the water cisterns of London as to defy the detection of any trace in its noxious condition.'[20] He had little to say about the purpose of this 'convenient expression,' but it is clear that the term was to remind the public of the danger of a sewage-contaminated water supply. Initially it would compensate for his failure to find the 'actual sewage contamination' (organic nitrogen) that would provide grounds for condemning the water. Its quantity was presumably so small as to be masked by the error that arose in the separate determination of inorganic forms of nitrogen.

With the launching of PSC in early 1867 Frankland had begun to integrate analytical processes with formats for presenting results. He was developing a strategy to make analysis a basis for social action. The strategy embodied the following principles: 1) that both negative and positive results of water analysis were untrustworthy (analytical nihilism); 2) that the onset of water-borne zymotic disease was unpredictable; 3) that decision-making bodies, having great and misplaced faith in the abilities of chemistry, would not take decisive action in the face of uncertainty; 4) that therefore, in attempting to protect the public, the scientist could not rely on normal democratic processes, but would have to pre-digest information.

Frankland by no means founded this practice of colouring facts in this way. Hassall, who was openly glad to see that Frankland's analyses were 'calculated to alarm the public mind,' had done something similar. But he did raise substantially the level of insidiousness, and it became necessary for analysts representing other interests to respond with similar tactics.[21]

The Invisible Enemy

If PSC was an analytical construct designed to serve a political function, so too was Frankland's depiction of the probable morbid poison of water-borne diseases, the germ, able to increase its numbers rapidly, resistant to the elements, and wholly undetectable. Despite the early work of Pasteur, the experiments of Hallier and Thiersch in Germany, and the hypotheses of William Crookes, Lionel Smith Beale, and Robert Angus Smith on the cause of the cattle plague of 1865, in the late 1860s germ theories remained in about the same state of speculativeness as they had been when John Snow had published the second edition of his cholera pamphlet in 1855. What had changed was the background of expectation. The concept of specific diseases, each caused by a unique morbid poison, was more widely accepted: if no less speculative, germs were more plausible. During the 1866 cholera Frankland had considered the possibility that the exciting causes of cholera and similar diseases might be 'the germs of organisms,'[22] and his superior, William Farr, who had already taken the step of positing and naming the particular cholera poison, had taken a similar view: what caused cholera was some extremely tiny particle 'at war with those that constitute man.'[23]

Neither felt a need to take a stand on whether the poison was obscure fermentation or belligerent cell; that is whether, from the modern viewpoint, the Liebig or the Pasteur model was the more accurate. There was no good way of choosing between these hypotheses, no consensus that the dichotomy was a real one, and in many respects it seemed not to matter. The Cattle Plague Commissioners had made it clear that the term 'germ' was to be used metaphorically, to represent a set of characteristics rather than an entity: 'the terms "germ," or "growth" are used because no better expressions can be found,' they wrote. 'They seem to imply an independent living existence of the poison, and on this point our knowledge is not yet sufficiently definite. Care must be taken that the terms used do not lead to erroneous conclusions.' Indeed, throughout the '60s and '70s the term 'germ' was an extraordinarily vague one; as J L W Thudichum pointed out as late as 1878, germs were eggs for some, seeds for others, 'shapeless ferments' for still others.[24]

The scheme of water analysis Frankland announced in the spring of 1867 could be justified alternatively in terms of germs or putrefying matter. Focusing on organic nitrogen made sense in terms of the zymotic theory since only nitrogenous organic molecules could acquire a dangerous form of putrefaction, but it also made sense in terms of a germ theory, since many felt that germs would require a congenial home of nitrogenous organic matter in which to 'vivify and develop' whilst in between human hosts.[25] In the next two years, however, Frankland did begin to recognize that living germs and putrefying matter differed in some important respects, and while for many years he would maintain that the identity of the agents of water-borne disease was not yet known, his analytical approach came increasingly to reflect the belief that living germs were morbid poisons.

By early 1868 Frankland had fully worked out his new scheme of water evaluation. He had introduced new techniques, new ways of interpreting results, and new formats for communicating conclusions to the public. The various aspects of this synthesis were made public in a lecture to the Chemical Society on 15 January, in testimony to the Water Supply Commission on 27 February, and in a Royal Institution lecture on 3 April. The January lecture detailed the new combustion process for direct measurement of organic carbon and nitrogen in water developed by Frankland and his student H E Armstrong. Their process was a modification of the procedure normally used to determine the proportions of the elements in organic matter. In it the sample was slowly evaporated with sulphurous acid, a mild reducing agent, which would destroy nitrites, nitrates, carbonic acid, and carbonates, leaving a residue in which any remaining carbon and nitrogen could be assumed to represent organic matter. The residue was then placed in a combustion tube in the presence of lead chromate, an oxidizing agent, and heated to combustion. From the combustion gases the analyst could compute the quantities of organic carbon and nitrogen, with the determination of organic nitrogen needing to be corrected only for ammonia. Though simple in principle, the process demanded exceptional skill from the analyst, took two days to complete, and required equipment and facilities beyond what many analysts possessed.[26]

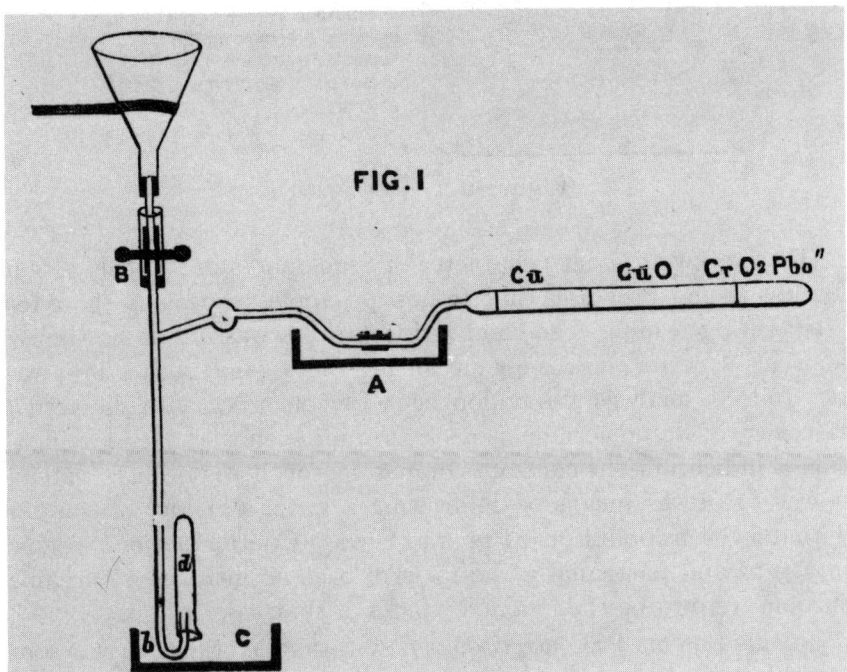

Figure 6.2 Edward Frankland's combustion apparatus for the analysis of the organic nitrogen and carbon in water. The simplicity is deceptive. The cost of the apparatus and skill needed to carry out the process meant that few chemists used it (*J Chem Soc* ns 6 (1868): 90).

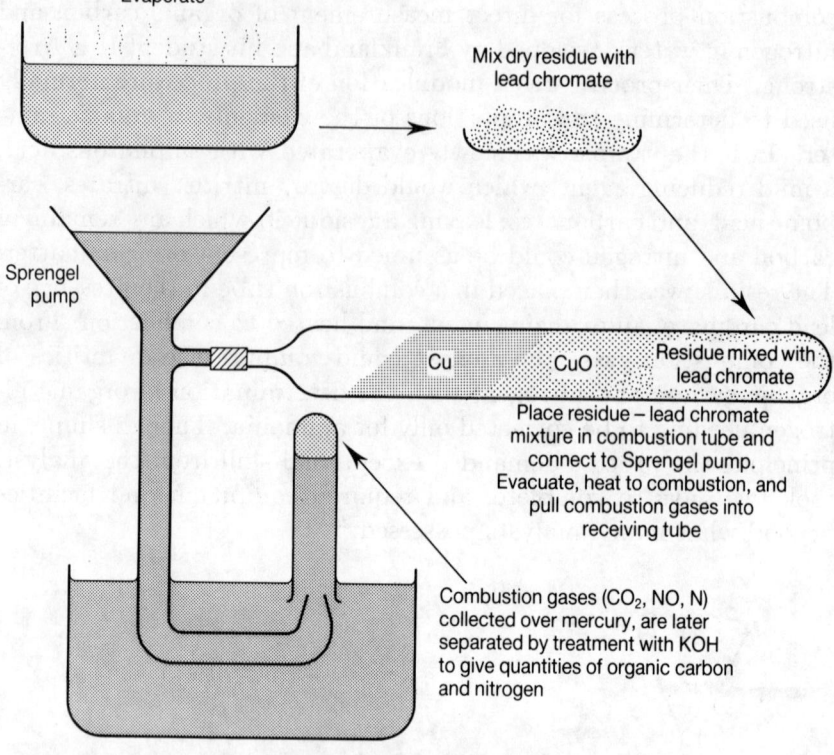

Evaporate

Mix dry residue with
lead chromate

Sprengel
pump

Cu   CuO   Residue mixed with
lead chromate

Place residue – lead chromate
mixture in combustion tube and
connect to Sprengel pump.
Evacuate, heat to combustion, and
pull combustion gases into
receiving tube

Combustion gases (CO_2, NO, N)
collected over mercury, are later
separated by treatment with KOH
to give quantities of organic carbon
and nitrogen

Figure 6.2   (continued)

Yet the revolutionary character of Frankland's analytical system lay less in new processes than in new principles, especially those for interpreting results. The combustion process was not to be simply another, if better, means for measuring the organic matter in a water. Indeed, analysis was no longer to be concerned with the actual discovery of water-borne poisons at all, nor even with the identification of indicators, the constant companions of those poisons. Instead it was to be the means of discovering a water's history of contact with dangerous pollutions. Previous Sewage Contamination, organic nitrogen, and particularly the ratio of organic nitrogen to organic carbon, were to be the indelible marks of that association.

In this scheme PSC acquired a new rationale. Since it was conceivable that living germs could survive even after the non-living sewage matter which brought them to the rivers or wells had decomposed, the presence even of such purified sewage (PSC) was a sign of immediate danger. As for the organic nitrogen and the N:C ratio, they yielded information about the source of the yet-to-be-decomposed organic substances in the water. By early 1868 Frankland had decided that it was important whether this organic nitrogen was of animal or vegetable origin for the agents of disease were likely to be associated only with animal contamination. He believed that each class of contamination would produce a characteristic nitrogen:carbon ratio, and, in general, that the worst contaminations, sewage for example, would show the greatest proportion of nitrogen. Hence this ratio might be more important than the actual quantity of organic matter present.[27] In practice Frankland only considered the N:C ratio in questionable cases; with regard to London's water, known to be contaminated with sewage, he habitually treated all organic nitrogen as actual sewage contamination and ignored the accompanying carbon.[28] Here too, however, Frankland was not claiming that organic nitrogen was harmful, or even that water having a particular N:C ratio indicated that harmful matter had contaminated the water. It showed only that the water had at some time past been contaminated with dangerous matter that might become actively harmful at any time.

Frankland's designation of a water's history as the primary concern of the analyst may seem obvious to modern readers. If we want to know whether there is anything bad in the water it may seem common sense to ask where it has come from and what kinds of things are likely to have gotten into it. In fact, Frankland's move was not at all obvious, but required the linking of two discrete traditions, the indicator approach of the 1850s and the concept of contingent contagionism.

Those who had used indicator arguments in the '50s, like Hassall and Hofmann, had been concerned with discovering a measurable entity that was present when the danger was present, but they were not much concerned with how the noxious matter had come to exist in the water. Indeed, they often assumed that the danger in water was simply a condition of foulness that water assumed when it contained a high concentration of putrescible matter and a low concentration of oxygen. Frankland's move away from this perspective occurred only gradually over the course of several years. In late 1866 he had still been looking at water quality in this traditional way, arguing that the key index of quality was putrescibility and the key index of putrescibility was the level of dissolved oxygen.[29] But as he became more of a contingent contagionist, convinced that the dangerous matter was a discrete substance that had entered the water at a particular time and place, Frankland became more concerned with finding the remains of substances that might have entered the water in the company of the dangerous matter. Increasingly, this would become the significance of the organic nitrogen measurement; it was a component of potentially germ-bearing sewage. Similarly, nitrates, nitrites, and ammonia would become important because they showed water had once been polluted and might still bear living germs. Thus no longer were indicators contemporaneous with the harmful substance; now they were to be regarded as fossil records of a dangerous event.

In early 1868 Frankland's conception of germs was still tenuous. In later years, he would continue to admit his ignorance of the nature of water-borne morbid poisons, but germs—germs with definite sizes and capabilities no less—emerged more and more prominently as a model of what morbid poisons might be like, a model which supported the conclusion that once-polluted water could never again be used safely. Vitality, for example, might be exactly the quality that would allow a morbid poison to maintain its virulence during the long flow between the sewage outfalls of upstream towns and the intakes of the London water companies. As an analogy he invited the Water Supply Commission to think of an egg floating down the Thames:

> if you were to break an egg and beat up the contents, and mix them with Thames water at Oxford, the organic matter so introduced into the Thames ... probably would be entirely destroyed and converted into mineral matter before it reached Teddington [near the companies' intakes]; but if you were to throw an egg in without being broken, it would be carried down by the stream and would reach Teddington with its vitality undestroyed.[30]

The reason the public could find no security in filtration was that germs might be so small as to pass through filters: 'I should not be prepared to say that after any amount of filtration we should be guaranteed from the presence of those minute germs, which being smaller in some cases ... than blood globules, would pass through the pores of the chalk ... like human beings pass through the streets of London.'[31] These examples illustrate the use Frankland was making of the germ concept. It offered an argument that was irrefutable because germs were hypothetical; it was sufficiently vague to be adapted to a wide range of rhetorical requirements, yet concrete enough to convey a vivid image—of a poison resisting the elements, of a poison able to slip between the pores of a filter.

New Analysis Meets Old Expectations: Frankland and the Royal Commission on Water Supply, 1868

For Frankland, the germ was more a symbol than a theory. It symbolized several characteristics of water-borne zymotic disease: the unpredictability of outbreaks yet the constant links to sewage pollution, the undetectability of the poison and its ability sometimes to resist the effects of dilution and oxidation, and the ability of the poison to increase itself in a suitable environment. It was not necessary to insist that these characteristics must inhere in some entity, though it was certainly convenient.

Frankland's testimony to the Water Supply Commission in February 1868 reveals the force of these arguments and the power of the germ as a symbol. His testimony confused the commissioners. 'Previous Sewage Contamination' gave them trouble, chiefly because they could not accept the underlying premises that no purification technique could be trusted and that chemical analysis could never show whether harmful materials were absent from water. If previous sewage contamination represented sewage that was no more, why then worry about it, they wondered. Their perplexity is not wholly Frankland's fault for he made his position clear early on:

> I consider that water contaminated with sewage contains that which is noxious to human health. There is no process practicable upon a large scale by which that noxious material can be removed from water once so contaminated, and therefore I am of opinion that water which has once been contaminated by sewage or manure matter is thenceforth unsuitable for domestic use.[32]

The dogma was clear, but did Frankland really mean it? The commissioners assumed he did not; that despite his strong statement, Frankland must, like any other chemist, use his analyses to determine whether a water was safe. Time after time Frankland made it clear that he rejected that conventional assumption. He blithely agreed that the nitrates he measured in London's waters (the main ingredient in the previous sewage calculation) were harmless. The Duke of Richmond took that response to mean that the water was good—if Frankland, a chemist, agreed that 'there is nothing in them [nitrates] that could be injurious to health, ... therefore the water is a wholesome water to drink?' 'I did not intend my statement to go so far as that,' Frankland replied. Nitrates might signal the presence of something dangerous.[33]

Frankland admitted that purification processes improved water

ANALYSIS of the METROPOLITAN WATERS in FEBRUARY 1868.   By Professor FRANKLAND, F.R.S., of the Royal College of Chemistry,

Companies.	Date and Place of Collection.	Temperature in Centigrade degrees.	Total Solid Impurity in 100,000 parts.	Organic Carbon.	Organic Nitrogen.	Nitrogen, as Nitrates and Nitrites.	Ammonia.	Total combined Nitrogen.	Previous Sewage Contamination. (estimated.)	Total Hardness.
THAMES.										
Chelsea - -	{ 4th February, Cab rank, Horse Guards }	7·2 C.	30·8	·326	·052	·340	·002	·394*	3,100	18·8
West Middlesex -	{ 4th February, Cab rank, Portland Road }	7·2 ,,	30·0	·357	·031	·321	·002	·354	2,910	18·4
Southwark and Vauxhall	{ 4th February, Barclay's Brewery }	7·7 ,,	32·2	·324	·032	·344	·001	·377	3,130	19·7
Grand Junction -	{ 4th February, Cab rank, Woodstock Street }	5·8 ,,	32·6	·329	·055	·343	·001	·399	3,120	21·1
Lambeth ·-	{ 4th February, Cab Rank, Lambeth Road }	7·2 ,,	31·2	·360	·045	·310	·001	·356	2,790	18·5
OTHER SOURCES.										
New River - -	{ 4th February, Cab rank, Tottenham Court Road }	6·6 ,,	30·8	·217	·026	·355	·001	·382	3,240	20·5
East London -	{ 5th February, Slater's Lamp Black Works, Old Ford Road }	7·2 ,,	34·4	·272	·037	·370	·003	·410	3,400	20·5
Kent . -	{ 7th February, Deptford Railway Station }	11·6 ,,	59·2	·081	·013	·564	·001	·578	5,330	30·0
Loch Katrine -	{ February 1867, Glasgow }	—	3·28	·256	·008	·031	·002	·041	0	0·3
Lancaster -	{ September, Lancaster }	—	3.54	·157	·001	·036	·001	·038	50	0·1
Manchester -	{ November, Manchester }	—	6·8	·242	·026	·001	·001	·028	0	2·7
Leicester -	{ November, Leicester }	—	23·7	·506	·020	·001	·001	·022	0	13·4
Preston -	{ December, Preston }	—	14·7	·515	·040	·001	·003	·044	0	6·7
Worthing -	{ November, Worthing }	—	36·7	·162	·000	·426	·000	·426	3,940	23·8
Column 1	2	3	4	5	6	7	8	9	10	11

* In the last month's report through a clerical error the total combined nitrogen in the Chelsea Company's water was given as ·339 instead of ·394.

The numbers in columns 4, 5, 6, 7, 8, 9, 10, and 11 all relate to 100,000 parts of the waters. The table is to be read thus: the Chelsea Company's water collected on 4th February at the Horse Guards' cab rank had a temperature of 7·2° C.; 100,000 lbs. of it contained 30·8 lbs. of solid impurity; the organic matter, constituting a portion of this impurity, contained ·326 lb. of carbon and ·052 lb. of nitrogen. This solid impurity also contained ·340 lb. of nitrogen in the form of nitrates and nitrites, besides ·002 lb. of ammonia, whilst the total amount of *combined* nitrogen in every form was ·394 lb. The above quantity of water supplied by the Chelsea Company had been, after its descent to the earth as rain, contaminated with sewage or manure matter equivalent to 3,100 lbs. of average filtered London sewage. By gradual oxidation, partly in the pores of the soil, partly in the Thames and its tributaries, and partly in the reservoirs, filters, and conduits of the company, this sewage contamination had been converted into innocuous inorganic compounds before its delivery to consumers. Finally, 100,000 lbs. of the Chelsea Company's water contained 18·8 lbs. of carbonate of lime, or an equivalent quantity of other soap-destroying ingredients.

REPORT for June 1868.

Figure 6.3 One of the first of Edward Frankland's monthly analyses of London's waters done according to the format that the water companies found objectionable. Especially troublesome were references to thousands of pounds of previous sewage contamination (*Report on the Analysis of the waters supplied by the Metropolitan Water Companies*, p 49, 1872).

and that whatever lowered PSC could be assumed to be making water safer. He even developed a scheme for classifying potable waters based mainly on previous sewage contamination.[34] He supported measures to make Thames and Lea water safer, and saw great room for improvement in filtration since the filters of some of the companies did a better job purifying the same water than those of others. Sewage treatment by upstream towns would help too. But while all these might improve the water, none of them changed the fact that London relied on sewage-polluted water, and that such water was inherently unsafe.[35]

What so exasperated the commissioners was that Frankland was advocating an approach to water analysis wholly incompatible with the practice of his colleagues, even those who admitted that morbid poisons were beyond the reach of analytical chemistry. The commissioners expected chemists to base their advice on the results of their analyses. Hitherto chemists had believed that even though they might not be able to isolate the causes of water-borne disease they could offer useful approximations based on the measurement of indicators. Frankland rejected even this limited assumption. Water with a bad history was to be allowed no redemption, no matter how pure it might appear. Eventually Frankland would liberate the concept of impurity from all analytical constraints. The Appendix demonstrates how, as he became familiar with the analytical characteristics of different kinds of lake, river, spring, and well waters, he modified his interpretive framework in such a way that any set of results could lead to the conclusion either that the water was safe or polluted, depending on what was independently known of its history of contamination.

This approach, in which the actual analysis was secondary, marked a sharp break with tradition. Most of Frankland's predecessors (and most of his colleagues) took the view (at least in public) that analysis was the final authority. This claim had been asserted and defended by generations of mineral water analysts and by promoters of the indispensability of chemistry, men like Brande. They might admit that analysis was not perfect, but maintained nevertheless that whatever his limitations, the chemist could offer some crucial bit of information, otherwise unavailable, that could significantly change the assessment of a water. If chemistry were to play this role of final arbiter, it seemed to follow that the chemist's assessment could not depend on other public sources of information such as knowledge of the source of the sample or of the effects on health of the water under examination. Indeed the power of chemistry seemed to stem directly from the chemist's ability to discover such secrets.

From such a perspective it might seem that Frankland was guilty of trading under false colours since analysis was not the main basis of his evaluations. In 1884 William Odling, then an analyst to the London water companies, condemned Frankland's practices as

> an abuse of chemistry, that a chemist ... should state and summarise the results of his analyses in such a fashion as to make it appear that the unwholesomeness, which he really infers on other grounds, is strictly deducible from the results of his periodical chemical examinations.[36]

To those with conventional notions of the role of chemical analysis, Odling's complaint was well founded. It is easy to get the impression from Frankland's reports that chemical analyses were the sole basis for his judgments of water quality, particularly if that is what one expects from a chemist and if one only reads one of the brief monthly reports.

In fact Frankland had recognized that potable water analysis presented a different kind of problem from many other problems of practical analytical chemistry, such as the evaluation of fertilizer, mineral ores, or foodstuffs alleged to be adulterated. In such cases there was reason to believe that statements made as to the purity or value of the material might be untruthful; the analyst was to validate or reveal the falsity of those statements. But such was not the case with potable waters. The circumstances of their origin were unavoidably public. There were cases in which clients concocted water samples to deceive analysts, but only a few. Hence in coming to a judgment on the quality of a water there was no reason not to take into account all the information one could assemble including descriptions of the circumstances of the site from which the water had been taken and details of its apparent effects on those who drank it.[37]

Frankland, then, was using chemistry to complement other sources of information. In his view the sort of expert needed was someone with comprehensive knowledge of epidemiology, of the nature of morbid poisons, and of the pollutions to which waters were subject, not someone with narrow expertise in analytical chemistry. Sound advising required the integration of all these sources of information. However heretical this view might be in the late 1860s, it did become the dominant perspective among water analysts and was a central tenet of an analytical protocol developed by the Society of Public Analysts in the early 1880s.

Frankland Becomes an Authority, 1867–68

During the years that Frankland was developing his radical views on water quality he was also gaining prominence as a sanitary scientist. In spring 1867 he was still not a specialist in water analysis. A year later he was rapidly becoming Britain's leading authority on water quality. It is not clear that Frankland meant this to happen though he did make the most of two fortuitous opportunities. These were the establishment of the Royal Commission on Water Supply in early 1867 (Frankland was appointed one of its consulting chemists) and the collapse and subsequent re-establishment of the Royal Commission on Rivers Pollution, with Frankland as one of the new commissioners.

The main task of the Water Supply Commission was the evaluation of the schemes put forward as alternative sources of supply for London. To assess the quality of the various alternatives, it chose Frankland and William Odling, his successor at St Bart's and colleague at the Royal Institution. They submitted three reports, one on the waters from Cumberland and Wales (November 1867) and two others on the Thames (July and September 1868). In two of these substantial portions appeared under Frankland's name alone—his views on water quality had already become too extreme even for a sympathetic colleague.

It is likely that Frankland's service on the Water Supply Commission led to his appointment to the new Rivers Pollution Commission in April 1868, and that the appointment had more to do with the Treasury's tightness than Frankland's expertise. Shortly after the first Rivers Commission disbanded in January 1868 the Treasury asked William Pole, secretary of the Water Supply Commission, whether the Rivers Commission's laboratory could be used by the Water Supply Commission. Pole was not sure that the laboratory would be appropriate and was instructed to check with his chemist (Frankland). In April 1868 Frankland was chosen for the chemist's spot on the three-member Commission and it may well have been that sharp minds in the Treasury saw a way to avoid supporting duplicate water laboratories by having a single analyst for both commissions.[38]

While Frankland's influence on the Water Supply Commission was small—his views were restricted to appendices and to the minutes of his evidence, and the Commissioners neither accepted nor fully comprehended them—his impact on the second Rivers Commission was enormous: he effectively became the Commission. The first (1865) Commission (of Rawlinson as engineer and chairman, Way as chemist, and John Thornhill Harrison as agriculturalist) had been instructed to report on six representative kinds of river pollution: the agricultural and sewage pollutions the Thames received, and the pollutions associated with the Lancashire cotton industry, the Yorkshire woollen industry, the metals trades along the lower Severn, and the mining and metals trades along the Taff in south Wales and along one of the Cornish rivers.[39] In almost two and a half years it had accumulated a mass of ill-ordered information, but completed only two of these investigations.

Along with Frankland the new Commissioners appointed in 1868 were Sir William Denison, an ex-army engineer and colonial administrator, and John Chalmers Morton, an agriculturalist and journalist. Denison died in 1870, having collaborated only on the Commission's first report. In the next four years Frankland and Morton published five additional reports: on the ABC patent sewage precipitation process (1870), on pollution by the woollen industry (1872), on pollution in Scotland (1872), on pollution from mining and manufacturing operations (1874), and on the nation's water supply (1874). Morton appears not to have played a strong role in directing the Commission's inquiries or in shaping its conclusions, and we may accept, as did many of Frankland's contemporaries (defenders and critics), that the Commission was for all practical purposes Frankland's mouthpiece.[40]

The second Commission was far better organized than the first had been. Its first report, though ostensibly concerned with pollution in the Mersey and Irwell basins, began with a thorough review of the state of the art in sewage purification and water analysis and presented the results of experiments undertaken to test widely held generalizations. With respect to the claim that rivers spontaneously became pure, Frankland showed that they did so only very slowly. With respect to sewage treatment by irrigation, a technology which royal commissions had touted for nearly two decades, Frankland determined the purification capacities of different soils and showed that any soil had a far higher capacity if one began with a small dose and raised the application rate slowly. With respect to water analysis

Frankland made clear why no process for determining organic matter could provide useful information about the most harmful contaminants in water and presented the combustion process as the only scientific means of water analysis.

On the other issues too—standards for effluents, processes for recycling industrial wastes, determinations of the effects on health of sewage polluted water—Frankland relied on experiment or systematic observation, where his predecessors had been content with speculation.[41] Besides offering authoritative answers to vexed questions, the six reports were also striking in the quantity and organization of information they contained. Frankland's assistants, mostly students in the School of Mines, had done more than 3200 analyses of river water, town water supplies, and urban and industrial effluents.[42] The Commission sent detailed questionnaires to each town or manufacturer it investigated on such matters as sewerage and sewage treatment, water supplies, and general sanitary conditions. Having collected information in this way the Commissioners were much more successful in making generalizations about river pollution in Britain as a whole. They were also able to ask more probing follow-up questions in local hearings. One reviewer aptly commented on the work, 'It treats of the subject in its entirety. No argument is left out, no proof is wanting. Each statement is carefully verified by experiment and observation and the whole work is filled with analyses and the most complete and minute details.'[43]

Hence regardless of whether he had set out to become a water expert, Frankland had become one. The Rivers Commission appointment was, if not a full time position, certainly a demanding one. It provided Frankland with the means (a salary of £800/year and an additional £700/year for the laboratory) to carry out thorough investigations on all questions of water quality. So superior were the Commission's reports and so closely was Frankland's name linked to them, that he became the leading authority on water questions.

Late in 1873 the Rivers Commission completed its work, having spent £39,625 since 1865, making the combined rivers commissions the most costly government-sponsored scientific research that had yet been undertaken.[44] The immediate effects of the reports were modest: a weak anti-pollution bill based loosely on its recommendations passed Parliament in 1876. The long-term effects were more far-reaching. Thorough, definitive, and superbly organized, the reports acquired an international reputation. Long extracts were translated and published by French and German public health agencies.[45] The Commission's opinions were taken seriously and its recommended effluent standards were widely adopted. The standards were used in law courts to define pollution despite the fact that they had not been included in the 1876 Act and hence had no legal standing.[46]

Most important was Frankland's emergence as a spokesman for those seeking to obtain uncontaminated water supplies. The 1870s and 1880s were a period of water reform: towns were buying out private water companies and acquiring new and purer supplies. Often their bills were opposed in Parliament and Frankland regularly appeared as an expert witness to defend such proposals. He was an effective witness: experienced, quick witted, prestigious, in command of a wealth of facts, familiar with continental science, able to explain himself clearly.[47] No previous sanitarian nor any contemporary could match his combination of activism and authority.

During these decades Frankland continued his monthly reports on the London water supply. Often he found reason to criticize one or more of the water companies: turbidity, a high PSC or organic nitrogen level, or the presence of 'living and moving organisms.' He stopped calculating previous sewage contamination at the end of 1876 but replaced it with a mode of comparing different waters that the companies found equally objectionable: Frankland's 'Table E' showed how many times more impure in terms of organic nitrogen were the waters the river-water companies drew from the Thames or Lea than the water the Kent Company pumped from deep wells.

Prior to the development of bacteriological culture methods in the mid 1880s, there were few significant changes in Frankland's approach to measuring water quality. An 1876 paper on 'Some Points in the Analysis of Potable Waters' presented improvements in the combustion process and reiterated the unacceptability of its competitors. An 1880 paper 'On the Spontaneous Oxidation of Organic Matter in Water' was a polemic denying the oxidative self-purification of rivers and an attack on the views of Charles Meymott Tidy, a physician, chemist, and later barrister who was Frankland's principal adversary on water matters. In the last decade of his life, Frankland's opposition to the water companies moderated and he was not a protagonist in the water controversies of the '90s. For two full decades, however, he was at the centre of questions about water quality, and the positions taken by other scientists, and the analytical processes that were developed or discarded, only make sense when Frankland's central role is understood.

1   A H Hassall, 'On "Living Organisms" in Potable Water,' *FWA* 1 (1872): 143–4.
2   *Lancet*, ii, 1856, pp 576.
3   G R Burnell, 'On the Present Condition of the Water Supply of London,' pp 169–77; J F Bateman, 'On the Present State of our Knowledge of the Supply of Water to Towns,' pp 62–77; *Lancet*, i, 1866, pp 212, 235; *Times*, 13 Aug 1866, 9d; 20 Aug 1866, 5f; *Builder* 23 (1865): 313. For the fullest account of the alternative schemes see R C Water Supply, *Report*.
4   But see Luckin, 'The Final Catastrophe,' pp 32–42.
5   *R C Water Supply*; William Farr, *Report on the Cholera Epidemic of 1866 in England*; *Report by Captain Tyler on the Quantity and Quality of the Water supplied by the East London Waterworks Company*; J Netten Radcliffe, 'Cholera in London especially in the eastern districts,' in *9th Annual Report MOPC*, pp 264–331; R C Rivers Pollution (1865), *First Report* and *Second Report*; *S C Thames Navigation Bill*; *S C East London Water Bills*; 'Correspondence between the Board of Trade and the East London Water Works Company with reference to Captain Tyler's Report'; *S C Lee Conservancy Bill*; Lancet Analytical Sanitary Commission, 'On the epidemic of cholera in the east end of London,' *Lancet*, ii, 1866, pp 157–60, 217–9, 273–66.
6   'Reports on the Examination of Thames Water,' *JRSA* 31 (1882–83): 88. On R Dundas Thomson (1810–64) see *DNB* 19, p 748; *Lancet*, ii, 1864, p 226.
7   Partington, *A History of Chemistry* IV, pp 500–1; W H Brock, 'Frankland, Edward,' *DSB* 5 (1972): 124–7.
8   H Letheby, 'Methods of Estimating Nitrogenous Matter in Potable Waters,' p 429.
9   Burnell, 'On the Present Condition,' pp 171–3. Cf GBH Medical Council, *Report of the Committee for Scientific Inquiries on the 1854 Cholera*, p 178.
10   Eyler, *Victorian Social Medicine*, pp 22–3, 35, 100, 141, 161, 199–200.
11   W A Miller, 'Observations on the Analysis of Potable Waters,' pp 120–24; discussion in E Byrne, 'Experiments on the Removal of Organic and Inorganic Substances in Water,' pp 6–53; Playfair in *GBH MWS*, p 78; C R C Tichborne, 'On the Nature and the Examination of the Organic Matter in Potable Waters,' *CN* 17 (1868): 147–9; T Spencer and others, *CN* 17 (1868): 192–3, 203, 215.
12   C B Fox, *Sanitary Examinations of Water, Air, and Food*, 2nd edn, pp 26–7.
13   E Frankland, 'Water supply of the Metropolis during the year 1865–1866,' pp 240–2. See also *CN* 13 (1866): 136.
14   Edward Frankland, 'The Purification of Water,' *CN* 14 (1866): 71.
15   But see Farr, *Report on the Cholera of 1866*, p xxv. Farr argued that

the reason not everyone who drank East London Company water came down with the disease was due in part to the fact that the poison was not evenly distributed in the water.

16 *S C East London Water Bills*, App 8, p 357.

17 *Ibid*, pp 373–4.

18 Edward Frankland, 'On the Water Supply of the Metropolis,' [RI], 114–9.

19 'Dr Frankland's Report on the London Water Supply in February,' *Lancet*, i, 1867, p 347, ii, p 493.

20 Frankland, 'On the Water Supply of the Metropolis,' p 118.

21 A Hassall, 'London Water,' *FWA* 1 (1872): 121; 'On "Living Organisms" in Potable Water,' *FWA* 1 (1872): 143–4. See also 'The Contamination of Water,' *Lancet*, 1867, ii, p 493.

22 *S C East London Water Bills*, App 8, pp 352, 373–4; R C Cattle Plague, *3rd Report*, pp vi–ix, 146–54, 156–9, 187–8; Foster, *A Short History of Clinical Pathology*, pp 53–7; Pelling, *Cholera*, pp 236–44; Crellin, 'Airborne Particles and the Germ Theory,' pp 49–60.

23 Farr, *Report on the Cholera of 1866*, pp lxv–lxxxi; Eyler, *Victorian Social Medicine*, pp 105–7.

24 R C Cattle Plague, 3rd Report, p. vi, fn; Thudichum in C T Kingzett, 'The Chemistry of Infection, or the Germ Theory of Disease from a Chemical Point of View,' *JRSA* 26 (1877–8): 311–20, esp p 318; 'The Cholera Poison,' *CN* 14 (1866): 84, 109; Gairdner, *Public Health in Relation to Air and Water*, pp 31, 72, 158–78, 191–2.

25 E Frankland, 'The Water Supply of London,' [*QJS*], p 315.

26 E Frankland and H E Armstrong, 'On the Analysis of Potable Waters,' pp 88–98. The early use of the process is described in letters from Frankland to Armstrong of 8 January 1867, 21 September 1867, 20 April 1868, undated (late 1868), in *Royal Society Misc. Mss.* v 10, #90–93, and in letters from J J Day to Armstrong of 23 November 1867, 14 December 1867, 4 April 1868, 6 May 1868, 21 June 1868, 27 March 1869, in *Imperial College, Armstrong Papers*, C241–C246. According to J Vargas Eyre, Armstrong was the real developer of the process working on 'the barest instructions' from Frankland (J Vargas Eyre, *Henry Edward Armstrong, 1848–1937, The Doyen of British Chemists and Pioneer of Technical Education* [London: Butterworth Scientific, 1958], pp 40–1).

27 RCWS, *Evidence*, Q 6289.

28 *Ibid*, Q 6405.

29 Edward Frankland, 'On the Effect of Temperature on Organic Matter in Water,' *CN* 14 (1866): 275.

30 RCWS, *Evidence*, Q 6372.

31 *Ibid*, QQ 6401, 6244.

32 *Ibid*, Q 6222.

33   *Ibid*, Q 6238.
     0
34   RPPC, 1868, *6th Report*, pp 16–17.
35   RCWS, *Evidence*, QQ 6235–6, 6246, 6376, 6390–92, 6381, 6396, 6417–18.
36   William Odling, 'On the Chemistry of Potable Water,' *CN* 50 (1884): 206.
37   See H Swete, 'On the Interpretation of Water Analysis for Sanitary Purposes,' *SR* ns 1 (1879–80): 182.
38   *PRO HO* 74 3, pp 425, 427, 433.
39   RPPC, *1st Report*, 1865, p 4.
40   *CN* 64 (1891): 222–23; 'London water supply,' *CN* 45 (1882): 180–1; C M Tidy, 'The Treatment of Sewage,' *Van Nostrand's Eclectic Engineering Magazine* 35 (1886): 3; *idem* in disc. of P Frankland, 'The Upper Thames as a source of Water Supply,' p 446; W Dunbar, *Principles of Sewage Treatment* trans. by H T Calvert (London: Griffin, 1908), p 23; George Rafter, 'Sewage Irrigation,' *USGS Water Supply Paper #3* (Washington, DC: GPO, 1897), pp 29, 45–8; Theophile Schloesing, 'Assainisement de la Seine. Épuration et utilisation des eaux d'egout. Rapport fait au nom d'une commission,' *Annales d'Hygiène Publique et de Medecine Légale*, 2nd ser 47 (1877): 207.
41   RPPC, 1868, *1st Report* (Mersey and Irwell Basins), pp 13–23, 28–39, 43–8, 63–70, 96–102, 112–7.
42   RPPC, *5th Report*, 33 (1874), p 51.
43   'Professor Frankland,' *Biograph and Review* 4 (1880); 336.
44   'Return of all royal Commissions issued from the year 1866 to the year 1874,' *P P*, 81, 1888, (426).
45   *Reinigung und Entwassering Berlins* (Berlin: Hirschwald, 1871), v 1–2; Prefecture de la Seine, *Assainissement de la Seine. Épuration et Utilisation des Eaux d'Egout. Documents Anglais* (Paris: Gauthier Villars, 1877).
46   T B Dudley, *From Chaos to the Charter*, p 188.
47   Cf House of Lords Record Office, [HLRO] *Minutes of Evidence*, v 5 (S C Cheltenham Corp Water Bill), pp 246–303.

7 Frankland and the Chemists, 1866–85

month after month, with a process of analysis that is faulty in the extreme, and with speculations that admit of no proof, we are either frightened or amused [by figures] ... showing that such water ... had been contaminated with sewage or manure matter equivalent to so many hundred parts of average filtered London sewage. Gentlemen, if this were not put forth in the garb of science, and, moreover, in a semi-official form, it would be regarded as a burlesque, and would excite nothing but ridicule.[1]

H Letheby

To make sense of Frankland's career as a water analyst we must view it in two contexts. The first is of water policy-making and of the participation of scientists in that process. The second is of the growth and changing structure of the profession of chemistry. During the period in which Frankland came into government service, government-by-experts was coming to the fore. According to Oliver Macdonagh and those who have followed him they came in through a common pattern.[2] Entering government initially to cope with crisis or as a result of tepid legislative initiative, inspector-experts expanded their mandates, discovered ever more that needed doing, and by the presentation of undeniable fact, prevailed upon the legislature to grant them expanded powers. At first glance Frankland's career seems to conform to this pattern. He had stepped into a marginal position and expanded its importance. He inspected, found things wanting, and reported, calling for major new legislative initiatives. Yet the divergence from this pattern is more important. Despite his official appointments, Frankland retained his independence as a consultant. And his policy recommendations were not so much the outcome of his investigations, but the result of principles of water supply he had adopted early in his career as a water analyst. His subsequent investigations and reports, from 1870 onward, were more important in maintaining his reputation as a well-informed expert than in advancing his views, which changed little until the mid '90s. His monthly reports on London's water had their greatest impact directly on public opinion through their publication in newspapers; those officials to whom they were directed ignored his recommendations on those occasions when they were not actively trying to moderate his extreme interpretations.

Frankland's stature as a chemist, his skilful exploitation of his official status, and his adroit circumvention of the limits of empirical investigation profoundly affected British water analysis for the remainder of the century. To understand how we need to understand the structure of British chemistry. For present purposes we can divide British chemists into three groups. First, there were several other eminent chemists, holders of professorships, recipients of government patronage with interests in sanitary matters—chemists, that is, who would seem to have been well enough placed to have acquired prominence in water matters equal to Frankland's but didn't. Second were other prominent chemists who for various reasons came to form a loosely knit faction in opposition to Frankland. Most important were William Crookes, James Alfred Wanklyn, and Henry Letheby and Charles Meymott Tidy, the latter two the principal consultants to the London water companies. Third was the large (and growing) mass of practicing chemists, often with training from the Royal College of Chemistry or another technical school, and working for the most part quietly in the provinces in industry and in private analytical practice, which might occasionally include water analysis. The great range of competence and ethical standards among these chemists led the leaders of the profession, including Frankland, to establish the Institute of Chemistry in the early 1880s, an organization which would certify competence and set professional standards.[3] The first two of these groups are considered in this chapter, while the public analysts, those of the rank-and-file chemists who took most interest in water analysis, are considered in the next.

Potential Rivals

During the years following the cholera of 1866 many of Frankland's colleagues were wrestling with the same problems of purification and analysis, and many of them came to share his belief that chemistry could neither discover the poisons of cholera nor guarantee any method of removing them. The range of opinion (and bewilderment) is evident in the three nights of discussion that took place at the Institution of Civil Engineers following a short paper on water filtration by Edward Byrne in 1867. At issue was whether water was harmful, what it was in water that was harmful (changing organic matter, disease germs, organic and inorganic poisons of some kind), how it acted (by predisposition, infection, or neither), how to analyse the organic matter in water (ignition and permanganate processes were both objectionable), and what, if anything, filters did to purify water. Well respected experts (including arch enemies Thomas Hawksley and Edwin Chadwick) could still be found who dismissed the notion of water-borne disease. There was a great variety of dogmatic opinion on particular points at issue, but no clear route to consensus, either in matters of analytical technique or water policy.[4]

Among the most prominent of Frankland's contemporaries taking an interest in such issues were W A Miller, William Odling, and Robert Angus Smith. Miller, professor of chemistry at Kings College, had been with Hofmann and Graham one of the three 'government chemists' to investigate London's water in 1851. He later served as consulting analyst to Britain's senior public health official, John Simon. In June 1865 Miller presented the Chemical Society with his 'Observations on Some Points in the Analysis of Potable Waters.'[5] Before joining Frankland as one of the consultant chemists to the Water Supply Commission, William Odling had gained experience in sanitary matters as medical officer to the Parish of St Mary's Lambeth. Trained as a medical man, he had studied chemistry with A S Taylor at Guy's Hospital and was among the most successful in combining medicine with chemistry. Odling became one of the leading chemical theorists in Britain and succeeded Brodie in the Oxford chemistry chair in 1872. In the early '80s he became one of the analysts for the London water companies and the most astute of Frankland's critics on water analysis.[6] Chadwick's ally Robert Angus Smith had been one of the Cattle Plague Commissioners who in 1865 had tentatively considered the implications of a germ theory of disease. In taking a serious interest in the work of Pasteur he was unusual among British chemists, and tried to take water analysis away from a narrow focus on the quantity of organic matter. All three testified to the Water Supply Commission on the state of water analysis.

These three chemists shared most of the elements of the Frankland–Brodie perspective. Miller agreed that chemists did not know exactly what in water was harmful or potentially harmful and agreed that both analysis and purification were untrustworthy, but he thought that if chemists would only adopt uniform procedure, it would be possible to correlate composition with health. On the key question of whether contaminated water could be safely consumed, Miller replied that 'in the majority of instances' it could though 'there may be cases in which danger is produced.'[7] Odling, similarly vague, saw drinking water purity as simply 'a practical question'—the purer the better, but water might still be acceptable even if it were not quite so pure. In 1860 and again in 1868 he expressed general satisfaction with the water of the London companies, yet admitted great inadequacies in water analysis.[8] Smith believed cholera to be caused by some sort of living germ, which if 'carefully nursed' would be visible through the microscope.[9] He still held that because not all types of organic matter were equally harmful, the population of microscopic creatures in a water rather than its chemical composition was the best indicator of the type of organic matter present, yet he retained the old and vague Hassallian standard; it was the number of organisms, their size, and disgustingness that remained the measure of water quality.[10] On the vexed issue of previous sewage contamination all three acknowledged that inorganic forms of nitrogen sometimes came from sewage but denied that they afforded grounds for calculation of Frankland's 'previous sewage contamination'. They all expressed great admiration for the combustion process—indeed Odling was its most ardent early defender—yet none of them followed Frankland in insisting on the 'utter untrustworthiness' of alternative processes.[11] Odling, like Miller and Smith, used the permanganate process, Wanklyn's new process, and Frankland's process, acknowledging that each offered useful information.

Each of these three was a chemist of stature, each held some manner of official appointment relating to sanitary chemistry and water quality; any one of the three might well have become the dominant figure in water analysis. Yet though each recognized much that was problematic in water analysis none of them saw the need for such thoroughgoing reformulation as Frankland did, and consequently none was able to provide the leadership that he did. Satisfied that science, however inadequate the existing state of its art might be, provided the best guide to better water, they saw no need for Frankland's manipulations of public opinion.

These three, however, remained on the margins of water matters. They analysed, consulted, testified, yet avoided being drawn into the maelstrom of controversy that surrounded Frankland. Such was not the case with William Crookes, Henry Letheby (and his successor Charles Meymott Tidy), and James Alfred Wanklyn. The careers of these men, at least in matters of water quality, came to be dominated by opposition to Frankland.

Crookes and *Chemical News*

As editor of the *Chemical News*, the profession's chief trade paper, William Crookes had a significant impact on the reception of Frankland's ideas. Crookes' pages were open to controversy, and through editorials and book reviews he took part. Crookes vacillated for several years before finally coming to an anti-Frankland stance on water matters. Although in 1865 he had been one of the developers of a germ theory of the cattle plague, he initially had little patience with Frankland's undetectable and unremovable water-borne cholera particles. In Crookes' view germs were air-borne: they were in sewer gas and they infected cisterns, cholera therefore had more to do with local unsanitary conditions than with the systematic distribution of bad water. Basing his conclusions in part on the 1851 Graham, Miller, Hofmann, study, Crookes maintained that 'there is no evidence, physical or chemical, to show that this [organic matter in London water] is otherwise than harmless.'[12]

By April 1867, however, Crookes had admitted that the East London company's polluted supply and poor management had caused the epidemic there, though he was relieved to learn that the fault had been the engineers', not the chemists'.[13] The spring of 1869 marked the high point of his support for Frankland. He defended 'previous sewage contamination': 'the phrase simply states an indubitable fact.' He recognized that Frankland was not denying that bad water might become pure, only challenging the claim that purification was inevitable. The burden of proof, Crookes argued, properly lay with those who put their faith in purification.

> We know as an absolute fact that tons upon tons of human excrement are thrown into the waters of the Thames and ... Lea before we drink them, and we have a right to look for absolute demonstration of the complete destruction of all this filth; and not only

of the filth itself—of the lifeless organic matter;—we must also be convinced that it is impossible, at all times and under all circumstances, for living matter—the low forms of life and their germs with which the processes of disease and putrefaction appear to be so closely connected—to retain their vitality.[14]

But growing differences caused a rift between the two in the early '70s. While in 1869 he had praised the combustion process as 'very greatly more accurate than any that has hitherto been suggested,' by 1872 he was urging a ban on its use in official investigations.[15]

The grounds for Crookes' opposition were threefold. The first was personal. The two disagreed about the best means of sewage treatment. Crookes favoured the sewage precipitation approach in which chemicals were added to sewage to coagulate suspended matters into a form that could be marketed as a dry fertilizer. Frankland regarded such schemes as unworkable, even downright fraudulent. In 1868–69 his Rivers Commission investigated the ABC process of the Native Guano Company in which Crookes was deeply involved as a director, publicist, chemical advisor, salesman, and even for a time as chairman. Frankland made an example of the process, hinting that Native Guano was trying to fool its manure customers by sending out doctored samples and even trying to fool his investigators by secretly diluting its effluent.[16] Throughout the 1870s *Chemical News* championed precipitation and attacked Frankland's alternative of land treatment as a rarely practicable technique advocated by muddle-headed idealists.[17]

The other grounds for opposition were more clearly professional. *Chemical News* was a mouthpiece both for professional chemists and for the British chemical industry. Industrialists rejected Frankland's claim that industrial wastes could be profitably recycled and maintained that the Commission's proposed effluent standards would be impossible to meet. Crookes took up their case: the Commission's effluent standards were harsh, inconsistent, and arbitrary.[18] Parliament did drop these standards from unsuccessful bills presented in 1872 and 1873. Much to Crookes' chagrin, the Rivers Pollution Prevention Act which passed in 1876 had no standards whatever.[19]

Finally, Crookes was concerned with the implications of Frankland's system for the practice of water analysis. The time, expense, and experience required to use the combustion process successfully meant that for practical purposes it was unavailable to most analytical chemists, who would only occasionally be analysing waters. According to Cornelius Fox, a leading medical officer of health, it would require six months for an ordinary medical man to learn the process while the apparatus would cost at least 13 guineas.[20] What concerned Crookes and others was that the effluent standards proposed by the Rivers Commission were expressed in terms of organic carbon and organic nitrogen, variables that could only be measured by the combustion process. If these standards became law water analysis would become centralized. Chemists would be forced either to adopt Frankland's costly techniques or abandon water analysis. The rank and file of analytical chemists who made up the *Chemical News* constituency did not use combustion, Crookes pointed out repeatedly. The issue was thus one of professional democracy: chemists looked upon Frankland's process as 'impracticable and fallacious'; in the name of professional unity he ought to give it up.[21]

These concerns were inter-related and the issue of the validity of Frankland's water analysis underlay all of them. If Frankland's analysis were discredited the rest of his water science would fall with it. Crookes was an able polemicist and rarely neglected an opportunity to make a scornful aside about Frankland's water science. A review of Pettenkofer's work on cholera or of an annual report by the Massachusetts State Board of Health was occasion to snipe at PSC or the Rivers Commission.[22]

The Wanklyn Affair

Crookes' criticisms were occasional; he had other polemics to occupy him. Such was not the case with James Alfred Wanklyn, for whom revenge on Frankland became a *raison d'être*. Wanklyn's criticisms of Frankland were the longest sustained, the bitterest, and yet the most substantive. Between 1868 and 1877 the relative merits of Frankland's combustion and Wanklyn's 'ammonia' processes dominated discussions of water analysis in Britain. In Wanklyn's view the dispute was over which process better measured the harmfulness of a water. To Frankland such a quest was futile; analysis was to shed light on a water's history. Nonetheless, the debate was conducted largely on Wanklyn's terms, it being assumed by most chemists that the process that more accurately determined the quantity of putrescible matter gave the better measure of harmfulness. Indeed, quibbling about parts per million of organic nitrogen or 'albuminoid ammonia' completely displaced the more important questions of what, if anything, these entities signified. Still, Wanklyn's

criticisms are important: his attacks on Frankland's character dam-
aged Frankland's credibility; his substantive criticisms of the combustion process were extensively used by others who knew little of
chemistry and whose reasons for attacking Frankland lay elsewhere.

In its early stages Wanklyn's career had paralleled that of Frankland, his senior by nearly a decade. A Lancastrian like Frankland,
he had been apprenticed to a Manchester doctor and later studied
with Frankland at Owens College in the early '50s, where he was
Frankland's assistant. Like Frankland he took advanced training
at Bunsen's laboratory at Marburg and with Frankland's assistance
secured appointment as a demonstrator to Lyon Playfair in 1859.
From 1863 to 1870 Wanklyn was a lecturer at the London Institution; he spent the rest of his career as public analyst to a number
of authorities, as a lecturer at St George's Hospital, and in private
practice as an analyst.[23]

In June 1867 Wanklyn described to the Chemical Society the new
'ammonia' process that he had developed with E T Chapman and
Miles H Smith. The process was based on the belief that albumin
was the dangerous material in water since albuminoid substances
putrefied unusually rapidly. Wanklyn believed that a definite proportion of a water's albumin was converted to ammonia when the
water was distilled with a caustic solution of potassium permanganate. The amount of this 'albuminoid ammonia' could be determined by the Nessler colour test, a common and well accepted
approach for quantitative determination of ammonia. This quantity
was to be the main index of water quality.[24]

Wanklyn's process was thus made public in the middle of the same
year that Frankland was developing the combustion approach. Both
were working in the Hofmann tradition and were deeply concerned
about the types of organic nitrogenous matter that might undergo
dangerous decomposition. Like Frankland, Wanklyn rejected existing evaporation techniques. He ridiculed Miller's call for uniformity
in analytical procedures and in the statement of analytical results,
arguing that to pursue uniformity on Miller's terms would merely
be to multiply comparable results of doubtful accuracy.[25]

Initially British chemists were not impressed and Wanklyn and his
collaborators quickly found it necessary to moderate their claims.[26]
The first Wanklyn–Chapman–Smith paper had reported trials on
natural waters. When Wanklyn and associates tried to verify the
process by using it on artificial solutions of pure organic compounds
they ran into some surprises. In early experiments pure albumen

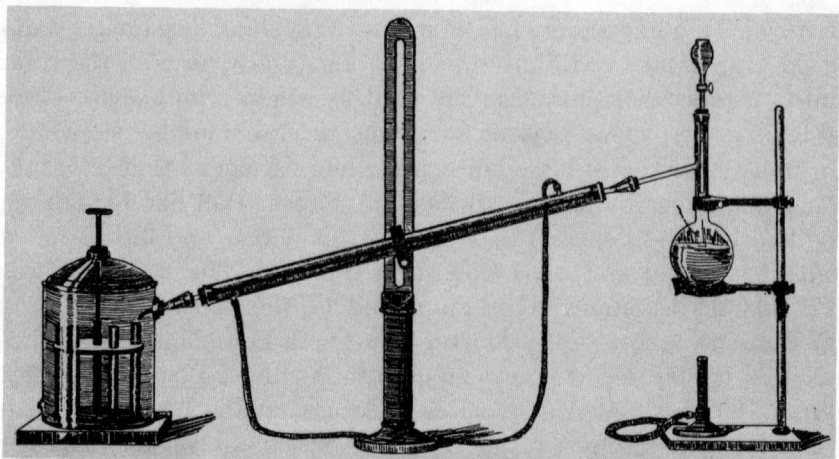

Figure 7.1 Wanklyn's ammonia process was for most British chemists the central technique for determining the salubrity of potable water during the '70s and '80s (From T B Stillman, *Engineering Chemistry* [Easton, PA: Chemical Publishing Co, 1897], p 75).

(from egg white) gave up two/thirds of its nitrogen as ammonia. But other nitrogenous compounds apparently gave up all, or a half, or a quarter, or even a seventh of their nitrogen as ammonia. When the empirical results were multiplied by the proper factor there was close correspondence between 'theory' and observation, and to Wanklyn this was proof of the soundness of the process: substances were breaking down in an orderly way and yielding a definite proportion of their nitrogen as ammonia. Yet he was able to offer a truly theoretical explanation in only one case. To other chemists the process looked arbitrary. Wanklyn's 'theory' seemed nothing more than selecting the nearest whole number which when multiplied by his empirical result would give a result close to the calculated level of ammonia. Moreover, the question remained of what trials on pure substances implied for natural waters. In Wanklyn's view there was nothing to worry about: natural waters could be relied upon to yield a definite and unchanging proportion (two-thirds) of their nitrogen as ammonia.[27]

It may seem that the rationale for the ammonia process was flimsy. It was. The so-called 'albuminoid ammonia' was not a natural substance but an artefact of the analysis. Nor was it clear whether its quantity bore any relation whatever to harmfulness, potential for harmfulness, or even putrescibility. (In practice sanitari-

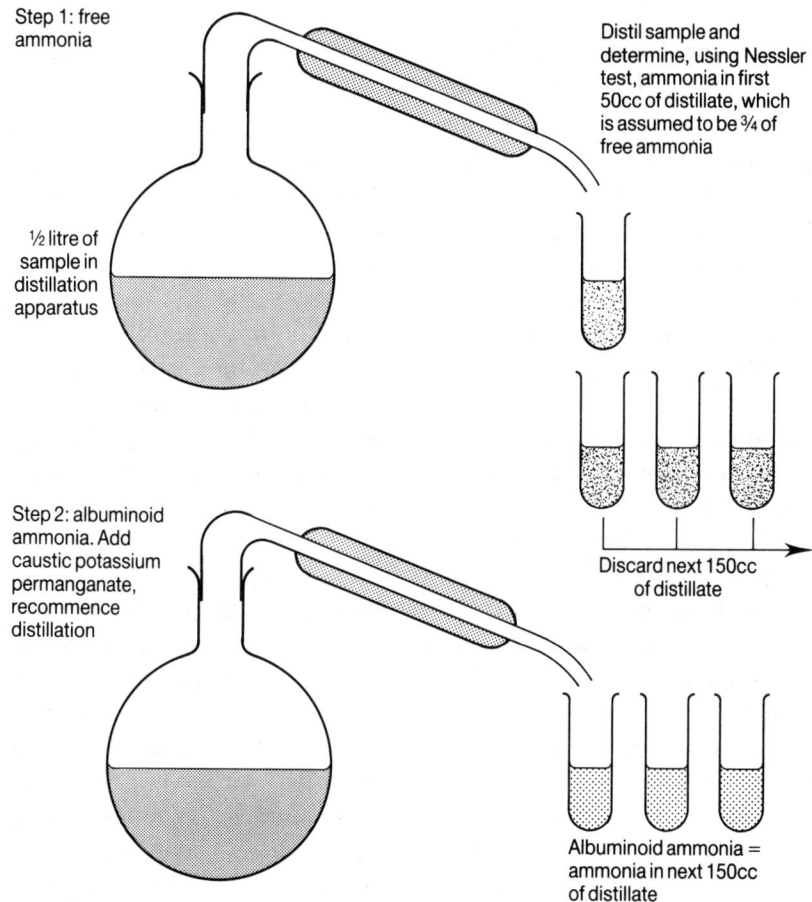

Step 1: free ammonia

½ litre of sample in distillation apparatus

Distil sample and determine, using Nessler test, ammonia in first 50cc of distillate, which is assumed to be ¾ of free ammonia

Step 2: albuminoid ammonia. Add caustic potassium permanganate, recommence distillation

Discard next 150cc of distillate

Albuminoid ammonia = ammonia in next 150cc of distillate

Figure 7.1　(continued)

ans would come to minimize these problems; as Fox put it, the yield of albuminoid ammonia was seen 'to keep pace with the purity or impurity' of waters.[28])

Indeed, it is not too much to say that the launching of the ammonia process was a disaster and that Wanklyn and his colleagues made fools of themselves in it. *Chemical News* published a savage review of Wanklyn and Chapman's handbook on water analysis, whose authors suffered from

an excess of scientific fervour, a sort of scientific *afflatus*, which forms a part of the unconscious poetry of these gentlemen's natures, and which impels them to burst forth in paeans at the Chemical Society whenever any one of them conceives a new idea. We may admire such gushing enthusiasm, but we cannot but regret that the scientific fame which these gentlemen are acquiring . . . should be sullied by the publication of raw, incomplete, and sometimes inaccurate results.[29]

Even had there been no Frankland and no combustion process to contend with, Wanklyn would have been in for rough treatment from his peers. But for him it was Frankland who came to symbolize the injustice that could be expected from the elite of chemistry.

Their dispute began in June 1867. On the same day that he formally presented the ammonia process to the Chemical Society Wanklyn testified about it to the Royal Commission on Water Supply. He said that he had been in frequent contact with Frankland, who 'had admitted . . . the extreme value of the [ammonia] process' and was testing it. He was familiar with Frankland's attempts to measure organic nitrogen through combustion and claimed that Frankland had now come to recognize that that approach was 'unsatisfactory to the last degree.'[30]

Given Wanklyn's bravado on June 20, the events of ensuing months must have been deeply humiliating. Frankland and Odling did test the ammonia process in their work for the Water Supply Commission, but in a way that was completely unfair in Wanklyn's view. They compared results obtained with the new and unverified ammonia process with those obtained with the new and unverified combustion process. Finding the two processes incompatible they rejected the ammonia process as unreliable. The rejection was not in fact quite so high-handed as it may seem—the combustion process was intended to measure organic nitrogen directly according to a simple and well-accepted principle while the ammonia process depended on the novel and somewhat dubious (and soon discarded) claim that the caustic permanganate would convert all organic nitrogen to ammonia.[31]

Matters came to a head in January 1868 when Frankland publicly introduced the combustion process in a Chemical Society lecture. He made short work of the ammonia process: it had been surpassed. The discussion took up the merits of the rival processes. Campbell, Odling, and Frederick Abel spoke against Wanklyn. Wanklyn maintained that in Frankland's own trials the combustion process had produced errors sometimes larger than the amount of organic nitrogen likely to be in the water. He and Chapman raised what seemed insuperable problems in Frankland's procedure. If combustion were to measure organic nitrogen accurately, it was necessary that all nitrates be destroyed during the initial evaporative step, yet Chapman still found nitrates in the residues of samples evaporated according to Frankland's instructions. Further, any volatile organic materials would be lost during evaporation. Any attempt to eliminate one of these errors would exacerbate the other. Efforts to ensure total destruction of nitrates such as adding more sulphurous acid would likely increase volatilization.[32]

Wanklyn's points were well taken: 'one of the most formidable pieces of criticism which we have ever met,' noted the *British Medical Journal.* Throughout the early 1870s others of Frankland's critics reiterated these objections and Frankland himself devoted great attention to resolving these problems.[33] But substantive criticism was not Wanklyn's style. The *ad hominem* attacks made on him during the early days of the ammonia process, and what he saw as the ruthlessness and arbitrariness of Frankland's rejection of the process, played upon his paranoia. Chemical objections became secondary; he came to believe that he was a victim of a conspiracy: Frankland was the head of a washed-up elite of 'chemists whose activity does not take the direction of ... original research.'[34] In 1871 he wrote a 'History of the Ammonia Process,' partly to warn 'younger chemists ... [of] the sort of reception which awaits them at the hands of their chemical brethren in this country, should they be so unfortunate as to make any notable advance in chemical methods.'[35]

Ironically, as Wanklyn became increasingly bitter and increasingly isolated, the ammonia process was becoming more widely used. In 1876 Frankland himself acknowledged that the ammonia process was 'now almost [as] generally used by analytical chemists as were formerly the incineration and permanganate processes.'[36] Using the process did not necessarily mean sympathy with Wanklyn, however; many who used it had little concern with the feud or even with the validity of the two processes. The ammonia process was adopted because it was easy to learn, easy and cheap to use, and because however weak its rationale, chemists saw it as a reliable way to distinguish good water from bad. So long as one did not attach too much significance to absolute quantities of free and albuminoid ammonia, but used the results to suggest further investigations or discover changes in the condition of a particular water supply, the ammonia test was useful. Thus, ironically, the test came to have exactly the significance Frankland thought water analysis ought to have—its results were to complement other information, not to dictate whether water was safe.[37]

By the early 1870s Wanklyn had attacked most aspects of Frankland's water science and had systematically taken positions opposing

Frankland's. He had allied himself with the London water companies, defending them from Frankland's criticisms, and reversing his own earlier position.[38] In 1872 he ridiculed 'previous sewage contamination' and accused Frankland of using it for political purposes. In 1872 and again in 1879 he began programs of alternative analyses of London's waters. By employing a chemist (Frankland) who defiantly used 'illusory and defective methods,' the government had shirked its duty; Wanklyn would selflessly perform that duty.[39] He repeated the analyses of the Cumberland and Welsh waters examined by the Water Supply Commission and found them no better than the Thames.[40] In 1878 he accused Frankland of having stolen his work in the years during which he had been Frankland's assistant and noted that Frankland's great discoveries had ceased when he had left Frankland's lab.[41]

However much social factors may have kept Wanklyn from fulfilling his potential as a chemist, it is hard to grant credence to the accusations or to sympathize with the accuser: the complaints are inconsistent, incoherent, and hysterical. Even as chemists were adopting the ammonia process as an easy and useful component in a broadly based empirical approach to water analysis, Wanklyn continued to insist that it was the ultimate in water analysis and that albuminoid ammonia should be regarded as the harmful material in the water (no matter what form this was presumed to take). As W H Brock has pointed out, Wanklyn still did not accept the germ theory of disease as late as 1906; it was too much a part of Frankland's system.[42]

In the long run Frankland prevailed. He, not Wanklyn, was internationally acknowledged as the greatest expert on water quality. Many analysts continued to use the ammonia process, but in the early 1880s a few, notably Charles Meymott Tidy (who would replace Wanklyn as Frankland's main adversary during the decade) did adopt combustion, if only to put themselves in a better position to attack Frankland's conclusions about water quality.

Letheby, Tidy, and the London Water Companies

The third great challenge to Frankland's water analysis came from a group of chemists employed by the London water companies as analysts, consultants, expert witnesses, and publicists, very much the sort of multiple role Pearson and Gardner had played in 1828 and Brande and Taylor in 1850–52. There was one difference: owing to the establishment of the position of a water analyst in the Registrar General's Office the companies now faced continual scrutiny of their supplies. They therefore had need of their own analysts to ensure that the public's knowledge of the water did not come solely from some minion of William Farr, no friend of the companies. Among the companies' chemists were men of high reputation in the world of science (it was of course precisely that reputation that made them useful to the companies) such as William Crookes, William Odling, and James Dewar. Others, like Henry Letheby and Charles Meymott Tidy, were of lesser stature, yet willing to tailor their statements to the companies' needs, no matter how far this might depart from current scientific consensus.

The first of these regular analysts was Henry Letheby, medical officer for the City of London and professor of chemistry at the London Hospital. In 1856 Letheby had succeeded John Simon as medical officer of the City of London and continued Simon's policies. In 1861 he began monthly analyses of the water of the New River, Kent, and East London companies, and by 1864 he was analysing waters of all the companies. Letheby continued these analyses for the rest of his life (he died in 1876). Ostensibly they were sponsored by the Association of Metropolitan Medical Officers of Health, but the expenses of analysis were met by the companies. However 'uninspired' and 'plodding' he may have been as a sanitary administrator, Letheby the water analyst was outspoken and aggressive.[43]

By the early 1860s Letheby had recognized specifically polluted water as a route of zymotic disease transmission and was campaigning to close the City's cesspool-polluted shallow wells. During the 1866 cholera he, like Frankland, found existing methods of purification and analysis inadequate. Even the best filtration would 'never be sufficient to   [secure] the complete removal of those subtle agents of disease, which even the most refined appliances of the chemist have failed to discover.' And worse, if living germs were responsible for cholera—and 'unquestionably' they were—oxidative purification, whether in rivers or by Condy's fluid (potassium permanganate), might not destroy these presumably resistant organisms. He was skeptical of chemistry: a chemist 'would be putting forth very dangerous propositions, ... if by relying on his science alone he ventured to dogmatise on so difficult a subject.'[44] His assessment differed from Frankland's in one striking particular however: whereas Frankland believed the water companies were distributing dangerously polluted water, Letheby believed contamination occurred in household cisterns, and was therefore the responsibility of the individual.

By the end of 1866 Letheby had begun to back away from the strong stand he had taken during the epidemic. He testified as a representative of the East London Water Company at several of the official inquiries on the epidemic.[45] Letheby insisted that not only had the company's water not caused the cholera, but that its customers had possessed a 'singular' exemption from the cholera that surrounded them.[46] Among his sharpest turnabouts was with regard to analysis. In February 1867 he affirmed that analysis did accurately reflect the sanitary quality of water.[47] Since the water had been analytically good during the epidemic it could not have been the source of the cholera.

Through 1867 and 1868 the split between Letheby and Frankland widened as it became clear how pervasively Frankland was using his monthly reports to serve the cause of water reform. Letheby and others of the companies' allies were particularly troubled by the 'previous sewage contamination' calculation, which they rightly saw as a means of undermining public confidence in the water. In an attack on Frankland in the *Saturday Review*, PSC was described as 'an imponderable and imaginary element ... entirely delusive.' Writing in the *Quarterly Review*, the engineer William Pole noted:

> the companies complain, and we think with reason [that] ... the Registrar's analysis [Frankland's] is calculated to produce needless alarm and groundless popular prejudice ... .The prominent reiteration, month after month, of the terrifying charge of an enormous 'sewage contamination,' must produce an impression in the great mass of the public (who know nothing of the doubtful and disputed reasoning on which the statement is founded, or the far-fetched and metaphorical interpretation it is intended to bear), that it refers to some well-ascertained and offensive and unwholesome present state of the water.[48]

From the presidency of the Institution of Civil Engineers, Thomas Hawksley likewise thundered against 'theoretical chemists' (Frankland) who used unwarranted phrases 'invented to frighten Her Majesty's subjects from the use of some of the purest and most harmless ... of waters the world can furnish.'[49]

Letheby took his stand against Frankland in April 1869 in an address to the Association of Metropolitan Medical Officers. His title was 'Methods of Estimating Nitrogenous Matter in Potable Water,' yet the focus of the address was his subtitle: 'On the Value of the Expression "Previous Sewage Contamination," as used by the Registrar-General in his Reports of the Metropolitan Waters.' Coming slightly a year after the Frankland–Wanklyn clash of January 1868, the paper occasioned another round of heated discussion and editorials and letters to editors. Letheby recounted the enormous improvement in Thames water in the previous two decades. Organic matter had been reduced from four to six grains per gallon to about one-half grain. Yet 'great as the improvement is, the public mind continues to be agitated and alarmed by vague fears ... unnecessarily excited by the persistent use of certain expressions of an improper kind by those who have taken upon themselves to report in a pseudo-official manner of the quality of the London waters.' As the 'custodians of the public health' medical officers should be particularly concerned about 'undue excitement and alarm of the public mind,' Letheby maintained.[50]

In the discussion Frankland replied that he had not sought the position of water analyst, but otherwise sidestepped water politics to deal with the scientific issues at hand. He upheld the propriety of his analysis. Some standard was necessary to replace the unreliable incineration (ignition) and permanganate methods, he insisted. He maintained that 'previous sewage contamination' was now an empirically verified construct: nitrates, at levels greater than could be accounted for by rainfall, were found only in waters contaminated with sewage or manure. He was strongly supported by B H Paul and also by Charles Heaton, a medical officer, and J M Rendel, an engineer. But most of the meeting was sympathetic to Letheby, and to Wanklyn, though Letheby was pressed to explain his close relationship with the water companies.[51]

An Infusion of Neutrality: The Office of the Water Examiner, 1871

Beginning in 1871, however, Frankland was no longer the sole government official having regular oversight of water quality. The Metropolis Water Act of 1871 (Parliament's response to the Royal Commission on Water Supply) called for establishment of the post of a water examiner who was to ensure that the filtration required by law was effectively carried out. The first occupant of the post was Colonel Francis Bolton, a military engineer with experience in signalling apparatus and lighthouses.[52] Bolton was not to be an analyst; he was simply an inspector working within the revolution-in-government tradition as it is usually understood. Bolton ieant well and he was industrious, but he was politically naive and knew little of sanitary chemistry.

NUMBER of occasions when MOVING ORGANISMS were found.																
—	1869.	1870.	1871.	1872.	1873.	1874.	1875.	1876.	1877.	1878.	1879.	1880.	1881.	1882.	1883.	1884.	1885.
Chelsea -	3	2	2	3	2	5	4	4	1	0	2	0	0	0	1	0	0
West Middlesex	0	0	0	0	0	0	0	0	0	1	2	0	0	0	0	0	0
Southwark -	8	1	4	1	2	5	5	7	5	3	0	0	0	0	3	0	0
Grand Junction	4	1	1	2	3	5	7	3	3	3	1	3	3	0	1	0	0
Lambeth -	5	0	4	6	3	4	5	4	1	1	0	2	1	1	1	1	0
New River -	0	0	0	0	1	1	0	0	1	0	0	2	0	0	0	0	0
East London -	4	3	3	1	0	2	0	0	0	0	2	0	0	1	0	0	0
Kent -	0	0	0	0	0	0	0	0	0	0	0	0	0	0	0	0	0
Colne Valley -	–	–	–	–	–	–	–	–	.0	0	0	0	0	0	0	0	0
Tottenham -	–	–	–	–	–	–	–	–	0	0	0	0	0	0	1	0	0

Figure 7.2 Among the least sophisticated but most effective of Frankland's propaganda techniques was the listing of the 'living and moving organisms' detected in the monthly water samples. Their presence ostensibly signified ineffective filtration (*15th Annual Report of the LGB*, app B, p 96).

On taking up the post Bolton found himself in the midst of the furor over 'previous sewage contamination' and the other Frankland stratagems. In December 1871 he received a petition from the Vestry of St Mary, Newington, complaining of their wretched water supply. The petition was studded with extracts from Frankland's reports, including references to 'living and moving organisms' that Frankland had found in the water. Bolton examined the water microscopically, and found in it paramecia, confervae, fungi, and organic debris. It deposited a brownish sediment on standing. Nevertheless he rejected Frankland's claim that the water was bad: E A Parkes' *Manual of Hygiene* set a maximum safe limit of three grains per gallon of 'organic or volatile matter' and the water in question contained only 1.9 grains. Misunderstanding Frankland's views on water analysis, Bolton insisted that the water was also safe according to Frankland's standards. He took issue with Frankland's listing of 'living and moving organisms,' arguing that microscopic life was ubiquitous in waters.[53]

Frankland was indignant. Bolton received a chilling rebuke from S J Smith, secretary of the Rivers Commission. Having pointed to Bolton's misreadings, errors, and reliance on obsolete analytical methods, Smith informed him that 'Her Majesty's Commissioners' had no objection to his stating his opinion but thought that

> in a matter of such importance . . . it would have been a proper course for the Water Examiner [Bolton] to have put himself in communication with them before using their experimental data and conclusions in a mutilated form to support his own opinions, and, as an engineer, review the work of a chemist: in so doing he has shown his limited acquaintance with physics and chemistry, and consequent misunderstanding of chemical language and results.[54]

Poor Bolton! As if holding one's 'own opinions' were not sin enough, he had the misfortune to be an engineer. Bolton sputtered, but there was little to say.

With the coming of Bolton, questions arose of what Frankland's future status was to be. Inasmuch as it was Bolton's responsibility to certify that London's water was being 'effectually filtered' and inasmuch as Frankland was analysing the water at government expense in a way that would show the adequacy of filtration, there was some thought that Frankland ought to submit his results to Bolton, who would then decide whether they represented effectual filtration. This would have made Frankland a glorified technician, working under someone who knew far less about water quality and analysis than he did, but it might have been attractive to those uncomfortable with Frankland's activism. Exactly what was proposed is not clear, but George Graham, the Registrar General, vigorously (and successfully) resisted any major change in Frankland's status. On 8 May 1872 Graham wrote to the Local Government Board, not to claim that Frankland was impartial, but to praise Frankland's sanitary activism:

> nothing has had so great an effect upon them [water companies] as Prof. E. Frankland's undisguised description of the impurities discovered, published by the authority of the government, and his comparison of what we are here compelled to drink, with what is supplied to Glasgow, Manchester, etc.[55]

Graham admitted that he himself knew nothing of the chemistry at issue. But he thought it was perfectly normal that Frankland would be criticized, for 'the higher the individual attacked may be, the greater glory will be their fate who demolishes his reputation.' The water companies might well find 'rival chemists glad to dispute the accuracy of the analyses,' but one ought not to be misled by them. He added an ultimatum:

> Of course upon receiving monthly reports from Prof. E. Frankland
> framed conscientiously on analyses which with his large experience
> and acknowledged learning he considers the best and truest that can
> be made, it is not wished that before publishing them I should expur-
> gate what may be likely to offend any company and their hired chem-
> ical advisor? It cannot be contemplated that the professor would
> submit to such treatment; still less can it be thought that I should
> publish garbled statements in a matter of such importance.
>
> The reports in future must be published as I receive them on the
> responsibility of Prof. E. Frankland, or not be published at all.[56]

Graham professed not to care which option was taken, but he had
left the LGB in a difficult position. Frankland's reports continued
to appear independently, under aegis of the Registrar General, until
1875 at which point the water analyst's position was transferred to
the LGB which was coordinating more and more aspects of public
health policy. In most respects this was a more suitable arrangement since the Board was responsible for most aspects of sanitary
policy, but it did mean that Frankland's monthly analyses and his
often inflammatory interpretations were published (without editing)
in Bolton's monthly report, leaving Bolton with the problem of how
to reconcile Frankland's views with his own.

The administrative arrangement was troublesome in other ways.
As water examiner Bolton was the closest thing to a public regulator
of privately owned public utilities. He was a specialized civil servant,
a subordinate in the LGB hierarchy. Frankland, on the other hand,
was an unsalaried consultant, paid directly for his analyses and not
subject to the bureaucrats. From time to time there was talk in
the LGB of getting Frankland to restrain his speculations about
the water danger, but it was realized that even the appearance of
censorship would be politically embarrassing.[57]

Matters were further complicated by Bolton's political naiveté.
He failed to appreciate the symbolic power that an official publication conferred on any partisan reports included in it. At the end
of 1873 Bolton began including reports of analyses by Letheby and
J K Bamber (the companies' analyses). As it came to appear that
Bolton's pages were open, more of those with anti-Frankland axes
to grind found excuses for the inclusion of their analyses. Always
there were good reasons: a need for more data to confirm, complement, or supplement Frankland. Always these programmes of
analyses were made under the sponsorship of some ostensibly non-

partisan body; just as the Association of Medical Officers of Health received Letheby's analyses (though the companies paid for them), the metropolitan vestries sponsored a series begun in 1879 by Frankland's nemesis J A Wanklyn and his associate William Cooper. But the intent of the programmes was invariably to dilute Frankland's influence; to give members of the public free choice to take truth from whichever official expert most closely matched one's ideology.[58]

By the later '70s Bolton's superiors at the LGB had become concerned about this situation and in 1879 they advised against including the Wanklyn/Cooper analyses. Bolton disagreed, citing Wanklyn's eminence as a water analyst. He liked the simplicity of Wanklyn's format, noted that the analyses were expressly for the vestries, and that besides, there was extra white space on the page.[59] The issue came to a head in spring 1883 when a question was raised in the House of Commons as to the official status of the analyses done for the water companies (the Association of Medical Officers had withdrawn its sponsorship in 1879) by Letheby's successor Charles Meymott Tidy. Tidy (1843–92) was so central in water matters during the '80s that it is appropriate to digress to introduce him here. Son of a Hackney medical practitioner, Tidy studied medicine and chemistry with Letheby, qualifying in 1864 and taking an M B degree from Aberdeen in 1866. On Letheby's death in 1876 he became professor of chemistry, public health, and medical jurisprudence at the London Hospital and medical officer for Islington. He also lectured on medical jurisprudence at the Inns of Court and toward the end of his life qualified as a barrister. From 1876 until his death Tidy was Frankland's principal antagonist. He was far more tactful than Wanklyn, and cleverer than Letheby. Not only was he a defender of the London water companies, but Tidy (like Wanklyn) opposed Frankland on almost every water issue: he believed chemical methods of water analysis were adequate to pronounce waters safe, doubted germs caused disease, and maintained that polluted rivers purified rapidly. He was an adept publicist, and a talented expert witness. Scientists did not take him seriously, among sanitarians he lacked Letheby's stature, yet he was effective and influential nevertheless.[60]

In November 1880 the Association of Medical Officers of Health had informed Tidy that he need no longer supply them with monthly analyses of the London waters.[61] Without the Association to legitimate the analyses it became even more important that they be included in Bolton's reports. In early 1881 the companies took steps to

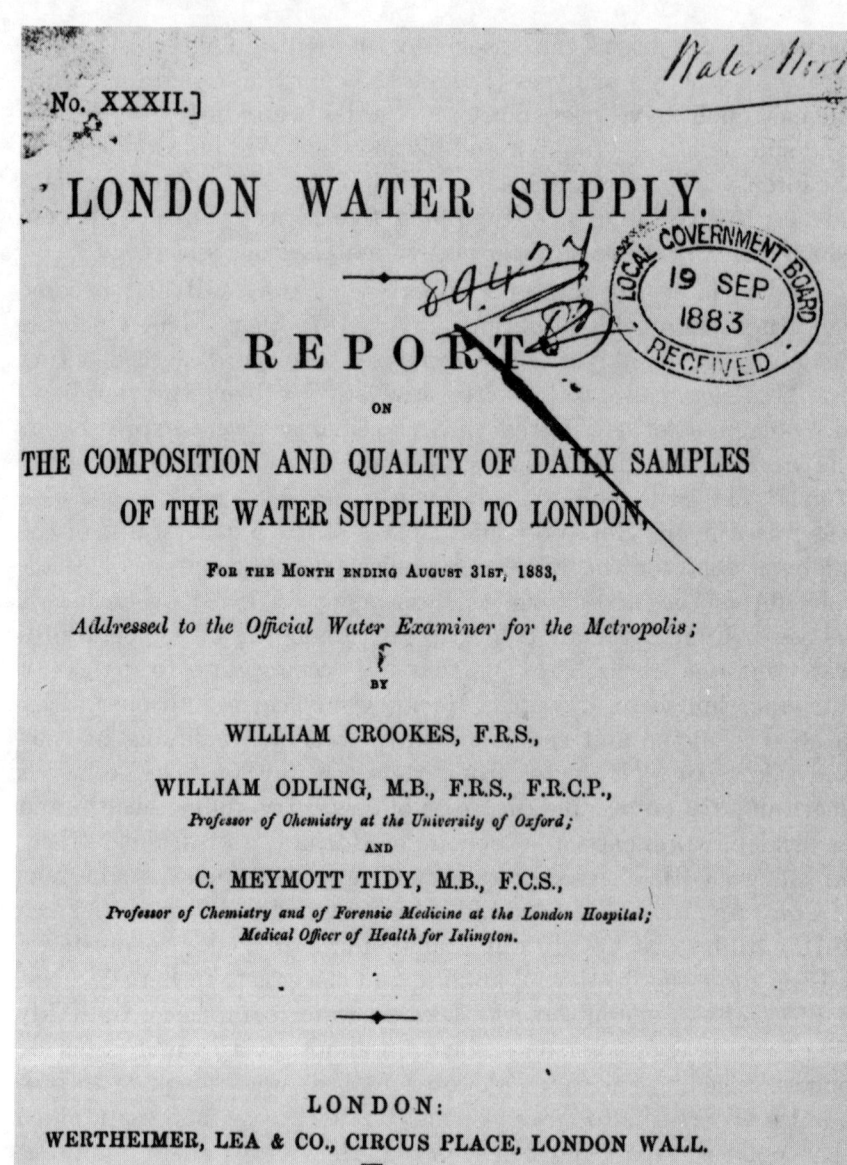

Figure 7.3 The printing of analyses done for the water companies in official LGB reports gave these analyses some claim to official status, which the analysts William Crookes, William Odling, and Charles Meymott Tidy attempted to expand, much to the distress of the LGB. As these examples show, even when they changed wording in response to LGB complaints, their reports still looked official (*PRO MH* 29 5).

TO THE

RIGHT HONOURABLE THE PRESIDENT OF THE LOCAL . GOVERNMENT BOARD.

November 1st, 1882.

Sir,

We submit herewith the results of our analyses of the 182 samples of water collected by us during the month of October, on the days and at the times indicated, from the mains of the seven London water companies taking their supply from the Thames and the Lea.

In Table I. we have recorded the analyses in detail of samples, one taken daily, from October 2nd to October 31st inclusive. The purity of the water, in respect of organic matter, has been determined by the Oxygen and the Combustion processes; and the results of our analyses by these methods are stated in Columns XIV. to XVIII.

We have recorded in Table II. the tint of the several samples of water, as determined by the colour-meter described in a previous report.

Of the 26 samples supplied by the New River Company, the whole were found to be well filtered, clear and bright.

Of the 26 samples from the mains of the East London Company, the whole were found to be well filtered, clear, and bright.

Of the 26 samples from the mains of the Chelsea Water Company, the whole were found to be well filtered, clear, and bright.

Of the 26 samples from the mains of the West Middlesex Company, the whole were found to be well filtered, clear, and bright.

Of the 26 samples from the mains of the Lambeth Water Company, one was "slightly turbid," and six were recorded as "very slightly turbid." The remainder were well filtered, clear, and bright.

Of the 26 samples from the mains of the Grand Junction Company, the whole were found to be well filtered, clear, and bright.

Figure 7.3 (continued)

strengthen the credibility of their analyses, commissioning William Odling and William Crookes, chemists of substantial reputation, to collaborate with Tidy. Moreover the reports were henceforth to be based on daily analyses of the waters of each company, not just the once-a-month samples that Frankland took. The three chemists took the offensive when the question arose in 1883 of the status and impartiality of the analyses they sent Bolton. Theirs weren't the partisan analyses, Frankland's were, they maintained. Further, a public report, like Bolton's, ought to present a balanced view.[62]

From the LGB viewpoint the problem was not simply one of the publication of the Crookes–Odling–Tidy analyses in Bolton's reports. What the Board received each month was a copy of a printed report, addressed to the Board but distributed to the public and the press as well. The format of this privately published report implied that the analyses were commissioned by the Board. It was addressed (in large capitals in a font similar to that used in official blue books) 'To The RIGHT HONOURABLE THE PRESIDENT OF THE LOCAL GOVERNMENT BOARD.' When LGB staff tried to persuade the three chemists to make it clear that their analyses were not official, the three responded with formats that met the Board's specific complaints yet conveyed the impression of being official in ingenious new ways. Their June 1883 report was addressed to the Board's secretary (in small letters), and admitted that the analyses were 'for the information of the Local Government Board' and had been 'made at the expense of the Water Companies.' Bolton, however, insisted the analyses be addressed to the companies. This Tidy refused to do, and beginning in August the cover page of the report (again with font and layout mimicking a blue book) announced that the analyses were 'addressed to the Official Water Examiner for the Metropolis.'[63]

Bolton and his superiors had better luck dealing with the Wanklyn–Cooper analyses. Bolton's September 1884 report did not include a Wanklyn–Cooper analysis; their contribution had not come in, Bolton maintained. His reports through January 1885 carried similar statements, even though the analyses were being received, in some cases several days before Frankland's (which were included) were received, and in all cases but one in time for publication. Somewhere in the LGB Wanklyn's reports were going astray; whether Bolton knew what was happening is not clear. By August Wanklyn and Cooper had stopped sending reports.[64]

These episodes indicate how pervasively politicized analyses of London's waters had become. Bolton's monthly reports, ostensibly the record of an impartial investigator appointed to safeguard public health, in fact represented the struggles of conflicting interests, each of which sought to gain power by gaining control of the icons of an official report. On one side was Frankland, possessing 'official' status, yet not being responsible to any official. On the other was the triumvirate of Crookes, Odling, and Tidy, all able polemicists and masters in the construction and manipulation of images of authority. In the middle was poor Bolton, holding a marginal position in the LGB bureaucracy, trying unsuccessfully to pretend that supervising London's filters could be based on straightforward empirical investigations. It was an impossible job. The financial and political issues involved were too large, the scientific issues shrouded in too much uncertainty.

The Companies' Critique

The reports of Tidy, Crookes, and Odling were clearly undertaken to undermine Frankland's credibility. Were it possible to regard their writings only as political propaganda it would be unnecessary to consider them. Yet the sorts of arguments they made were equally those of the science of water analysis. There was no independent fount of truth, free of the taint of politics, which might then be prostituted into the service of some vested interest. Instead, the issues of water analysis were inextricably mixed with social and political conflict. The science developed as a dialectic between those who were trying either to prove the sure and certain safety of public water supplies or to incite public action to make those supplies far more trustworthy than they currently were. Hence the work of Tidy and his colleagues deserves scrutiny not because it was disinterested, but because in the period before the development of bacteriology, they put the case that analysis and purification could be trusted more thoroughly and thoughtfully than had any others.

Shortly after taking over the analyses Tidy began to solidify the scientific foundations of his attack on Frankland. In 1879 he re-opened the vexed issue of which process was best for the determination of organic matter, delivering a long paper on the subject to the Chemical Society in early February.[65] Like most of Tidy's presentations, this one was a rhetorical masterpiece. It presented little in the way of new or significant scientific findings, and served a larger strategic end in providing a foundation for claims and arguments Tidy would make in subsequent years. Setting himself up at the outset as an impartial authority, Tidy based his criticism of the combustion process on Wanklyn's arguments and his criticism of the ammonia process on Frankland's. At the level of chemical principles, both sets of criticisms were unanswerable. Having used Frankland and Wanklyn to neutralize one another, he turned to the oxygen-absorbed (permanganate) process, which Letheby had long championed and which he had inherited. Frankland and Wanklyn had represented the process as naive in conception, meaningless in results. Tidy admitted as much, but claimed that he and Letheby had greatly improved the process. It had been objected that the process failed to distinguish between harmless and harmful organic matter, so Tidy began the practice of taking two readings, treating the oxygen absorbed after one hour as representing the oxidation of the dangerous highly putrescible, animal organic matter, with the three-hour reading representing relatively unobjectionable vegetable organic matter.[66]

Throughout the paper Tidy maintained a tone of superiority. To the argument that combustion was too tricky he retorted that if chemists could not do it they ought not to be chemists (it might help weed incompetents from the profession). He made much of his own medical training, which gave him an authority in such matters of life and death that those trained only as chemists could not pretend to have.[67] The most important of Tidy's stratagems was the stance he took toward the rival analyses he and Frankland were doing of the London waters. Though highly critical of both combustion and ammonia, Tidy ended up supporting Frankland against Wanklyn. He based his support on his own trials on the combustion, ammonia, and permanganate processes. Tidy found a significant concordance between combustion and permanganate and on this basis condescended to endorse combustion, however objectionable it might be in theory. He still favoured the permanganate process as the more prudent for its errors lay in the direction of yielding false positives (which would induce the analyst sometimes to condemn safe waters), while combustion might give false negatives in the case of volatile organic matters that would be lost during evaporation.[68]

These moves enabled Tidy to represent his disagreements with Frankland in a much different light than had Letheby. For Letheby the conflict had been as much about which analytical process was better as about whether London's water was safe. Endorsing combustion allowed Tidy to focus on interpretation. He soon adopted combustion in his analyses (keeping the oxygen process as well) and attempted, unsuccessfully, to have his analyses and Frankland's done on duplicate samples so that there might be no equivocation about changes in composition. The strategy was thus to remove any pretense that his differences with Frankland could be ascribed either to their competence or to their samples. If both used the same methods, and Frankland continued to come up with inflammatory conclusions, it could mean only that Frankland was an irresponsible rabble rouser. Hence while Tidy might claim that the analyses were 'not made . . . in any spirit of antagonism or opposition,' he had simply found a far stronger basis for opposition. (In fact, as W C Young would point out in 1895, the levels of impurities discovered by the companies' chemists were generally higher than those on which Frankland based his condemnations of London's waters.)[69] The 1879 paper on analysis set up Tidy's next production, a similarly lengthy paper on 'River Water' presented a little over a year later. Here he challenged Frankland's claim that rivers did not self-purify to any significant degree and Frankland's belief that the morbid poisons of water-borne diseases were resistant living germs.[70]

A lengthy and acrimonious exchange ensued. Frankland's exhaustive and sometimes sarcastic response appeared in the *Journal of the Chemical Society* as a research report (on the grounds that it was the first public presentation of self-purification experiments done by the Rivers Commission). A second 'River Water' by Tidy in March 1881 added nothing new. W Noel Hartley of the College of Science in Dublin and Charles Folkard, a Frankland student, published lengthy papers supporting Frankland, as did Frankland's son Percy, in a savage 1884 attack on 'The Upper Thames as a Source of Water Supply.'[71] In all these forums Tidy was thoroughly taken to task. His meagre experiments, numerous unwarranted assumptions, and self-serving concepts of morbid poisons were repeatedly pointed out. Nonetheless, Tidy had made himself a celebrity, the champion of all those who felt Frankland was too unrelentingly (and perhaps unrealistically) hard on existing water supplies. All the while Tidy was taking a beating in the learned societies, he and Odling and Crookes were taking their case to the public. Their monthly reports became a forum for denouncing Frankland and extolling the unexceptionable yet constantly improving London waters. As in the early '50s and the mid '60s, the fierceness of the battle reflected a new round in the struggle for control of the supply. Again there was grumbling in the vestries about high cost and poor service. In 1879 the government had called for public takeover and a Select Committee had taken up the problem in 1880. The companies were willing to sell; the political problem remained how much they would receive for their fixed assets and in recompense for future dividends, and there would be no resolution of this problem for another two decades.[72]

In the main, Tidy, Odling, and Crookes followed the strategy of ignoring all the unanswered questions (e.g. those of the identity and characteristics of the morbid poisons of water-borne diseases) and returning the debate to the validity of Frankland's inferences and the propriety of his methods of presenting his results. With a confidence rarely seen since the days of Brande, they reaffirmed a simple and absolute faith in the reliability of chemical water analysis. When they wrote that 'mere chemistry, however refined, has not been able to distinguish a difference, or even to establish a presumption of a difference' between the water taken from the headwaters of the Thames and that taken from near the companies' intakes (contaminated with sewage from upstream towns) they were, in effect, asserting that no difference existed.[73] With their assumption of the unassailability of chemistry, the three went on to chide Frankland for intentionally misleading the public. In May 1883 they brought up Frankland's periodic allusions to 'moving organisms' in the water. They called upon him to state what kinds of organisms these were and noted that Frankland himself had stated (in a lecture to the Royal Institution 22 years earlier, no less) that such organisms were safe.[74] In August they took issue with Frankland's 'Table E.' Here Frankland compared the organic contamination in the water supplied by the river-water companies with what the Kent Company took from deep wells. The table used the Kent's level as unity and hence represented how many times dirtier the river-derived waters were than what London might have if it converted to the deep wells that Frankland now advocated. Crookes, Odling, and Tidy complained that comparing waters from different types of sources was unfair, and noted that the organic matter in Thames-derived water was roughly the same as that in Glasgow's pure Loch Katrine supply.[75]

The campaign of Crookes, Odling, and Tidy and the contemporaneous analytical programme of the Society of Public Analysts that will be considered in the next chapter were the last great hurrahs of chemical water analysis. On strictly chemical grounds the points the three chemists made were defensible—using chemical methods Frankland could not prove that there were significant differences in waters from various places along the Thames; he was guilty of inappropriate comparisons and misleading use of terms. Yet what the three chemists did not fully realize was the extent to which determinations of water quality were escaping (and indeed had escaped) the province of chemistry. They were defending an empty orthodoxy.

Frankland: The Triumph of the Symbolic

It is as well to take stock here, to assess the transformation Frankland had wrought on the community of analysts. In 1866 most analysts would have looked upon water analysis as an empirical issue. They might have admitted that their processes were not yet very good for making fine qualitative distinctions or even terribly accurate quantitatively. Yet they did believe that in measuring organic matter, 'animal' organic matter, or hydrogen sulphide in water they were measuring the harmful matter, or at least approximating it as closely as possible. They might admit that this matter did not always lead immediately to acute disease, yet it was still possible to conceive that one was measuring the harmful substance: according to contemporary medical theory one might believe that bad water only debilitated health, acted, that is, as a predisposing cause. Or one might regard the effect of bad water as cumulative and as having no effect until a threshold was reached. Or one might argue that the water would become lethal only when epidemic conditions were present. Underlying any of these interpretations, however, was the belief that the analyst really was measuring the harmful matter—its harmfulness simply might not be quickly, conveniently, or obviously manifested.

This is the empiricism: one solves the problem of whether the water is harmful by measuring the harmful material in it. In such a perspective claims about health effects were subject to the findings of chemistry and such an outlook was a natural outgrowth of chemists' long-standing claims that their science provided the ultimate authority in matters of water quality. If now we regard this claim as over-reaching its warrant, or even as unfounded, we must remember that it was regarded at the time as a prudent claim: even if not all organic matter was harmful, it was organic matter (at measurable levels) that was harmful. Hence the analyst could only go wrong by making false positives, by advising caution when there might be no need for it.

Taking this empiricist perspective involved minimizing or conveniently overlooking Hofmann's observation that what one could measure was really very different from what one ought to be worrying about. It was different in category; it was process not substance. It was characterized not by its quantity but by its transitoriness: the matter which underwent the dangerous process might constantly be present, yet the deadly process might only occasionally occur in it. Such a perspective was not logically incompatible with the empiricism of traditional chemists, for they could still maintain that their processes identified and quantified an essential antecedent to the development of water-borne disease. But if it did not exclude empiricism, Hofmann's new perspective did strain it: if the important distinctions were really unmeasurable qualitative distinctions, knowledge of the amount of organic matter was virtually irrelevant. If all nitrogenous organic matter in water were to be accepted as capable of undergoing pathogenic putrefaction false positives would become so frequent that it would become questionable whether analysis was of any service at all.

The final blow to this philosophy came from Frankland's discovery of false negatives during the 1866 London cholera. On precisely the occasion when water analysis was most needed—the occasion of an epidemic—it failed. The cholera germs, whatever they were, were so minute that a water which would be regarded as safe according to any set of chemist's standards might actually be lethal. Hence empiricism might lead not to prudence but to folly. But what to do then? Frankland had pretty well falsified traditional conceptions of morbid poison but he had nothing—at least nothing analytically tangible—with which to replace them. Hence the campaign of what might best be called counter-analyses that he embarked on in the following two years. They were counter analyses in the sense that they were effective because they were presented in an environment (and in a form) in which they were viewed in traditional empiricist terms. All the while the conclusions Frankland was coming to derived from a few prudent and simple principles about what waters could be trusted—the main one being 'don't trust any water unless you know where it's been.' What Frankland was doing then was to reverse the direction of argument, but he did so behind the scenes. Now chemistry was to be subject to epidemiological demonstrations, not the other way round. And since the phenomena of concern were epidemiological, not chemical, chemical information was really irrelevant. Yet it was, as Frankland astutely realized, the only form of information policy makers were likely to pay attention to. Brande had done his job well; chemistry was the unquestioned source of authority in such matters, and besides, Frankland was a chemist.

In removing chemical analysis from centre to periphery, in using it mainly to legitimate conclusions arrived at through other rationales and to persuade others to accept those conclusions, Frankland was liberating himself from certain modes of accountability as well as from certain assumptions. The initial reactions of colleagues and competitors, Letheby for example, convey a sense of shock, outrage, and exasperation—Frankland had broken unwritten rules in an unexpected way. By using *reductio ad absurdum* arguments about undetectable, resistant germs, he was no longer subjecting himself to the arguments chemists could make, no longer was he accountable to the profession. But Frankland had not so much broken rules as changed them, and Tidy, Odling, and perhaps Wanklyn had by the early '80s become skilled at playing the game according to Frankland's rules. Frankland's rules were much looser. They permitted one to arrive at an opinion of a water's quality by whatever route and on whatever basis one chose, and to defend it by whatever means seemed most effective. On the one hand the adoption of this sensibility by at least the elite London chemists can be taken as a sign of maturity: no longer were they perpetuating an instrumental rationality by taking refuge in simplistic and arbitrary chemical standards which for years had been becoming increasingly fraudulent and hypocritical. By the same token, the realization that the proper end of water analysis was not knowing parts per million of organic nitrogen but securing better water, i.e. that water analysis must be subject to water policy, meant that it was hard to take the statements of analysts at face value. A statement by Tidy that London's waters were safe, or by Frankland that they were unsafe, had immediately to be translated into the political message it symbolized, either that existing arrangements were adequate or that great change was needed.

As it turned out this abandonment of empiricism was to be shortlived. In the mid '80s empiricism returned to water analysis in two forms. One of these was bacteriology, which engendered as much naive optimism as chemistry once had. The other was a more modest, and ultimately more successful empiricist programme. It emerged from the reaction of rank-and-file water analysts to the controversies over analytical methods between Frankland and Wanklyn and Frankland and Letheby. Especially important are the responses of two emerging groups of public health officials: the local medical officers of health (mandated by the sanitary legislation of 1875), and the local public analysts, appointed under the adulteration acts to ensure that consumers could be assured of safe and pure foods and drugs. Neither medical officers nor public analysts were specifically required to inquire into water quality, but a great many did and found the processes of the London chemists wholly unsuited to their needs.

1   H Letheby, 'Methods of Estimating Nitrogenous Matter in Potable Waters,' p 432.

2   For the most recent review see Roy M MacLeod, *Government and Expertise in Britain* (Cambridge: Cambridge University Press, 1987).

3   A Chaston Chapman, *The Growth of the Profession of Chemistry*; Colin Russell, N G Coley, and G K Roberts, *Chemists by Profession*.

4   E Byrne, 'Experiments on the Removal of Organic and Inorganic Substances in Water,' pp 6–53.

5   W A Miller, 'Observations on the Analysis of Potable Waters,' 117–32; *CN* 11 (1865): 269–70, 283–5; RCWS, *Evidence*, q 7102. On Miller see *DNB* 13, pp 429–30.

6   H B Dixon, 'William Odling, 1829–1912,' *Proc. Royal Society of London* 100A (1921–2): i–vii; W H Brock, 'William Odling,' *DSB* 10 (1974): 177–9.

7   RCWS, *Evidence*, qq 7082, 7074, 7094, 7106–14.

8   *S C Serpentine*, qq 734, 739, 763, 2142, 2143–6; RCWS, *Evidence*, qq 6459, 6462, 6472.

9   RCWS, *Evidence*, qq 7209, 7236, 7242, 7211; R Angus Smith, 'On the Examination of Water for Organic Matter,' (Manchester Lit. and Phil.), pp 37–78; same title, slightly different version, *CN* 19 (1869): 278–82, 304–6; 20 (1869): 26–30, 112–115.

10   Smith, 'On the Examination,' (Manchester), p 43.

11   *BMJ*, i, 1868, pp 331, 378–9, 391, 413, ii, pp 71, 86.

12   'London Water Supply: Past, Present, and Future,' *CN* 15 (1867): 37–8, 49. On Crookes see E E Fournier D'Albe, *The Life of Sir William Crookes* (London: Unwin, 1923).

13   'London Water,' *CN* 15 (1867): 190.

14   'Dr Letheby on the Methods of Water Analysis and on 'Previous Sewage Contamination,' *CN* 19 (1869): 231–2.

15   *Ibid*, 'A Bill to Amend the Law relating to Public Health,' *CN* 25 (1872): 145.

16   RPPC, 1868, *Second Report*; Fournier D'Albe, *Life of Crookes*,* pp 257–70.

17   *CN* 28 (1873): 121–2, 191, 216; 29 (1874): 63, 124, 156, 166. Cf Crookes in discussion of C N Bazalgette, 'The Sewage Question,' *MPICE* 48 (1875–6): 163–6.

18   'The Pollution of Rivers Bill,' *CN* 28 (1873): 37–9; *S C Pollution of Rivers*, House of Lords, 1873.

19 'The Rivers Pollution Bill,' *CN* 31 (1875): 221.

20 Cornelius B Fox, *Sanitary Examinations of Water, Air, and Food*, 2nd edn, p 54. Cf Kenwood, *Public Health Laboratory Work*, pp 106–7.

21 'The Pollution of Rivers Bill,' *CN* 28 (1873): 37.

22 'Review of Pettenkofer, *Cholera: How to Prevent and Resist It*,' *CN* 31 (1875): 139-40; 'Review of the Fourth Annual Report of the Massachusetts State Board of Health,' *CN* 27 (1873): 311–2.

23 W H Brock, 'James Alfred Wanklyn,' *DSB* 14 (1976): 168–70.

24 Wanklyn, Chapman, and Smith, 'Water Analysis: Determination of the Nitrogenous Organic Matter,' pp 445–54.

25 Cf E T Chapman, 'The Relation between the Results of Water Analysis and the Sanitary Value of the Water,' *CN* 16 (1867): 275.

26 D Campbell, 'A Note on Messrs Wanklyn, Chapman, and Smith's Method for Determining Nitrogenous Organic Matters in Water,' *CN* 16 (1867): 139; comments in 'Chemical Society discussion, 6 February 1868,' *CN* 17 (1868): 80.

27 Wanklyn, 'Verification of Wanklyn's Water Analysis,' pp 591–5; Wanklyn and Chapman, 'On the Action of Oxidizing Agents,' pp 161–72; comments of Campbell and Frankland in 'Chemical Society discussion, 6 February 1868,'* p 80.

28 Fox, *Sanitary Examinations of Water, Air, and Food*, 1st edn, p 50.

29 'Review of Wanklyn and Chapman's *Water Analysis*,' *CN* 18 (1868): 151–3.

30 RCWS, *Evidence*, qq 5420–5437.

31 Wanklyn to *CN*, *CN* 17 (1868): 60.

32 'Chemical Society discussion, 6 February 1868,'* pp 79–80.

33 'Water Analysis by Modern Methods,' *BMJ*, i, 1868, p 331; E Frankland, 'On some Points in the Analysis of Potable Waters,' pp 88–9.

34 Wanklyn, 'Wanklyn and Chapman's Water Analysis,' *CN* 18 (1868): 165.

35 Wanklyn, 'History of the Ammonia Process,' *CN* 24 (1871): 10–11.

36 Frankland, 'On Some Points in the Analysis of Potable Waters,' p 847.

37 A H Allen, 'On Some Points in the Analysis of Water, and the Interpretation of the Results,' *The Analyst* 2 (1877): 61–5.

38 Compare *RCWS*, qq 5401, 5446–51, 5482–85 with *RCWS*, appendix AK, p 95.

39 J A Wanklyn, 'The Registrar General's Reports on the London Water,' *CN* 25 (1872): 169–70; *idem*, 'Report on the condition of the Water Supply of London,' *CN* 26 (1872): 239.

40 Wanklyn to *BMJ*, i, 1871, p 153.

41 Wanklyn, 'Dr Frankland's Researches,' SR 8 (1878): 174–5.

42 Brock, 'Wanklyn,' p 169; SR 6 (1877): 15–16, 96.

43 Lambert, Sir John Simon, pp 213–16; *DNB* 11, p 1010; *RCWS* qq 3861, 3909.

44 Letheby, 'City Pumps,' *CN* 4 (1861): 260–2; *idem*, 'Composition and Quality of Metropolitan Waters in July, 1866,' *CN* 14 (1866): 83.

45 'Report by Capt. Tyler to the Board of Trade,' p 5.

46 *S C East London Water Bills*, Appendix 3, pp 315–6.

47 *Lancet*, i, 1867, p 185; *S C East London Water Bills*, q 2230; 'Report by Capt. Tyler to the Board of Trade,' pp 18–9.

48 *Saturday Review*, 10 October 1868, p 480; [W Pole], 'The Water Supply of London,' *Quarterly Review* [American edn] 127 (1869): 245fn.

49 Hawksf(1871–2): 345. See also *Our Water Supply. A Discussion for and against the fitness of the Thames and River Water for Domestic Use*, reprinted from the Surrey Comet (London: Trounce, 1880).

50 Letheby, 'Methods of Estimating the Nitrogenous Matter in Potable Waters,' p 429. Cf 'Water Analysis,' *BMJ*, i, 1869, p 402; [B H Paul], 'Water Analysis for Sanitary Purposes,' *BMJ*, i, 1869, pp 427–8, 495–7, 543–4, 1869, ii, pp 32–3.

51 'Processes of Analysis of Potable Water,' *BMJ*, i, 1869, p 379; 'The Value and Meaning of Existing Reports on Potable Water,' *BMJ*, i, 1869, pp 432, 574.

52 'Col Sir Francis John Bolton,' *MPICE* 93 (1887–88): 497–501.

53 'Copy of any Reports to the Board of Trade by the Water Examiner,' pp 4–14.

54 'Copies of a Letter from the Royal Commission on Rivers Pollution to the Board of Trade,' pp 1–4.

55 G Graham to LGB, 8 May 1872, *PRO MH* 19 67, #26162/72.

56 *Ibid.* Cf G Graham to LGB, 29 May 1872, *PRO MH* 19 67, #30440/72.

57 Minute on Frankland's ms report for January 1876, *PRO MH* 29 2, #8084/76; minute on Frankland's ms report for October 1876, *PRO MH* 29 2, #67673/76; unnumbered memo, Thomas to Dalton with minutes, misfiled in *PRO MH* 29 5, probably from summer 1884.

58 A Hardy, 'Water and the Search for Public Health in London,' 275–6; 'Reports on the Examination of Thames Water,' *JRSA* 31 (1882–3): 87–8.

59 Wanklyn and Cooper to LGB, 22 October 1880, *PRO MH* 29 3, #92868/80.

60 *DNB* 19, pp 864–5; *CN* 65 (1892): 143; *J. Chem. Soc.* 63 (1893): 766–8; J H Balfour Browne, *Forty Years at the Bar* (London: H Jenkins, 1916), p 97.

61 *SR* ns 2 (1880–81): 224; Luckin, *Pollution and Control*, p 61.

62 Crookes, Odling, and Tidy to LGB, 17 May 1883, *PRO MH* 29 5, #52429/83.

63 Letters and reports, *PRO MH* 29 5, #57540/83, 67506/83, 88334/83, 116365/83. On confusion about the official status see *SR* ns 3 (1882–3): 440; *Hansard's Parliamentary Debates* ser 3, v 267, c. 1808, v 278, c 318–9.

64 *PRO MH* 29 7, #90942/84, 112129/84, 116767/84, 118242/84, 3501/85, 4685/85, 8771/85.

65 C Meymott Tidy, 'Processes for Determining the Organic Purity of Potable Waters'.

66 *Ibid*, pp 66–85.

67 *Ibid*, pp 51, 88–9.

68 *Ibid*, pp 57, 76.

69 *Ibid*, pp 89–94; W C Young, 'A Comparison of the results obtained by Dr Frankland and the Companies' Analysts,' pp 159–64.

70 C Meymott Tidy, 'River Water,' (1880), pp 267–322.

71 C Meymott Tidy, 'River Water,' (1881), pp 113–4; W N Hartley, 'The Self-Purification of Peaty Rivers,' *JRSA* 31 (1882–83): 469–84; Charles W Folkard, 'The Analysis of Potable Water,' pp 57–115; Percy Frankland, 'The Upper Thames,' pp 428–53.

72 'The Metropolitan Local Authorities and the Water Supply,' *Lancet*, ii, 1881, p 770; G Phillips Bevan, *The London Water Supply*, p 37; David Owen, *The Government of Victorian London, 1855–1889*, pp 136–40; A Mukhopadhyay, *Politics of Water Supply*, ch 2.

73 'London Water Supply,' *CN* 45 (1882): 181. Cf 'The London Water Supply,' *Lancet*, ii, 1881, p 519.

74 'The London Water,' *CN* 47 (1883): 241.

75 'The London Water,' *CN* 48 (1883): 63.

8 Water Analysis and the Working Sanitarian

I am not conscious of ever having made a mistake in water analysis.[1]

Cornelius B Fox

In May 1884 Alfred Ashby, medical officer of health for Grantham, complained to the Society of Medical Officers of Health of 'the wretched wrangling, during the ten years past, over these rival processes, and ... the ridiculously exaggerated importance that has attached to them.' This wrangling was the cause of 'the discredit [in]to which water analysis has fallen among engineers and sanitarians.'[2]

It is easy to share Ashby's view that the London chemists had become so preoccupied with deprecating one another's laboratory prowess that they had lost sight of the key question of what water analysis was to be for: the pursuit of precision had come to overshadow questions of sanitary significance. The community of sanitary chemists had become so caught up with whether Frankland's or Wanklyn's process was the better, that the possibility that the best answer might be neither, might instead be something as radical as correlating analytical and epidemiological results, was rarely considered. Frankland and Wanklyn checked the accuracy of their processes on solutions of pure organic compounds of known strength, yet each claimed that the key determination he chose to make also had great significance for distinguishing safe from harmful water. In retrospect their attempts to measure a few parts per million of what was in Wanklyn's case an experimental artefact seem ludicrous. It is not so much that they were using chemistry to detect what turned out to be biological, but that they kept egging on one another for so long without stopping to ask whether they were taking water

212

analysis in a useful direction. How, we may wonder, could this be tolerated by the larger community, and especially by those actively seeking to safeguard the public from water-borne disease?

When we take into account the reasons analyses were commissioned and the uses they were put to, the rationale is clearer. Chemists of the stature of Frankland, Wanklyn, Letheby, and Tidy were regularly called upon to act as expert witnesses, both in legal actions involving pollution and before Parliament in the assessment of alternative sources for town supply. In neither context was there likely to be a clear-cut right answer; in such cases issues involving water quality were entangled in larger matters of law, policy, finance, and politics. We have already seen that these had become prominent contexts for water analysis beginning in the late '20s, yet we have not yet fully addressed the implications of working in such a context. Expert witnessing called for a set of attributes— in demeanor, authority, willingness to speculate—quite unlike those needed either in pure science or in front-line public health work. In the former uncertainty was admissible, while the latter placed a premium on thoroughgoing analysis of individual cases. The expert witness on the other hand had the job of persuading lay decision-makers that his claims and the perspectives of those he represented were sounder than his opponent's. As matters of water composition became more important and as methods of analysis became more sophisticated and their rationales more complex, the importance of the external symbols of credibility—credentials, assurance, coolness on the stand, the finality with which one could demolish an opponent's argument—all these became more and more important. And one's credibility hung to a great degree on the apparent credibility of one's analytical process. The chemist who worked at an order of magnitude smaller than his rivals and whose process had been more rigorously tested, was more likely to be believed, regardless of whether his measurements contributed anything of import.

This adversary context favoured the manufacture of certainty, however flimsy that certainty might later be shown to be. The expert who could deliver certainty on some matter, no matter how small and useless, was in a stronger position than one who could make informed, prudent, and plausible assessments of the central matters at hand, yet provide no certainty. It was far easier for the former to claim greater territory for his certainty than for the latter to recover from an honest admission of doubt. To the vultures of the parliamentary bar, doubt was vulnerability.[3] One may wonder how much use all this would be in the pursuit of public health; with a community of analysts bent on showing up one another's incompetence using analytical processes incapable of indicating anything certain about the presence of dangerous microbes, it might seem that decision-makers were in effect operating in ignorance. It is true that the links between an analyst's claims about the safety of a water and any investigation which could reasonably be regarded as warranting such a claim are sometimes tenuous indeed; moreover public health was often not the main *agendum* either in pollution litigation or in efforts to obtain purer water supplies. Better water was often sought by industrialists for industrial use and litigation undertaken to protect property values. But regardless of the motives of the promoters of better water and regardless of the means their hired scientists used to fight their credibility battles, better water and better health were often the outcome.[4]

To those on the front lines of public health work this sort of performance was much less useful. The health officers responsible for investigating outbreaks of epidemic diseases, both those in central government like John Simon and his staff of medical inspectors at the Privy Council (and later at the Local Government Board), and those serving local governments as medical officers of health, were rarely much concerned with the great questions of water supply policy. What they needed were quick and easy techniques for distinguishing fluctuations in the quality of service on a single street, or perhaps even for a single house. Frequently epidemics of waterborne diseases were the result not of an ongoing contamination, but of some unusual combination of circumstances.[5] The Caterham typhoid of 1879, for example, was started by a workman engaged in digging a new well who had the disease.[6] Sanitary engineers likewise were out of sympathy with the practices of the London chemists: 'the extreme refinements of modern analysis has seemingly led chemists to set up a standard of purity in excess of the necessities of the case,' one complained.[7]

After 1875 the dominance of the London consultants began to wane, as this kind of immediate and on-site approach became increasingly important. The great Public Health Act of 1875 mandated the appointment of a medical officer of health by each sanitary authority. In many cases these positions were for part-time officers only, but it was becoming increasingly common for their holders to have had specialized training in public health. One could earn a diploma in public health medicine at the University of Dublin after

1871, or at Oxford or Cambridge after 1875. In addition, a sort of British Association for sanitary matters, the Sanitary Institute of Great Britain, was founded in 1877. It took it upon itself to certify the competence of local sanitary officials. After 1875 towns were also required to employ a public analyst, an office established to fight food and drug adulteration. Both the medical officers of health and the public analysts founded professional organizations, the Association (later Society) of Medical Officers emerging out of the older Association of Metropolitan Medical Officers in 1875, the Society of Public Analysts being founded in 1877. In 1891 a third organization, the British Institute for Public Health, was founded, its membership restricted to holders of the diploma in public health. Water analysis was officially within the mandate of neither medical officers of health nor public analysts, yet in many cases holders of these offices took an interest in water quality as did their professional organizations.[8]

Relations between these local sanitary professionals and the London elite were not good. Front line sanitarians knew at first hand the inadequacies of the ammonia and combustion processes for distinguishing disease-causing waters. They were also well placed to see how contradictory were the opinions of different chemists on the same source of water. In 1875 the chemist J Carter Bell, later public analyst for Cheshire, wrote to the *Chemical News* to report such an outrage. A water sample from a newly dug well had been sent to an unnamed chemist for analysis. Finding no putrescent organic matter, this analyst had pronounced it satisfactory. A year later a sample had been sent to a commercial analytical firm, which reported that it contained flocculent organic matter and stated that it was dangerously contaminated. A third analyst, a medical man using the ammonia process, concurred. A fourth, a chemist, judged the water safe on the basis of chlorine and oxidized nitrogen levels. A fifth, relying mainly on the ammonia process, broke the tie by declaring the water safe. Bell found the situation scandalous. Each chemist had used a different set of processes, he noted. He suggested that a jury of the Chemical Society ought to select and enforce a set of standard methods.[9]

Alfred Ashby had a similar story. Having found analytical evidence of 'sewage pollution' in a well water from Newark-on-Trent, he requested the magistrates to order it closed. The well's proprietors countered with an analysis by Charles Graham of University College which showed the water to be good, and faced with conflicting scien-

104   PRACTICAL HYGIENE.

2. Usable Water.

Character or Constituents.			Remarks.
Physical characters : Colorless, or slightly greenish tint; transparent, sparkling, and well-aërated ; no suspended matter, or else easily separated by coarse filtration or subsidence ; no smell ; taste palatable.			In some usable waters, such as peat waters, the color may be yellow or even brownish. In some also the taste may be flat or only moderately palatable.

Chemical Constituents.	Grains per gallon, 1 in 70,000.	Centi-grammes per litre, 1 in 100,000.	Remarks.
1. Chlorine in chlorides.....*under*	3.0000	4.2857	This may be much larger in waters near the sea, deep well waters, or waters from saline strata.
2. Solids in solution : total..*under*	30.0000	42.8571	
Solids in solution: volatile, *under*	3.0000	4.2857	The solids may blacken, but no nitrous fumes should be given off.
3. Ammonia, free or saline, *under*	0.0035	0.0050	This may be greater in deep well waters.
Ammonia, albuminoid ...*under*	0.0070	0.0100	This may be larger in upland surface waters, peat waters, etc., when the source is chiefly vegetable.
4. Nitric acid (NO_3), } in nitrates.. ..}.....*under*	0.3500	0.5000	The amount of nitrates varies greatly, so that an average is of doubtful value.
Nitrous acid (NO_2), } in nitrites......}	nil.	nil.	
Nitrogen in nitrates .....*under*	0.0790	0.1129	
Total combined nitrogen, including that in free ammonia ....*under*	0.0819	0.1170	
Total nitrogen, including that in albuminoid ammonia ..*under*	0.0876	0.1252	
5. Oxygen absorbed by organic matter within half an hour, by permanganate and acid, at 140° F. (60° C.).....*under*	0.0700	0.1000	The oxygen absorbed may be greater (about *double*) in upland surface waters, peat waters, etc.
Do. in fifteen minutes, at 80° F. (27° C.) ....*under*	0.0210	0.0300	
Do. in four hours, at 80° F. (27° C.) ....*under*	0.1050	0.1500	
6. Hardness, total ....*under*	12.0°	17.3°	
Hardness, fixed ....*under*	4.0°	5.7°	
7. Phosphoric acid in phosphates..	traces.	traces.	
Sulphuric acid in sulphates.............} *under*	2.000	3.0000	In some waters the amount may be larger.
8. Heavy metals—Iron	traces.	traces.	
9. Hydrogen sulphide, alkaline } sulphides................}	nil.	nil.	

Microscopic characters : same as No. 1.

A water such as the above will in most cases be usable, but it will be improved by filtration through a good medium.

Figure 8.1 Writers of late nineteenth century public health manuals sometimes tried to give their readers sample analytical profiles of various qualities of waters, all the while advising that these were not to be taken seriously since so many factors had to be taken into account when judging waters (*Parkes' Manual of Hygiene*, 6th edn, p 104).

WATER. **105**

3. *Suspicious Water.*

Character or Constitnents.			Remarks.
Physical characters: Yellow or strong green color; turbid; suspended matter considerable; no smell, but any marked taste.			Where the impurity is mostly vegetable, the color may be very marked in usable water.
Chemical Constituents.	Grains per gallon. 1 in 70,000.	Centigrammes per litre. 1 in 100,000.	
1. Chlorine in chlorides..........	3 to 5	4 to 7	In some cases the chlorine may be greater.
2. Solids in solution: total	30 to 50	43 to 71	
Solids in solution: volatile	3 to 5	4 to 7	
3. Ammonia, free or saline	0.0035 to 0.0070	0.0050 to 0.0100	
Ammonia, albuminoid	0.0070 to 0.0087	0.0100 to 0.0125	
4. Nitric acid (NO₃), in nitrates ..	0.35 to 0.70	0.5 to 1.0	
Nitrous acid (NO₂), in nitrites..	0.0350	0.0500	
Nitrogen in nitrates and nitrites.............	0.0870 to 0.1661	0.1243 to 0.2373	
Total combined nitrogen, including that in free ammonia.....................	0.0871 to 0.1718	0.1247 to 0.2455	
Total nitrogen, including that in albuminoid ammonia.....	0.0879 to 0.1726	0.1255 to 0.2465	
5. Oxygen absorbed by organic matter within half an hour, by permanganate and acid, at 140° F. (80° C.)	0.0700 to 0.1050	0.1000 to 0.1500	
Do. in fifteen minutes, at 80° F. (27° C.)....................	0.0350 to 0.0700	0.0500 to 0.1000	
Do. in four hours, at 80° F. (27° C.)....................	0.1500 to 0.2800	0.2000 to 0.4000	
6. Hardness, total ....*above*	12.0°	17.0°	
Hardness, fixed ....*above*	4.0°	5.7°	
7. Phosphoric acid in phosphates	heavy traces.		
Sulphuric acid in sulphates .... } *above*	2.000	3.000	This may sometimes be larger.
8. Heavy metals—iron	traces.	traces.	
9. Hydrogen sulphide, alkaline sulphides	nil.	nil.	
Microscopic characters: Vegetable and animal forms more or less pale and colorless; organic debris; fibres of clothing, or other evidence of house refuse.			

A water such as the above ought to excite suspicion: its use ought to be suspended until inquiries about it can be made ; if it must be used, it ought to be boiled and filtered.

Figure 8.1 (continued)

tific testimony, the magistrates threw out the case. Ashby then had samples analysed by August Dupre and Otto Hehner, prominent members of the Society of Public Analysts. Both condemned the water. Still the magistrates were unconvinced, and sent a sample to an 'independent' analyst, Frankland. He too pronounced the water unfit—'soakage from drains or cesspools'—and finally they closed the well. Ashby saw all this as unnecessary. In some marginal cases there might be room for disagreement, he admitted, but this had not been such a case. And as local medical officer he had 'unusual facilities for judging' and 'the advantage over any chemist who may have to form an opinion from arbitrary standards alone.'[10]

Ashby's complaint about distant consultants who used their prestige to pronounce on matters about which they really knew little reflects the increasing impatience with centralized water analysts of local sanitarians. In 1876, when Frankland and other senior chemists (including Wanklyn) were attempting to form the Institute of Chemistry as a means of ending this kind of scandal, the *Sanitary Record* was scornful. Any lack of trust in analytical chemistry was due to contradiction by those at the top, not to the incompetence of those at the bottom.[11]

The Liberation of the Medical Officers

By the late '70s local sanitarians had gone beyond complaining and were actively seeking a foundation for a decentralized water analysis. That search is manifested in three distinct endeavours. The first was the attempt to develop a flexible approach to water quality assessment especially suited to the diverse needs of medical officers of health. It is well represented in two of the main manual/text books for health officers, those by Edmund Parkes and Cornelius Fox. The second was the attempt to revive the sort of microscopy Hassall had done as the mainstay of water assessment. It is exemplified by J D MacDonald's *Guide to the Microscopical Examination of Drinking Water*. The third is the attempt in the late '70s and early '80s of the Society of Public Analysts to develop standard methods and interpretations for water analysis.

Although Edmund Parkes was the better-known sanitarian, Cornelius Fox was the more outspoken.[12] Fox's *Sanitary Examinations of Water, Air, and Food* (1878, 2nd edn 1886) was written by a medical officer for his colleagues. Fox put a premium on speed in analysis. He claimed to be able to do a rudimentary water analysis (free and albuminoid ammonia, chlorine, nitrates and nitrites, and total solids) in about 40 minutes, in contrast to the 'two or three months' that might elapse before one received a verdict from Frankland.[13] Fox also emphasized the importance of including all evidence in forming an opinion. He revived a great range of qualitative tests—the odour of a water when heated, or the appearance of a residue when ignited—that had been discarded by more quantitatively minded chemists. The great weakness of the London chemists was their dogmatic faith in a particular parameter, he asserted, be it albuminoid ammonia or organic nitrogen: relying on the ammonia process alone might lead one to regard rain water as unsafe and fresh urine as safe.[14]

The key question, as Fox recognized, was not the one Frankland and Wanklyn were arguing about, of whose process was more accurate. Much more important was knowing whether the processes would lead the analyst to judge the quality of the water correctly. Leaving aside the particular results obtained, would an analyst using one of the processes reach the same conclusion as one using the other? Fox analysed 93 samples by both processes and in all but one case reached the same verdict; he could conclude only that Frankland and Wanklyn had been leading sanitarians on a quest for ultimate precision that was both futile and needless:

> Apart from these very warm controversies ... such disputes ... do retard the progress of sanitary science, and lead the public to imagine that the whole question, whether a water is or is not pure, is a 'toss-up'; this remark being generally clinched with the further reflection that it is universally acknowledged that doctors differ.[15]

But the main distinction between Fox and the London chemists had to do not with the acceptability of particular processes, but with the question of what analysis was for. Fox wrote that his intention was 'to treat [water matters] as a physician who studies them in connection with health and disease.' His book was to be as free as possible from chemical jargon—'from technicalities and all cloudy and chaotic surroundings,' as he put it. Analytical methods were to be simple enough that medical officers could learn them on their own, and analysis was to be only one among many tools for answering specific questions about water quality. Suspecting that a privy was contaminating a well the health officer was to look for specific signs of that contamination—high chlorine or nitrate levels, a peculiar odour, or a high level of albuminoid ammonia.

Thus, if analyses were to be useful it would be in terms of testing epidemiological hypotheses and the analyst would have to know a great deal about the circumstances of the water in question, including as much as possible about its origins—soil, drainage, what the land was used for. He knew of 'chemists [who] ... would rather know nothing about the sample' but held that 'an opinion so formed is worth very little .... the medical officer will always do well to obtain every item of information about it [the water] that it is possible to get.' No longer could analytical results be understood as providing ultimate and authoritative answers that would displace other forms of evidence. They were simply one type of information the health officer could use in coming to a diagnosis.[16]

In some ways this perspective resembled Frankland's. Frankland too recognized that analysis might be used to test hypotheses, even if it could not be expected to mirror nature in all particulars. For example, he used analysis to arrive at tentative generalizations about the oxidative capabilities of soils and the seasonal changes in rivers. Yet the kind of hypothesis-testing Fox had in mind, in which a local sanitarian would use certain carefully chosen analytical operations to track down the origin of a particular disease outbreak was not the sort of enterprise in which a distant London consultant, however elite, would be of much use.

Other sanitarians, though they did not go into the detail Fox had in re-orienting water analysis, shared many of his views. They returned to older and simpler processes, such as the ignition test for organic matter and the various approaches for comparing the colour of a water. They did not try to refute the theoretical objections that had been raised to such processes, but maintained that in the right circumstances these methods were good enough, and that they were easier to learn and to use.[17]

The Return of Microscopy

Among the approaches that began to come back into fashion were microscopical examinations. These were undertaken to look for signs of contamination, both life forms characteristic of sewage polluted environments and actual traces of sewage, such as bile-stained meat fibres. In Fox's view such methods held great potential for liberating local sanitarians from the irrelevant precision of the London consultants. With microscopy, local sanitarians could aspire to a similar level of precision. Moreover, it would be a field which they would have to themselves, since it was imperative that microscopic analyses be done on fresh samples.[18]

In the 1850s microscopical examinations of waters had been undertaken in connection with campaigns to end London's reliance on polluted river water. Like Frankland's monthly announcements of the 'living and moving organisms' in the companies' waters, this use of microscopy had been a way of making the idea of impurity more vivid. By the late 1850s most analysts had abandoned microscopical methods. Hassall, though he continued to rely on microscopy in establishing a reputation as Britain's leading food analyst, acknowledged the superiority of chemistry in water matters. In his journal, *Food, Water, and Air* (begun 1871), he endorsed Frankland's views on water and recognized that Frankland was doing with chemistry exactly what he had tried to do with microscopy: deliberately setting out 'to alarm the public mind.'[19]

In 1874 Jabez Hogg, a London ophthalmologist and microscopist, revived the old Hassallian programme of trying to force London's water-drinkers to confront the disgusting impurities which the microscope alone could reveal. Hogg even allied himself with Samuel Homersham, the engineer who in 1852 had commissioned microscopical studies by Lankester and Redfern, and who was still advocating an alternative water supply for London. Hogg was a master at the rhetoric of disgust. He was especially fond of the term 'noxious'—in the water there were 'filariae and larvae of the most noxious kinds' and 'small fish, eels, and numerous noxious animals.' The idea that water was a 'breeding ground' for various life forms Hogg also found loathsome; he had a knack for making ordinary biological processes morbid and menacing. Life itself was disgusting; one was not to drink lake water because 'living organisms find a congenial habitat in the mud and water, and rapidly increase and multiply.'[20]

It might be expected that arguments like Hogg's would be taken increasingly seriously as the prospect that zymotic diseases were caused by bacteria or 'germs' became more and more likely. Hogg did find the image of 'contagious cells' to suit his needs well, and he attributed a number of specific diseases to various forms of microscopic life, including bacteria. Yet in 1874 bacteria were still not the primary concern and Hogg's arguments were not taken seriously; he was recognized as an extremist. For Hogg the whole microscopic population was abominable—if a species were not actively harmful he was sure it either predisposed one to disease or facilitated the development of something that was harmful. Indeed, so thoroughgoing was Hogg's condemnation of minute living things that he cut off all opportunity to develop microscopic analysis into a means for making distinctions among waters: if all life forms were as malign as they could be, what was the point of making any distinctions at all?

The local sanitarians who were beginning to take an interest in microscopical water analysis in the late '70s were unconnected with the Hogg/Hassall polemics. They did, however, run up against the other problem that had plagued microscopists in the '50s, of whether there existed any set of microscopical distinctions that would tell anything one didn't already know. The most sober of the microscopical assays of the '50s had been done by the Bristol microscopists James Brittan and Robert Etheridge in connection with an 1854 cholera outbreak at Sandgate. They had begun their report by listing microscopic invertebrates commonly found in waters, along with the habitat of each. Then they reported the organisms found in the Sandgate water, noting that 'without exception, the organisms found ... are such as would be detected in waters yielding a considerable amount of decomposing vegetable and animal substances.' On this basis they concluded that the water must have considerable 'organic matter in a state of disintegration and decomposition.' They went so far as to suggest which samples probably had the highest concentrations of organic matter, judging from the 'characters of the organisms found, and their known peculiarities as to habitat, and food, etc.'[21]

The structure of the Brittan/Etheridge report is striking. They were forthright in stating their interpretive standards at the outset (unlike Hassall who sometimes appears to be assigning species to habitats as he goes along). They applied their standards to the water and drew the appropriate conclusions. Yet however rigorous their procedure, the most Brittan and Etheridge could claim to be doing was to be using biological procedures to measure chemical variables. It was, after all, a chemical entity, decomposing organic matter, that sanitarians were interested in and there was no reason to suspect that that entity could not be measured more directly, easily, and accurately through chemical techniques. Had the biologists been able to make qualitative distinctions among types of decomposing matter, their work might have been helpful. But they did not even try to make such distinctions. Brittan and Etheridge had perhaps demonstrated the feasibility of microscopic analysis, but they had shown no reason why it should be adopted.[22]

In the late '70s and early '80s microscopists faced the same problem. The range of conceptions of the nature of the harmful substances in water had shifted over the years—concern with decomposing organic matter had given way to concern with certain sorts of putrefaction and eventually to fears of specific, quasi-living, disease germs. But the actual disease-causing entities were still for the most part unknown and invisible to the microscopist. The most microscopy could offer was to demonstrate the likelihood of sewage contamination.

The foremost authority on the microscopic analysis of potable waters during the period was J D MacDonald, MD, FRS, professor of Naval Hygiene at the military medical college at Netley, and later Inspector General of Hospitals and Fleets for the Royal Navy. MacDonald's *Guide to the Microscopical Examination of Drinking Water* (two editions, 1878 and 1883) was the main source for the discussions of water microscopy that Parkes and Fox included in their manuals.

MacDonald's claims for microscopical water analysis were modest. He thought that the 'known habitat of certain organisms detected should enable us, in a general way, to determine whether the water had been taken from a river, stream, lake, pond, or well source. Indeed, if we were more perfectly acquainted with the natural history of the forms occurring in a sample of water even in the absence of more definite information, we would have little difficulty in forming a conclusion as to the source from whence the water was derived.'[23]

Though a water's source was recognized as an important factor in its assessment, the analyst was rarely in the position of having to discover that. But even so limited a goal demanded a much greater knowledge of habitat than was currently available, MacDonald pointed out. In fact, his book was closer to a traditional taxonomic guide, listing species and providing instructions for distinguishing them from one another, than it was to a handbook for microscopical water analysis. MacDonald included information on habitat in some cases, but he did not make a point of doing so in all. This was probably wise, since there were sometimes disagreements among microscopically minded sanitarians as to which organisms indicated which habitats. For example, A Wynter Blyth, following Hassall, insisted in 1874 that paramecia indicated bad water, yet Parkes wrote a few years later than 'subsequent observations have not, however, proved the relation between paramecia and animal matter in the water to be sufficiently constant to allow the former to be used as a test of the latter.' Similarly, ciliated forms indicated sewage to Blyth, while Parkes noted that there were many different kinds of these, and that they showed nothing more than the presence of 'vegetable or animal organic matter.'[24] Reluctantly, he admitted that microscopical analysis was of limited utility: 'So many are the objects in water that the observer will be often very much at a loss, first to identify them, and secondly, to know what their presence implies.'[25] Fine distinctions were rarely possible. There were exceptions—on occasion a microscopist might discover a clear trace of sewage, one which signified contamination more convincingly (and graphically) than could any purely chemical parameter.[26]

Some British sanitarians, especially those also involved in food analyses where microscopy was central, did make a practice of including microscopical examinations in their water analyses, but it is likely that they did so not so much for what microscopy could show but because it was one means of characterizing water that was available to them. Notably in Germany, France, and the United States the correlation of microscopic aquatic species with aspects of water quality was taken more seriously than it was in Britain. Independent traditions of diagnostic stream ecology developed in each of these countries. It was the German tradition, embodied in the *saprobeinsystem* of R Kolkwitz and R Marsson, that reignited the interest of British analysts in biological approaches in the first decade of the twentieth century.[27]

The Initiative of the Public Analysts, 1878–84

Among the most persistent advocates of microscopy in water analysis was George W Wigner.[28] Wigner is not one of the better-known analytical chemists, but he had been working in sanitary chemistry at least since the late '60s when he was one of the founders of the ABC Process, the most famous and long-lived of the sewage precipitation processes. In February 1878 Wigner read a paper on water analysis to the newly formed Society of Public Analysts, an organization restricted to those who had an official appointment as a public analyst. The society he addressed was not wholly typical either of learned or of professional societies. It had been founded in 1877 shortly after the office of public analyst had been made mandatory for urban sanitary authorities. While it was concerned with the advancement of knowledge and with the welfare of the nascent profession of the public analyst, the society spent much of its time trying to keep up with the rapidly evolving practices of food-and-drug adulterers and with the legal and technical complications encountered in bringing such villains to justice. In a profession in which consensus on technical matters had profound legal implications, matters of standard methodology were of great import.[29] Wigner shared many

of the views of Fox, Parkes, Ashby, and Charles Cassal, a cadre of
sanitarians, many of them young, who were critical of the domina-
tion of water analysis by Frankland and Wanklyn. Like Fox, he held
that an analyst ought to base an opinion on as wide a range of information as he could; like Ashby and Cassal, he was distressed at the
frequency with which water analysts contradicted one another, and
the apparent arbitrariness of their opinions of a water.[30] In Wigner's
view, the time had come to designate a wide-ranging set of qualitative and quantitative tests to be included in every analysis and to
translate the results of each test into a common numerical scale so
that at a glance anyone could compare results of two samples simply
by comparing the sums of the results of all the tests done on each.
The summing of the individual results would automatically ensure
that all processes were represented in the conclusion; the common
point system would ensure that all chemists would be weighting the
results in the same way and thus would arrive at the same conclusion about any given water. It was an ingenious and wonderfully
simple idea. The only questions were whether analysts could agree
on the weighting and whether they would be willing to substitute
a simple formula for the traditional freedom of drawing whatever
conclusion one chose on the basis of one's consummate expertise.
On both points Wigner was a bit naive.

and microscopic appearance.   But taken as an outline the table stands this way :—

5 grs. total solids ... ... = 1	Taste, decidedly offensive ... ... = 6
1 gr. loss on igniton ... ... = 1	Smell, flat rain water ... ... = 2
1 gr. chlorine calculated as chloride of sodium = 1	Ditto urine ... ... = 6
·0200 gr. free ammonia ... ... = 1	Colour, pale yellow ... ... = 2
·0010 gr. albuminoid ammonia ... ... = 1	Ditto yellow green ... ... = 4
·1000 gr. nitrates ... ... = 1	Ditto urine yellow ... ... = 6
·0020 gr. nitrites ... ... = 1	Ditto opaque yellow in 2-ft. tube ... ... = 9
·0100 gr. oxygen absorbed ... ... = 1	Microscope, bacteria ... ... = 3
5 degrees total hardness ... ... = 1	Ditto, other similar growths in greater
Traces of lead ... ... = 6	quantity ... ... = 4
Ditto copper ... ... = 6	Ditto, few living organisms ... ... = 6
Heisch's sugar test ... ... = 6	Ditto, animal remains ... ... = 12
Taste, good ... ... = 0	Ditto, urea and urates and muscular fibre... = 18
Ditto, slightly saline ... ... = 1	Suspended matter, traces ... ... = 2
Ditto, decayed leaves ... ... = 2	Ditto heavy ... ... = 4
Ditto, flat rain water ... ... = 2	

Figure 8.2 The first version of George Wigner's formula for evaluating
waters, presented to the Society of Public Analysts in 1878. The concept
of a uniform scale bothered Wigner's colleagues, and even after massive
revision the scale was not adopted (*The Analyst* 2 [1878]: 215).

In the initial version of the scale, Wigner used the quantity of one thousandth of a grain per gallon of albuminoid ammonia as a base—this was to equal one degree of impurity. The rationale was that this albuminoid ammonia was both the 'most injurious factor' discovered in the analysis and the smallest quantity it was useful to consider. All the other determinations and observations were then assigned values in accord with the degree of danger or unsuitability they posed in comparison with this albuminoid ammonia—nitrites were awarded one point per every two thousandths of a grain; free ammonia received a point for every two hundredths of a grain, and so forth. Wigner also awarded various numbers of points for total solids, weight lost on ignition, chlorine, nitrates, oxygen absorbed from potassium permanganate, hardness, and traces of lead and copper. Each determination was essential for a thorough analysis, he argued, hence each ought to figure in the classification of a water.

Wigner realized that while his views on the appropriate weighting of various substances might be questioned, standardization was feasible for quantifiable parameters. He faced a greater problem incorporating physical and microscopical observations for these were qualitative. He stipulated that a good-tasting water was to be given no points, one with a 'slightly saline' taste one point. A taste of 'decayed leaves' or 'flat rain water' warranted two points, a 'decidedly offensive' taste received six as did a water which emitted a urinous odour when heated. Colour was also important, worth as many as nine points in the case of a water which was an 'opaque yellow' in Letheby's two foot tube. The microscopical evidence was valued mainly in terms of its mass, along with the certainty with which it indicated sewage contamination. The presence of bacteria warranted three or four points, depending on their quantity. 'A few living organisms' received six points, 'animal remains' twelve, and traces of urea and partially digested muscular fibres eighteen. Waters with scores under 35 were to be regarded as satisfactory, those scoring between 35 and 55 were 'second class,' those between 55 and 75 were 'suspiciously dangerous,' while any sample exceeding 75 was to be regarded as 'sewage.' To ensure greater unanimity in making the qualitative determinations, Wigner hoped to make solutions of standard colour, taste, and smell, and sets of standard microscope slides.[31]

The expectation underlying the scheme was that a water impure according to one test would be impure according to others; the tests would overlap and amplify one another. Yet as Wigner himself knew, there were instances in which this was not the case—the danger in a water would show up in only one of the analytical operations done on it or not at all. Wigner admitted that he had examined waters recently contaminated with urine in which the analyst's only clue was the presence of solid urea or urates in the residue. In such cases there would be little albuminoid ammonia and possibly little free ammonia.[32] Hence the summing process, which was intended to ensure that each measurement figured in the conclusion, might actually hide the significance of that single measurement which represented the dangerous impurity. In such cases Wigner considered that 'the water should be condemned on the result of that examination alone.' But such a stance simply re-raised the issue of the utility of a numerical scale. If it remained essential to consider the result of each test independently and then to treat certain danger signs as valid reasons to ignore point totals, why then do the summing at all?

The Wigner scale did not arouse much interest until 1881 when a committee of the Society decided to institute a programme of regular and uniform analyses of the public water supplies of all towns in England. Making up the committee were Wigner, John Muter, August Dupre, A Wynter Blyth, Otto Hehner, Bernard Dyer, and Charles Heisch, all experienced analytical chemists who had managed to avoid the polarization in water analysis issues of the early '70s. Three of the members, Dupre, Hehner, and Heisch, had even at one time or another taken up problems in the methodology of water analysis.

Water analysis was not a part of the public analyst's job and the committee's existence probably reflects members' professional interests more than their specific concerns as public analysts. Public analysts had private practices too, and as an editorial in *The Analyst* pointed out, water analysis would normally be 'a fairly remunerative part of an analyst's practice.'[33] In announcing its decision to undertake the analyses, the committee made it clear that it was adopting the perspective of Col. Bolton, the metropolitan water examiner, and all who were bewildered by the proliferation of contradictory analyses in incommensurable formats. They envisioned a time 'when the same water sent to every Public Analyst in England will be returned with the same opinion, just like an analysis of milk or butter.' As Heisch recognized, this change would require a different attitude on the part of water analysts; they would have to get used to 'sinking their individual opinions and working loyally on the lines laid down by the majority to secure that which all must consider of the utmost importance—uniformity of results.'[34]

Ignoring the central issues of the Frankland–Wanklyn dispute, the committee chose processes likely to be the 'most rapid and reliable for such systematic analyses.' These were the oxygen absorbed (permanganate) process and Wanklyn's ammonia process, and several others Wigner had listed: physical and microscopical analysis, measurement of phosphates, nitrates, chlorine, hardness, and total solids. They issued a detailed set of instructions for these processes, the first appearance in Britain of anything resembling a set of 'standard methods' of water analysis. These were sent to all public analysts and given out free to anyone who requested them. So great was the demand that within the first year the Society had commissioned second and third printings.

The January 1881 number of *The Analyst* carried the first set of analyses of the waters of forty towns done according to this system. Even at the outset there was tension as to what, if any, interpretation of the results the committee should make. Much of the dissatisfaction with water analysis in Britain was with the extreme interpretations made by such men as Frankland and Wanklyn. Initially, the Society elected to offer no opinions whatever on the results it received each month; their value was to lie solely in the comparisons they allowed one to make of the waters of different towns. Since the returns were detailed those 'in the habit of collating such results' would be able to 'form a fair judgment for themselves.'[35] Thus the committee had accepted half of Wigner's proposal—uniformity of method—while ignoring the other half—uniformity in interpretation. It was not a satisfactory compromise: a central concern, for both Wigner and the committee, was that analyses be meaningful to the general public. For this reason they had rejected metric units in favour of the more familiar grains/gallon (1:70,000). Yet in leaving the facts uninterpreted, they were only returning responsibility for interpretation to other experts who alone could make sense of them.

In June 1881 Wigner presented the Society with a revised version of his scale. He assumed his colleagues would agree with him as to the value of uniformity in interpretation and he hoped they would be able to agree on a system itself. Wigner had fiddled with his scale a good deal during the previous three years, leaving out the taste test, for example, and changing the number of points assigned for other constituents. More importantly, he had tried out the system on several sets of analyses to ensure that his point totals actually corresponded to the opinions experienced analysts would be likely to give. He also tried to verify that the valuing system would clearly signal important changes in water quality, such as the discharge of a sewer into a river.

Despite these revisions and validations (and Wigner's willingness to compromise), when it came time to endorse the scale the membership balked. The discussion never even got to the details, it was the principle of uniformity itself that was called into question. The main objection was that it was erroneous to think that all one needed to know of a water was contained in analytical returns. Several speakers held Frankland's view that the most important factor in interpreting an analysis was knowledge of the water's history. Some northern towns, they pointed out, got moorland water that was heavily laden with organic extracts from peat. Such waters would receive high scores on Wigner's scale, though they were generally judged unobjectionable, being completely free of sewage contamination. Likewise, waters from deep wells frequently contained large amounts of nitrates and even nitrites and ammonia. These too were usually considered excellent waters, yet on Wigner's scale they might yield scores as high as 70 points, far above the fitness range. It seemed unfair to compare cities with such waters with other towns which might have lower scores, yet poorer water.

There was also concern that even though it was a good idea to consider all types of analytical evidence, it was important to be able to condemn water if it proved unsatisfactory in only one category, regardless of its point total. J W Tripe, medical officer and public analyst for some of the east London vestries, was bothered that microscopical evidence, no matter how damning, warranted no more than ten points. Some of the bacteria observed through the microscope might well prove to be the carriers of typhoid fever or some similar disease and yet in Wigner's scheme their presence was still not enough to condemn the water.[36]

The Society postponed action on the proposal but allowed Wigner a trial: *The Analyst* would publish the scores of the samples analysed as part of the programme of uniform analyses, but the scores were, however, to be printed on a separate page from the analytical results themselves. For nine months, from July 1881 to March 1882, Wigner listed the score of each water for that month, its score for the previous month, and its average score for January–June 1881. Thus anyone, no matter how untrained in chemistry, could at a glance see whether his water was improving or deteriorating and how it compared to the waters of other English towns. There were some striking vacillations in quality. The water at King's Lynn fell from 110 to 48 in a month and was back to 110 a month later; Darlington's water rose from 39 to 96 in a month.[37]

In February 1882 the Society returned to the question of a common scale and voted 25 to 11 against adopting the Wigner scale. They also decided that Wigner scores would not be published in *The Analyst* unless they were submitted by the chemists who had done the analyses. In the face of this opposition, Wigner and Muter, who were co-editors of *The Analyst*, decided to end the experiment.

In April 1883 Dupre and Hehner presented a modification in which standards would be calibrated for each region, and two months later Muter offered another approach to the standards problem,[38] but neither suggestion generated much interest. The fight to establish a chemical formula for good potable water had been lost. The Society's program of uniform analyses ended in November 1882, at the end of two years. Wigner and Muter noted that the analyses had been made 'at considerable cost ... by over fifty analysts ... without any payment from the Water Companies or Corporations who have supplied the water, but simply for the purpose of disseminating ... knowledge.' Nearly a thousand had been done, 'the largest series of uniform analyses of water supplies which have ever been published by any private body of analysts.' But while they insisted that the goals of the project—'to draw public attention to the character of the water' and 'to give facilities which were not then available for judging of the relative qualities'—had been achieved, it is clear that the membership was losing interest. The last of the tables included analyses of only 19 supplies outside London; at the start of the year there had been 39.[39]

It was the unwillingness of chemists to give up the professional's prerogative of independent judgement that made Wigner's initiative ultimately impracticable. The Society had begun its analyses, and Wigner had taken up the cause of uniform interpretation, out of the conviction that some chemists were using the trappings of professionalism as an excuse to justify their arbitrary interpretations and their use of processes not sanctioned by the profession as a whole. Wigner and the SPA had hoped to stop this by insisting that chemists use a standard set of processes and in Wigner's case by insisting on a common set of interpretative rules. Yet even during the two or three years when the attractions of uniformity seemed most compelling, chemists did not give up the conviction that independent judgement was an essential part of responsible professionalism.

The conflict was expressed most clearly by Dupre and Hehner in their defense of local standards. Universal standards had two negative consequences, they pointed out: they 'weaken[ed] the feeling of personal responsibility of the analyst,' and they gave 'a spurious belief in the possession of knowledge to the ignorant.' Good chemists would know when to discard standards, but that only showed that good chemists did not need standards, for they interpreted analyses 'according to the circumstances of every particular case.' There was no escape from the onus of judging, and no expedient way to abolish contradiction:

> what we wish to impress on our fellow analysts is this—by all means take into consideration and, on suitable occasions, make use of such general standards as have been laid down by chemists of high ability and large experience; but use these standards cautiously and with discrimination, and judge every case on its own merits. Analysts who lack either the ability or the experience to stand on their own legs, and slavishly adopt standards laid down for them by others, have no business to meddle with water analysis at all, and the sooner they leave such work to their more experienced brethren the better it will be for themselves and for the credit of water analysis.[40]

The Mallet Report

In terms of the distribution of authority among British sanitary chemists the most serious challenge to the dominance of Frankland and Wanklyn in the pre-bacteriological era was that mounted by the Society of Public Analysts between 1878 and 1882. They declared independence from the London chemists and insisted that the average public analyst was not only competent to analyse water, but the right person for such work, since public analysts could better know the circumstances of the analysis and could guarantee results according to a common format.

On the other hand, the most serious scientific challenge to Wanklyn and Frankland was an 1882 report on the accuracy and utility of the main processes for the analysis of organic matter in water, sponsored by the United States National Board of Health and carried out by a team of chemists and medical men under the direction of Professor J W Mallet of the University of Virginia. For nearly fifteen years Frankland and Wanklyn and their partisans had exchanged accusations. They had come up with unanswerable arguments explaining why the other's process could not possibly be accurate, yet no large-scale, systematic comparison of the two (or three, if we count Letheby's oxygen-absorbed approach) processes had been done to determine whether these arguments were valid. American and continental water scientists found the British scene hard to comprehend. They recognized that water quality was taken more seriously in Britain than elsewhere but were baffled by the vehemence of a controversy about how to measure an unknown and unmeasurable entity. Mallet's report emerged from this background. It was the first systematic attempt to assess exactly what the British chemists had accomplished in the development of analytical processes, the first investigation to provide definitive answers to the controverted questions. There were five of these: first, were the processes replicable; second, did any of them permit one to distinguish between the different sources of contamination waters might be exposed to; third, how accurate were the processes in measuring what they claimed to be measuring; fourth, did any of them have the capacity to allow the distinction of water known to cause disease from water known to be safe; and fifth, were biological or microscopical means of analysis any more satisfactory?

Let us take these in order. Especially in connection with the combustion process, the problem of replicability was one of the most troublesome questions for British water analysts, and the one with the most direct implications for the structure of the profession. Even if the combustion process gave accurate results in Frankland's lab, could it be successfully used by others? And if it could not what were the reasons? Mallet faced the problem directly. Three experienced analytical chemists, W A Noyes of Johns Hopkins, Dr Charles Smart of the U S Army, and Dr J A Tanner of the U S Navy, independently examined portions of a split sample. The three chemists had trained themselves in the use of the processes together to ensure as much uniformity as possible in their operations and they each analysed the sample on the same day to control for possible alteration of the substances in the water. The results were unexpected. The simple permanganate process of Letheby and Tidy gave the best results, but the ammonia process, widely viewed as giving concordant results, turned out to be not much better than the notorious combustion process, which was supposed to be too tricky to yield concordant results.[41]

The second and fourth questions were crucial. Did any process of analysis permit one to distinguish different types of contaminations, including waters which there was strong reason to think had caused disease outbreaks? Surprisingly, this fundamental issue had rarely been directly addressed by British analysts. Frankland, of course, had admitted as early as 1866 that it was impossible for chemical analysis to distinguish water that caused disease from water that didn't. While most sanitarians probably agreed with Frankland, no one had gone to the trouble of systematically collecting data to discover how well analyses could distinguish bad waters—here too the debate had been conducted through a series of assertions and denials. The results were not encouraging. The quantities of free or albuminoid ammonia or of the oxygen absorbed from permanganate could really tell the chemist very little about what was in the water, Mallet maintained, though the analyst could obtain some useful information from each of these processes by 'watching the progress and rate of the reaction,' rather than paying attention to the quantity finally determined.[42] As for the most important question of whether any of the tests allowed one to distinguish harmful water, none did. 'No one could, with these figures to guide him, refer a water of unknown origin to one or the other of the two classes [waters known to be safe and waters known to cause disease] on the evidence afforded by chemical analysis, using any ... of the processes in question.' There was one possible exception to this generalization. Waters known or suspected of being harmful did have significantly higher levels of nitrates and nitrites than safe waters. This was not wholly unexpected; we may recall that in 1867 Frankland had made the claim that such compounds, along with ammonia, constituted 'previous sewage contamination' inasmuch as they were the oxidation products of sewage or similar materials. The idea had remained controversial throughout the '70s and Frankland had quietly dropped 'previous sewage contamination' from his analytical returns in the beginning of 1877. Yet by 1883 Ashby and Hehner were reviving it, and maintaining that Frankland had not gone far enough—nitrates, etc. did not reveal previous sewage contamination so much as they demonstrated quite recent sewage contamination.[43]

What was unexpected was that these substances alone should be the peculiar signal of danger. In Mallet's view the link between high nitrates and dangerous pollution was bacteria. Citing the 1877–78 nitrification experiments of the French bacteriologists Theophile Schloesing and Achille Muntz, he speculated that the reason nitrates were associated with pathogenic water was that the microbe responsible for nitrification was probably also pathogenic. This was by no means implausible. Like many British scientists, Mallet had a vague belief in some sort of germ theory. Yet so little was known about the types of germs, and methods of detecting, isolating, and identifying microbes were still so rudimentary that simply to make a tentative association between disease and microbes was quite reasonable.[44]

It was the third question of which process most accurately measured organic matter (or something similar), that was the focus of so much of the Frankland–Wanklyn conflict. Mallet tried the processes on aqueous solutions of known strength of pure organic compounds and chose substances such as salts of butyric and valerianic acids that were exactly the sorts of things generated in putrefaction.[45] He found that both Frankland and Wanklyn were accurate in their criticisms of the other's process. On average the combustion process measured barely half the organic carbon present, but recorded an average of 118 per cent of the organic nitrogen.

In the case of the ammonia process on the other hand, 24 per cent of the albuminoid and 28 per cent of the free ammonia were generated on average. Contrary to Wanklyn's claim, the process did not accurately indicate the amount of putrescent matter. As for the permanganate process of Letheby and Tidy, the pure substances absorbed only about 13 to 16 per cent of the oxygen needed to oxidize them. Contrary to the claims of these chemists, the test did not discriminate particularly putrescible substances.[46]

All in all the combustion process supplied the closest approximation of organic matter. It was not so much that combustion was accurate, but that its competitors were more inaccurate. Yet Mallet denied that combustion could be regarded as a way of truly 'determining' organic carbon and nitrogen: 'it is a method of approximation, involving sundry errors, and in part a balance of errors.'

To answer the fifth question, Mallet commissioned H Newell Martin and E M Hartwell, both of Johns Hopkins, to classify samples on biological grounds. By itself, this biological method, even when supplemented by bio-assay tests in which waters were injected subcutaneously into rabbits, proved no better than the chemical methods in discriminating safe from deadly waters. Of 19 safe waters Martin marked four as suspicious, Hartwell seven. Of nineteen waters known to have caused disease, Martin marked none as 'dangerous' and two as 'suspicious,' while Hartwell marked none as 'dangerous' and one as 'suspicious.'[47]

Even in this brief summary the thoroughness of Mallet's approach should be clear. In his conclusions Mallet was unsparing. He saw his task not so much as choosing which of several problematic approaches was the best—the way the issue was framed by British chemists—but of laying out as fully as possible the errors to which each was subject. He had no loyalty to the state of the art; he saw no need to put the best face forward and insist that whatever problems there might be, the processes of water analysis were good enough. Yet however inadequate analytical processes, both biological and chemical, might be, analysts faced the necessity of coming to some decision about water quality on the basis of the methods available to them and Mallet recognized this. He divided his recommendations into three parts: the general problem of how waters should be evaluated, ways analysts could better utilize the existing processes, and substantial improvements to eliminate the more egregious sources of error. But even these improvements would not make the processes any more capable of detecting the deadly forms of pollution.

Most important are Mallet's 'General Conclusions.' In the tradition of Frankland he maintained that measurements of organic matter or its derivatives did not allow one to distinguish safe from unsafe water. Chemical tests had to be integrated with, and even subordinated to, knowledge of a water's history: there could be no general standards for distinguishing water quality of the sort Wigner was trying to formulate. In only two limited contexts would analysing the organic matter in water help. One was to detect unusual cases of gross pollution such as might occur through a leaking sewer. The other was to chart variations in the quality of an urban water supply. In such cases all three processes should be used.

In most respects, then, Mallet's perspective resembled Frankland's. Both were skeptical that water analysis could yield the information most important to the authorities responsible for public health. Yet in a footnote Mallet distanced himself from the Franklandians:

> it will not do merely to throw all doubts on the side of the rejection of a water, as has been more or less advocated by writers on water analysis [Frankland], for there are often interests of too serious character involved in such rejection to admit of its being decided on, save upon really convincing evidence of its necessity.[48]

However infelicitous the prose, the meaning is clear. The burden of proof ought not to rest wholly with those who advocated a particular water, where Frankland hoped to place it. To do so would be to put the advocates in an untenable position. Aside from the philosophical impossibility of proving a negative (that the water would do no harm), they would be forced to draw far stronger conclusions from water analysis than it could conceivably supply. However untrustworthy analyses were, cities needed water, and in Mallet's view just because one could not prove a water to be safe was no reason to think it unsafe. Unless there were strong reasons to think a water was unsafe it could be used.

A short version of the Mallet report was published in Britain in the *Chemical News*, the weekly organ of the profession, and the report was widely commented on in other journals.[49] To modern readers Mallet's report is apt to seem modern and common-sensical in its continual recourse to experiment to answer questions as they arise. It seems to clear away the mass of heated argument that had characterized British discussions of water analysis since the 1866 cholera. One is tempted to ask why no British chemist or group of chemists had produced anything similar in scope during that period. Two factors are important. The first has to do with the complexity of the sort of study Mallet carried out. It is not clear who in Britain would have sponsored such a study, though it could be argued that the project was not fundamentally different from collaborative researches organized by the Local Government Board or the British Association for the Advancement of Science. Other institutions—the Society of Medical Officers of Health, the Sanitary Institute of Great Britain, the Chemical Society, or the Society of Public Analysts— might also conceivably have undertaken such work. In fact the LGB did sponsor such an investigation in 1881, a short report by R D Cory showing that the ammonia, combustion, and permanganate processes were incapable of distinguishing waters contaminated with the excreta of typhoid sufferers.[50] (Indeed, a sample contaminated with typhoid stool would have shown up as excellent according to Wanklyn's standards.)

The second factor is more important and has to do with the social context of British water analysis during the period and the contrasting context in which Mallet worked. In Britain processes of water analysis had been developed and adopted or rejected in connection with disputes for political control of water supplies—the cases of Frankland and Tidy are exemplary in this respect, where social and political concerns informed the selection of analytical methods and the presentation and interpretation of analytical results. There was no neutral ground on which processes might be compared and evaluated. Men like John Simon and William Farr were hardly in the position to provide an umbrella of unimpeachable credibility for scientific research projects; they were far too ready to exploit science to rationalize their own social reforms. It was a system with some built-in constraints. In such a context conflict was more important than cooperation, credibility was more important than utility, political achievement was more important than scientific progress.

This context was foreign to Mallet. The National Board of Health, which sponsored the tests, was a new agency. It was controversial, but that had nothing to do with its evaluation of water analysis or similar matters. Simply by virtue of the fact that he was an American chemist Mallet was outside the ongoing conflicts that had produced such disunity among the British chemists. He could look upon the analytical processes as products to be judged, not recognizing the extent to which they reflected Wanklyn's thirst for revenge against Frankland, Frankland's crusade to liberate his countrymen from water-borne disease, the attempts of Letheby, Tidy, and Odling to represent the London water companies by countering Frankland's sensationalism, or the rivalries between the provincial chemists and the London consultants over what kind of institution water analysis was to become and who would be the analysts. In short, the devastating criticisms of Mallet (and Cory) were fine so long as one did not have to practice as a water analyst.[51] All this Mallet could ignore, yet all these factors had produced the products Mallet was evaluating, and would continue to shape water analysis even in the era of bacteriological analysis that was about to begin as Mallet's report was published.

1   Fox, *Sanitary Examinations of Water, Air, and Food,* 2nd edn, pp 215–6.
2   A Ashby, 'The Fallacies of Empirical Standards in Water Analysis as told by the Story of a Polluted Well,' *SR* ns 5 (1883): 533.
3   Hamlin, 'Scientific Method and Expert Witnessing,' pp 485–513.
4   J A Hassan, 'The Growth and Impact of the British Water Industry,' pp 531–47.
5   J H Timms, 'On Water Analysis for Sanitary Purposes,' *SR* ns 3 (1882–3): 216–7.
6   R Thorne Thorne, 'On an Extensive Epidemic of Enteric Fever at Redhill, Caterham, and Adjoining Places,' in *9th Annual Report of the LGB. Report of the Medical Officer for 1879*, pp 75–92, and Buchanan's comment, p viii; Same title, *JRSA* 27(1878–9): 871–2.
7   G J Symons, 'On the floods of England and Wales during 1875 and on Water Economy,' *MPICE* 45 (1875–6): 13.

8   Fox, *Sanitary Examinations*, 1st edn, p 2; 'Water Analysis,' *SR* 10 (1879): 27–8; *SR* ns 5 (1883–4): 46. On the SPA see Chirnside and Hamnence, *The Practising Chemists*, pp 10–12.

9   Fox, *Sanitary Examinations*, 2nd edn, pp 179–83; J Carter Bell, 'Water Analysis,' *CN* 32 (1875): 246–7. On Bell (1839–1913) see S Knecht, 'Bell, James Carter,' *The Analyst* 39 (1914): 1–2. For similar complaints see 'Provincial Physician to SR,' *SR* 8 (1878): 78; 'Previous Sewage Contamination,' *SR* ns 1 (1879–80): 23; 'The Fallacies of Water Analysis,' *SR* ns 6 (1884–5): 456; A W Scatliff, 'A Case of Well Pollution undetected by Chemical Analysis,' *TSIGB* 11 (1890): 241–4.

10   A Ashby, 'Water Analysis,' *The Analyst* 6 (1881): 108–9. On Ashby (1884–1922) see C Ainsworth Mitchell, 'Ashby, Alfred,' *The Analyst* 47 (1922): 49.

11   *SR* 4 (1876): 232–3. Contrast with C E Cassal, 'Hygienic Analysis,' *TSIGB* 7 (1885–6): 273, 276; 'review of Fox, *Sanitary Examinations*,' *Lancet*, ii, 1878, p 663. See also Chapman, *The Growth of the Profession of Chemistry*, pp 5–8; Bud and Roberts, *Science versus Practice*, pp 117, 158–60.

12   On Parkes, see M Pelling, *Cholera*, pp 70–3; Brockington, *Public Health in the Nineteenth Century*, pp 253–5; on Fox (1839–1922) see *Munk's Roll of the Royal College of Physicians*, v 4, p 274.

13   Fox, *Sanitary Examinations*, 1st edn, pp 139, 143. For other works in this tradition see Kenwood, *Public Health Laboratory Work*, pp 29–83; George Wilson, *A Handbook of Hygiene and Sanitary Science*, 4th edn (London: Churchill, 1879), pp 165–72. Cf Fox in *SR* 1 (1874): 199–200.

14   Fox, *Sanitary Examinations*, 1st edn, pp 14–17, 45–7, 2nd edn, pp 6–24, 48, 128.

15   Fox, *Sanitary Examinations*, 1st edn, pp 57–63; 2nd edn, p 64.

16   Fox, *Sanitary Examinations*, 1st edn, pp 79–80, 133–59; 2nd edn, pp x, 192–3; Parkes, *Manual of Hygiene*, 6th edn, I, p 99.

17   C A Cameron, *A Manual of Hygiene, Public and Private* (Dublin: Hodges and Foster, 1874), p 58; A H Church, *Plain Words about Water* (London: Chapman and Hill, 1877), pp 21–6; Parkes, *Manual of Hygiene*, I, pp 81–99; Louis Parkes, 'Water Analysis,' *TSIGB* 9 (1887–88): 377–8; W Lauder Lindsay, 'The Estimation of the Quality of Potable Waters,' *BMJ*, ii, 1876, pp 783–5; 'Fallacies of Water Analysis,' *SR* ns 6 (1884–5): 406.

18   Fox, *Sanitary Examinations*, 1st edn, p 130. On the tradition of amateur microscopy see M Pelling, *Cholera*, pp 152–3.

19   A H Hassall, 'On Living Organisms in Water,' *FWA* 1 (1872): 143–4; 'London Water,' in *Ibid*, p 121; 'The Registrar General on the Water Supply of London,' in *Ibid*, p 119.

20   J Hogg, 'River pollution,' p 581; [Hogg], *A Microscopical Examination of Certain Waters*, pp 7–25; Homersham and Hogg in discussion of C

W Folkard, 'The Analysis of Potable Water,' pp 86–90, 94.

21 J Brittan and R Etheridge in T E Blackwell, 'Report on the Sandgate Cholera,' pp 18–22.

22 See also Hassall, 'Report on the Microscopical Examination of different waters (principally those supplied to the Metropolis) during the cholera epidemic of 1854,' p 282, and comments p 47 in GBH Medical Council, *Report of the Committee for Scientific Inquiries on the 1854 Cholera*; Hassall, 'Report to the GBH on the Microscopical Examination of the Metropolitan Water Supply,' pp 12–14. A few did claim biological investigations might clear up things in cases where chemical results were ambivalent (W L Scott, 'On the Microscopical Examination of Water,' *Monthly Mic J* 18 [1879]: 237–40).

23 J D MacDonald, *Guide to the Microscopical Examination of Drinking Water*, 2nd edn, p 1.

24 Parkes, *Manual of Hygiene*, I, p 73; A Wynter Blyth, *A Dictionary of Hygiene and Public Health* (London: Charles Griffin, 1876), s.v. water, p 636. See also Fox, *Sanitary Examinations*, 2nd edn, p 168; J Lane Notter, 'On the Value of Chemical and Microscopic Analysis in Determining the Influence of Drinking Water in Originating or Propagating Specific Disease,' *SR* ns 1 (1879–80): 87-90; *idem*, 'The Filtration of Potable Water,' *SR* ns 2 (1880–1): 161–4.

25 Parkes, *Manual of Hygiene*, I, p 74.

26 Fox, *Sanitary Examinations*, 2nd edn, p 167.

27 On France see Gerardin, 'Alteration, Corruption, et Assainissement des rivieres,' pp 5–41, 261–91. For the United States see G H Parker, 'Report on the Organisms, excepting the Bacteria, found in the waters of the state, July 1887 to June 1889,' in Massachusetts State Board of Health, *Report on Water Supplies and Sewage* (Boston: Wright and Potter, 1890), pt 1, pp 582–620; Baker, *The Quest for Pure Water*, pp 390–414. For Germany see R Kolkwitz and R Marsson, 'Grundsatze fur die biologisch Beurtheilung des Wassers nach seiner Flora und Fauna,' *Mitthielungen aus der Koniglichen Prufungsanstalt fur Wasserversorgung und abwasserbeseitigung zu Berlin* 1 (1902): 3–72.

28 On Wigner see *DNB* vol 21, pp 197–8; and *J Chem Soc* 47 (1885): 345–6.

29 Ernst W Steib, in collaboration with Glenn Sonnedecker, *Drug Adulteration in 19th Century Britain* (Madison: U of Wisconsin Press, 1966), p 181; Dyer and Mitchell, *The Society of Public Analysts*, pp 18, 23, 92–3, 119; Chirnside and Hamnence, *The Practising Chemists*, pp 15–20.

30 G W Wigner, 'The Outbreak of Typhoid Fever at Baxenden and Accrington,' *SR* 7 (1877): 262–4.

31 G W Wigner, 'On the Mode of Statement of the Results of Water Analysis,' pp 214–16; cf G W Longstaff, 'Water Analysis,' *SR* 7 (1877): 402.

32 Wigner, 'Mode of Statement,' p 213.

33 'Water analysis,' *The Analyst* 3 (1878): 247.

34 'Notes of the Month,' *The Analyst* 6 (1881): 12–13; Heisch, 'Annual Address to the Society of Public Analysts,' *The Analyst* 7 (1882): 13.

35 'The Analyses of the Public Water Supplies of England,' *The Analyst* 6 (1881): 17–18.

36 G W Wigner, 'On the Valuation of the Relative Impurities in Potable Waters,' pp 111–25.

37 'The Public Water Supplies of England,' *The Analyst* 6 (1881): 168, 192, 208, 228.

38 Dupre and Hehner, 'On District Standards in Water Analysis,' pp 53–8; John Muter, 'On the most simple ... mode of expressing the Results of Water Analysis,' pp 93–8; *idem, A Manual of Analytical Chemistry*, pp 162–63. On Muter (1841–1911) see *The Analyst* 37 (1912): 77–80.

39 'To Our Readers,' *The Analyst* 8 (1883): 1.

40 Dupre and Hehner, 'On District Standards in Water Analysis,' pp 53–4.

41 J W Mallet, 'Determination of Organic Matter in Potable Water,' *CN* 46 (1882): 64.

42 *Ibid*, pp 74–5.

43 *Ibid*, pp 91–2; A Ashby and O Hehner, 'On So-Called "Previous Sewage Contamination",' *The Analyst* 8 (1883): 58–62; Fox, *Sanitary Examinations*, 2nd edn, pp 95–101; A Dupre, 'Presidential address to the section on Chemistry, Meteorology, and Geology,' *TSIGB* 9 (1887–88): 355–7.

44 Mallet, 'Determination of Organic Matter,' p 101. Cf H S Carpenter and W G Nicholson, 'A Method for the Examination of Water Biologically,' *The Analyst* 8 (1882): 94–6.

45 J W Mallet, 'Determination of Organic Matter,' p 63.

46 *Ibid*, pp 72–3.

47 *Ibid*, p 101.

48 *Ibid*, p 109 fn.

49 'The Sanitary Requirements of Water Analysis,' *SR* ns 4 (1882–3): 346–7; 'Water Analysis,' *SR* ns 4 (1882–3): 364–5; 'Special scientific investigations on sanitary subjects in America,' *SR* ns 4 (1882–3): 403–4; 'Water Examination,' *JRSA* 31 (1882–3): 215; Fox, *Sanitary Examinations*, 2nd edn, p 64.

50 R D Cory, 'On the results of the examination of certain samples of water purposely polluted with excrements from fever patients, and with other matters,' in *Eleventh Annual Report of the LGB. Report of the Medical Officer for 1881*, pp 127–65; 'The Sanitary Requirements of Water Analysis,' *SR* ns 4 (1882–3): 396–7.

51 A Dupre, 'Presidential address,'* p 361; Cassal, 'Hygienic Analysis,'* p 276.

9 Counting the Countless: The Temptations of Quantitative Bacteriology, 1880–90

The method of gelatine plate culture is excellent, if it is required to determine which of several samples of water contains more organisms capable of growing in gelatine.[1]

E E Klein to G Buchanan

In the late autumn of 1885 Colonel Francis Bolton, examiner of the London water supply and the official responsible for seeing that the water was effectually filtered, thought he finally had a way to gauge filtration success; there was a prospect for actually determining how well the companies' filters removed 'the living organic matter ... [the] portion of impurities which is now considered the dangerous one.'[2] The process Bolton had in mind was the culturing of water bacteria in a solid gelatine–peptone medium, on a glass plate. Developed over the previous five years by the German bacteriologist Robert Koch, the technique was currently being employed in a study of the London waters by Percy Frankland, a chemist–bacteriologist and Edward Frankland's second son. Percy Frankland was culturing samples of water taken before and after filtration, thereby documenting the effect of filters in removing bacteria. Bolton believed the process would allow the companies' engineers to monitor their filters on a day-by-day basis and quickly recognize when repairs were needed.[3] In this hope Bolton was disappointed. The LGB Medical Officer, George Buchanan, objected that Bolton's plan involved so many questionable assumptions that to rely on plate cultures to safeguard the metropolis was no less unwise than relying on any other single factor.

241

The conflict between the skeptical Buchanan and Bolton, desperately seeking a scientific foundation for his oversight of water quality, is a microcosm of a conflict about the utility of bacteriology that took place in the late 1880s and early 1890s. By the early '80s most British water analysts accepted that bacteria or similar organisms probably were the exciting causes of water-borne disease.[4] The incorporation of bacteriology into their analyses, however, took place in three distinct though overlapping phases. The first phase, underway by the mid '80s, was simply counting the colonies which grew on glass plates. Little attention was given to distinguishing species. The focus of the second phase, underway by the end of the '80s, was the conditions under which certain species could survive. The hope was to find laws of bacterial existence and hence to define more precisely the circumstances in which disease outbreaks might occur. The third phase, which peaked in the mid '90s, was to use new determinative techniques to check waters for the presence of the microbe thought to cause typhoid. Bacteriology in all its phases was looked to to resolve the political disputes over water quality which had for so long fed on the chemists' endless supply of uncertainty, yet bacteriology turned out to afford ample uncertainties of its own.

Worrying About Germs, 1878–86

By 1880 British sanitarians and water analysts were becoming increasingly concerned about their inability to measure water-borne germs. Although many British sanitarians in the '70s were doubtless aware of the spontaneous generation debates, of Lister's ideas about wound sepsis and its prevention, and of the identification of the anthrax organism, most of the talk about germs was loose talk—one could talk about germs of diseases without having to give up talking about products of putrefaction or obscure fermentations. In the late '60s and early '70s it was not clear that germs were bacteria, and they are better understood as something akin to point atoms of morbific force.[5] Furthermore, all the alternative causes of zymotic disease were tightly bound up with one another, and people might be talking at one and the same time on multiple levels of explanation and about various classes of causes—predisposing or exciting; necessary or sufficient. And finally, since it was not clear that the choice of morbid poison held out any significant practical implications for the conduct of epidemiological investigations or disinfection campaigns, there was little to prevent the loose talk from remaining loose. The move from vague talk of germs to disciplined discourse about bacteria—the reification of the germ theory into bacteriology—occurred not overnight as the result of any discovery, but slowly, as the boundaries of a metaphor became ever more circumscribed. And as germs became more tangible, sanitarians slowly learned to reinterpret old conceptions and rules of thumb in bacterial terms. Sewer gas, for example, might be conceived not as pathogenic in itself but as a vehicle which transferred germs from sewage to air.[6] Even in the early '80s there remained a great deal of barely bounded speculation about the nature of germs. Nevertheless, many sanitarians were coming to share Edward Frankland's perspective that germs were extraordinarily elusive and that it was wise not to put quite so much confidence in water analyses.

Such discussions took place in a variety of forums. Germs remained a central issue in legal and Parliamentary inquiries into questions of pollution, sewage treatment, and the acquisition of new municipal water supplies. They also came to be a central issue in discussions about water-purification technologies. Prior to 1884 they were not discussed with reference to water analysis as much as one might expect: water analysts remained too much caught up with the possibility of refining their chemistry or with finding some combination of chemical processes that might allow one to guarantee that a water was safe.[7]

It is worth briefly examining some of these sites of germ-discussion to recognize how far removed from the context of analysis discussions of water-borne germs were. An example of the first of these contexts is the consideration by a Commons select committee of the Cheltenham Corporation water bill in March 1878. The bill was promoted by Cheltenham to allow it to purchase the private water company which was then supplying it, to avoid imposition of a supply drawn from the polluted River Severn, and to obtain a new water supply of high purity. Much of the testimony dealt with the threat the Severn posed to the health of Cheltenham residents, i.e., with the question of whether it was possible or likely that germs put into the river by upstream towns would survive long enough to infect the town. Frankland was one of the expert witnesses for Cheltenham. He maintained that analysis did not allow one to distinguish the important 'physiological impurities [living germs].' He told of his experiments during the 1866 cholera epidemic. He insisted that the scientific world now accepted the germ theory, noted that the research of Pasteur and Koch had confirmed it in the case of anthrax, and declared that no member of the Royal Society would think of denying the existence of germs. In cross-examination Sir Edmund Beckett, one of the ablest members of the parliamentary bar, pressed Frankland to admit that the cholera germ was nothing but a theory; Frankland termed it an undiscovered fact. 'Have you got any evidence yet of the germ theory?' Beckett asked. 'You have been preaching this doctrine for ever so long ... trying it on, in committee after committee.'[8] Medical witnesses testifying for Cheltenham also took the Frankland line. William Thursfield, E Thomas Wilson, Thomas Wright, and Alfred Hill maintained that analysis was not to be trusted, that no distance of flow would necessarily remove all 'infective particles' or 'medical infection,' or 'centres of infection,' or 'living matter.'[9]

Similar arguments about undetectable poisons had been made as far back as 1850 and the claims Frankland made about the characteristics of germs were basically the claims he had made in 1868. Such claims were not significantly strengthened by the research on anthrax or by any other bacteriological discoveries, and they would remain only plausible speculations until there were convincing demonstrations of the existence and nature of germs of water-borne diseases. To those like Beckett, who argued the opposing case, Frankland's claims were no less outrageous in 1878 than they had been a decade earlier. As late as 1886, Charles Meymott Tidy, Frankland's nemesis, was still doubting the reality of germs: 'People talk about germs very freely, ... as though these things had been got hold of,' Tidy observed to a select committee on the condition of the River Lea. He demanded that the so-called cholera germ be put on the table before him before he would acknowledge its existence.[10]

A second context in which there was increasing talk of germs was with regard to water purification, both natural purification in rivers and artificial purification in filters. Especially important are the discussions following Tidy's March 1880 paper on the self-purification of rivers and William Anderson's 1882 paper on the 'Antwerp Water Works,' presented at the Institution of Civil Engineers. Tidy's paper was part of his campaign to discredit Frankland's water science. As he typically did when raising the germ issue, Tidy bemoaned its speculativeness. When talk turned to germs we were leaving 'the region of direct experiment, ... plung[ing] headlong into theories ... diving deeper and deeper into mere speculation when we discuss the laws governing the life of organised bodies, the very existence of which at present is unproved.'[11]

In fact Tidy wielded speculations as ably as anyone. He tried to make the image of germs as exceedingly delicate at least as plausible as Frankland's image of exceptionally hardy germs. Germs were 'so low in the scale of life that they [would] .... very soon suffer complete destruction by the bursting of their envelopes owing to the powerful endosmic action of the water in which they are immersed.'[12]

In light of the discoveries of the next few years that the germs of cholera and typhoid were in fact fairly fragile, Tidy's speculations can seem prescient, just as Frankland's were in 1866–68. But one could equally argue that in light of what was known of the hardiness of the spore-forming anthrax bacillus (the most important disease germ known at the time), Tidy's belief in fragile bacteria was wholly unwarranted. What is striking about the 1880 discussion (and the 1878 Cheltenham hearings) is the continued hypotheticality of germs. Frankland and Tidy paid little attention to the growing output of bacteriological science; they continued to make self-serving assumptions and to manipulate hypotheses drawn from them; it was scholastic disputation in its most barren form. While Frankland made an effort to ground some of his positions in experiment—he found, for example, that bacteria (species unstated) survived perfectly well in aerated water—his bacteriological experiments were exceedingly crude.

Two years later, when bacteriological aspects of water filters were discussed at the Institution of Civil Engineers, there was at least agreement about the proper terms of the question. William Anderson, an engineer, was involved with the chemist Gustav Bischof and with Frankland in a scheme for filtering Antwerp's water through Bischof's patent 'spongy iron' filters, which were supposedly germicidal. Frankland believed that a germicidal medium was necessary since germs might be 'too subtle' (too tiny) to be removed by ordinary filter media. He himself used such a filter to protect his family from the London water he regularly condemned.[13] Those who spoke agreed that the effect of spongy iron on microbes was the relevant question, and that answering it depended on the availability of means for detecting germs. Samuel Homersham and Jabez Hogg claimed that cultures of filtrate showed that spongy iron did not, in fact, kill all bacteria. Bischof replied by questioning the validity of gelatine cultures, while Anderson cast doubt on the bacteriological skills of Hogg and Robert Angus Smith, who had done the cultures. This brief exchange was a hint of what was to come in the next fifteen years: water analysts became as expert at raising doubts about one another's bacteriological technique as they had been at criticizing one another's chemical processes.[14]

On the few occasions during these years when 'germs' made their way into discussions on water analysis they again produced more heat than light. In 1881 Charles Folkard presented a paper on 'The Analysis of Potable Waters with Special Reference to Previous Sewage Contamination' at the Institution of Civil Engineers. Folkard, an assistant in the School of Mines, dealt with the topic in an elementary fashion, reviewing the major processes for an audience of engineers.[15] He closed with an image of what germs might be, much like the one Frankland had presented to the Water Supply Commission in 1868. One could plausibly imagine germs able to withstand any process of purification one cared to think of: whether filters (germs might be small enough to go through '1000 abreast'), chemical precipitation (germs might not be affected owing to their 'great vitality'), or sewage irrigation (germs might go through fissures into tile drains); science could supply no anodyne to any of these awful possibilities. In the discussion Tidy attacked Folkard's one-sided speculations with his usual plea for positivism: 'one could no more analyse a water for the germ of typhoid, than one could analyse the brain for an idea. Not only ... did the Author speak of germs as though they were tangible, but he had fixed the conditions of the life of a thing the very existence of which had never been proved.'[16] Folkard retaliated by accusing Tidy of pigheaded empiricism: Tidy based his conclusions on analysis of 'four thousand samples,' done by 'a process admitted by nine-tenths of the analysts of the present day to be worthless.'[17] Leaving aside the initiative of the Society of Public Analysts, this bitter and sterile exchange exemplifies British water science of the early '80s. Chemical analysis had been taken as far as it seemed useful to go (and possibly further); the detection of disease germs was seen as the way of the future but there were as yet no germs to detect nor any means of detecting them.[18]

Blundering Toward Enlightenment: Robert Angus Smith, 1882–83

All the while Tidy and Frankland were fighting round after round of their interminable duel to define the characteristics of disease germs, others were actually beginning to perform rudimentary bacteriological water analyses. These efforts belonged to two traditions. One was the enterprise of Robert Angus Smith, a scientist who defies classification. The other was bacteriology proper, mainly a continental science, but also the extension of proto-bacteriological initiatives into water analysis undertaken by a number of people from 1870 on. Here we consider Smith.

We have seen Smith before. He belonged to no mainstream research programme, being interested in deep and difficult questions about distinctions between chemical and physiological levels of explanation with regard to putrefaction and oxidation. Beginning with his 1848 study of the Thames biota he had made a number of attempts to correlate pathogenicity with microscopic life. In the mid '60s he was recommending storing a sample to see what grew in it, a rudimentary culturing method that was to be the central component in analysis.[19]

Among British sanitarians Smith was unusual in being fully conversant with the perspectives of Pasteur and Koch and he saw the implications of their work for water analysis. At the end of 1882 Smith presented the Manchester Literary and Philosophical Society with an account of his experiments using gelatine to culture the bacteria in water. Smith was unsure what to make of his results. He found that the waters most often complained of seemed to have the most 'points' (colonies), but he did not attach much importance to the number of colonies; indeed since the cultures were done in test tubes it was hard to count the colonies, and the depth of the tube made it likely that some aerobic forms would not develop. Nor was he interested in the appearances of different species. All the same he predicted that some such bacteriological process would supercede chemical water analysis.[20]

Smith's paper was well received, yet water analysts did not rush to adopt his methods. The *Sanitary Record*, the main public health weekly, called the approach promising and felt that it merited further exploration but was not yet practical.[21] But the problem was mainly Smith. He was, as his biographer admits, 'a fringe chemist,' holding a respected, if unique, station in the scientific community. Smith kept himself apart from the places (London) and contexts (Parliamentary select committees and the law courts) where water analysts made their reputations,[22] and focused on problems that others ignored. And in any case British analysts were preoccupied with other matters: the SPA standardization campaign, the reports of Mallet and of Cory, and the struggles of Frankland and Tidy for the hearts and minds of Londoners. Bacteriology figured in none of these.

The Coming of Plate Culture, 1884

Bacteriology proper was neither well nor widely practised in Britain, but it was practised. While one may think only of Lister (who was not in fact a bacteriologist), one could mention a number of others active during this period: Alexander Ogston (1844–1929) and William Watson Cheyne (1852–1932), students of Lister, J Burdon Sanderson (1828–1905), a pathologist, E E Klein (1844–1925), a histologist and physiologist, E Ray Lankester (1847–1929), mainly a botanist, and German Sims Woodhead (1855–1921), initially a pathologist. One might also mention T H Huxley and John Tyndall, who took an interest (and in the case of Tyndall did substantial research) on bacteriological questions, but were more fully involved elsewhere. Many of these men were at early stages of their careers in the early '80s and they did not yet function within any single disciplinary community. None, save Burdon Sanderson and Klein, had been much involved with the public health questions that concerned water analysts. The British bacteriologists whose interests came closest to these matters, D D Cunningham (1843–1914) and T R Lewis (1841–1886), were in India, where they were experts on cholera.[23]

By 1884 a few others besides Smith had sought a way to discover something about the bacteria in water. In studies commissioned by the Medical Office of the Privy Council in the early 1870s, Burdon Sanderson had tried to determine the 'zymotic' properties of waters, their capacity to induce decomposition in a sterile medium of Pasteur's solution. Since all his samples induced decomposition, it was not at all clear what the results meant; if 'microzymes' were everywhere it was hard to see how they could be the agents of occasional diseases.[24]

Charles Heisch, lecturer in chemistry at the Middlesex Hospital, was working along similar lines. In 1870 he reported that one could get some kind of minute 'fungi' or 'germs' (in particular a butyric ferment) to grow in contaminated water simply by adding a nutrient medium of sugar. Pure sugared water was not affected, Heisch claimed; nothing grew until a germ had been provided through addition of a small quantity of sewage. Among the few who took notice of Heisch's paper was Frankland, who was initially enthusiastic, hoping to have found 'a much nearer approach to the supposed morbific matter.' Yet Frankland soon decided that Heisch's interpretation was unwarranted. The sewage did not provide the germs, but only inorganic nutrients necessary for the growth of air-borne spores. These germs were ubiquitous; Heisch's test showed nothing.[25]

The belief that germs were everywhere and almost impossible to distinguish from one another is one factor that kept sanitarians from following up the initiatives of Burdon Sanderson and Heisch. Bacteria were known of, there was concern about them, but they seemed inaccessible to science. In retrospect we can see that both Frankland and Heisch were naive about the demands of sterile technique. Tyndall's spontaneous generation research, carried out during the next few years, made these criteria clearer, yet great problems in making cultures and distinguishing types remained. The crudeness of these early attempts makes even clearer how revolutionary were the new methods that became available in the early 1880s of using solid media and organic stains.

Although the German bacteriologist Robert Koch had demonstrated the use of solid media for culturing bacteria in London in 1881, it was only in 1883 that he published an improved version of the process and only in 1884 that British water analysts began to take much interest in it, the same year that Koch began teaching short courses in bacteriological technique.[26] That summer London hosted the International Health Exhibition, and among the exhibits was a working bacteriological laboratory set up by William Watson Cheyne, Lister's associate at King's College. The *Lancet* published a lengthy description of the plate culturing method being demonstrated there, noting that by such a process 'the numbers and nature of the organisms present in a sample of water may be estimated and ascertained.'[27]

That article, along with a similarly detailed description the following summer by C J H Warden, an Indian Army surgeon who had studied with Koch, stressed the simplicity of the method. The analyst mixed a drop of sample water (perhaps diluted with sterile distilled water if it was believed to be very badly contaminated) with a molten mixture of sterile gelatine and filtered meat broth, and then poured it onto a level glass plate of about five by six inches in area, which was immediately covered. In a day or two colonies appeared and by counting the colonies, the analyst could calculate the number of bacteria per unit of water (usually a cubic centimetre). If he knew the appearances of different species grown in such a medium he could also make tentative conclusions about the species present. Samples of interesting colonies could also be taken for the preparation of pure cultures.

In fact the simplicity of the technique was its greatest problem; it tempted those with little or no bacteriological training to try the process, yet if accurate and useful results were to be obtained, meticulousness and skill were required at every stage. 'Success in bacteriological work is largely dependent upon close attention to details: the methods are simple but their correct performance is often from this cause difficult of attainment,' wrote Warden.[28] He tried to include all those details. He listed the equipment one needed, explained how to sterilize it, gave rules for preparing the medium, collecting samples, estimating the bacterial population (necessary to know how much water to culture), counting the colonies, and so on. He tried to describe the fine movements of sound technique. In mixing the water with the gelatine the analyst was to hold the tube of medium 'in the left hand in a slanting direction, remove the [cotton] plug with a twisting movement, and place it between the first and second fingers of the same hand, and allow the water from the sterilized pipette to flow down the side; immediately replace the plug, and move the tube to and fro several times in order to thoroughly mix the contents.'[29] All stages were described in similar detail.

It may seem that the wiser thing would have been to tell would-be bacteriologists to get expert instruction. Yet as Warden pointed out, in Britain, unlike Germany, there was nowhere medical officers could learn such techniques, hence the need for his article. And despite the daunting demands for methodological rigour he had laid out, Warden ended his article by suggesting that bacteriological examination ought to be carried out on a widespread basis 'for the examination of water supplied by public companies. ... [and] for ascertaining the relative value of domestic filters.' He did not say who was to do these examinations, nor under what auspices they would be done, nor how their accuracy was to be guaranteed, nor how they were to be interpreted—all crucial questions.[30]

During the next few years a great many sanitarians took up plate culturing in its simplest form of counting the colonies that grew in a culture of one cc of a sample. A bacteriological boosterism set in as sanitarians tried to show how up-to-date they were. Much of this work was probably of poor quality: in such operations one gets results even with poor technique; what requires experience is knowing when to reject them. Moreover, the newness of the science made it hard to spot bad work. The distribution and properties of bacteria do vary greatly, but it was not yet clear how much they varied; bacteriologists could not be sure what results were reasonable. Hence far from removing the obscurity from water quality, bacteriology brought in its own obscurity. As had happened with chemists, bacteriological water analysts tended toward one of two poles; they tended either to be generally trusting of bacteriological demonstrations or to be highly skeptical of them. Percy Frankland exemplified the former, Gustav Bischof the latter.

The Transformation of Percy Frankland

Educated at the Royal School of Mines and at Wurzburg (PhD, organic chemistry, 1880), Percy Frankland (1858–1946) became a demonstrator under his father at the School of Mines and assistant in the water lab on his return to Britain. He apparently learned of plate culturing through the 1884 Health Exhibition demonstration, though at some point in the mid '80s he trained at Koch's laboratory. In 1888 he was appointed chemistry professor at Dundee, and in 1894 moved on to Birmingham where he spent most of his career. Aided by his wife Grace Toynbee Frankland, Percy Frankland became one of the leading late nineteenth century British bacteriologists, and an expert on water bacteriology.[31]

As his father's assistant during the early '80s, Percy Frankland took up the cause of water quality reform with vigour and even belligerence. In 1883 he published a popular article in the *Nineteenth Century* on 'The Cholera and Our Water Supply,' and in March 1884 delivered a major paper to the Royal Society of Arts on 'The Upper Thames as a Source of Water Supply.' In the former he spoke of 'numerous theories and apologies' that contaminated river water became safe after some period of flow, and claimed that these had been 'framed' to 'soothe the conscience of the river-polluter on the one hand, and of the purveyor of polluted water on the other.' The germ theory could be taken as proved, he argued, and since germs were exceedingly hardy, the only safe water was that which came from deep wells.[32] He even managed to slip in a plug for Bischof's spongy iron filters.[33] All these were positions his father had taken over the years. In the latter paper, he attacked Tidy *ad hominem* and sarcastically. Tidy's 'romances' had 'probably done more to check and paralyse the prevention of river pollution than anything else.'[34]

In mid 1885 Percy Frankland began using plate cultures to measure the bacteria-removing capacities of various filter media. In June he presented the results of small-scale experiments on several media to the Royal Society. Initially all the materials were highly effective, but they lost their effectiveness at various rates. He confirmed his father's fear that many domestic water filters as they became clogged with organic matter simply served as sites for bacterial multiplication. Spongy iron was an exception.[35]

At this time, bacteriology was still of limited significance in his perspective. He called the plate culture process an 'exceedingly beautiful and ingenious test for ascertaining the number of individual organisms present in a given water,' but recognized that the means of distinguishing different types of bacteria were as yet of 'little value.' Since the number of bacteria was a meaningless measurement, and since it was recognized that pathogens were only rarely in water anyway it was not clear that the results had any broad significance. Chemical water analysis remained superior and it remained necessary to secure water that had never been contaminated or that had percolated through thousands of feet of rock. Tidy's water science remained the object of a sneering aside.[36]

It was about this time that Percy Frankland began a series of plate culture tests on the waters the companies supplied to London before and after they were filtered. In the three years the series lasted, these tests became as politically volatile as the 'previous sewage contamination' calculations had been in the late 1860s. Moreover, they led Percy Frankland to change his views on London's waters. He soon discovered that the deep well water of the Kent company, which his father had long touted as exemplifying the pure water London should have, contained few microbes when it emerged from the ground, but vastly more on leaving the water works. By contrast, waters taken from the sewage-polluted Thames and Lea showed 'strikingly' fewer microbes after filtration. He first measured the rate of removal at around 86% and subsequently found it consistently in the 95–99% range.

From October 1885 on, the results were published in Bolton's monthly reports on the London waters. Initially there was alarm at the discovery of any microbes in the water. Germs were still so new, so malign, so mysterious that even to announce that only a few were in the water was to court hysteria from an ill-informed public. In early December 1885 two letters about interpreting the bacteria counts appeared in newspapers and trade periodicals. One was from Bolton and Percy Frankland. They wrote to reassure: the bacteria found in the water were not necessarily derived from sewage nor dangerous. The results showed the effectiveness of the filters. The other was from Crookes, Odling, and Tidy, the companies' analysts who, true to form, saw the issue in partisan terms. They were highly suspicious of the new measurements and looked upon the bacteriological measurements as another phase in Edward Frankland's nearly twenty year old campaign to deceive a gullible public. They declared that 'the attempt from the biological standpoint to condemn London water will ... be as conspicuous a failure ... as the attempt to condemn it from the chemical standpoint.'[37] They too would soon counter with their own plate culture measurements, they announced in early February.[38] It is easy to understand their skepticism. Edward Frankland had recently shown an interest in water bacteriology and thus far all Percy had written on water questions had followed his father's views.[39] It is also likely that they were right, that the filtration studies were initially undertaken in the expectation that they would embarrass the companies.

Bischof, Frankland, and Buchanan: The Value-ladenness of New Knowledge

Along with the water companies' chemists the officers of the Local Government Board were alarmed at the filtration studies, but for different reasons. The Board's medical officer, George Buchanan, feared plate culturing would lead to irresponsible complacency, not undue alarm. He wrote to Bolton and Frankland urging them to exercise greater caution in interpreting plate culture results. Since gelatine–peptone was not a proven medium for all microbe species there was no warrant for Percy Frankland's claim that he was demonstrating a water's 'relative freedom from organic life.' Nor was the alternative phrase 'average reduction' acceptable, for it was not proved that filtration affected all microbes in the same way.

Percy Frankland and Bolton nominally accepted Buchanan's conditions, yet doubtless much to his despair, they brought the whole affair into the open. In the third of the monthly reports (December 1885), Bolton published extracts from Percy Frankland's reply to Buchanan, in which Frankland defended the techniques. And Bolton continued to display what was either an utter insensitivity to the nuances of language or a bold disregard for Buchanan's wishes. He referred to the tests as nothing less than a measure of 'the efficiency' of the filters. Unwilling to let the matter drop, Buchanan called in Edward Emanuel Klein, a Slovakian emigré histologist and bacteriologist and regular consultant to the LGB.[40] Buchanan asked Klein four questions: whether gelatine–peptone was the best medium, whether it was the preferred medium for water-borne pathogens, what the relation was between the microbes (numbers and species) that grew on the plates and those actually in the water, and finally whether water company engineers would be able to master the technique. Klein confirmed Buchanan's doubts. He would not say that any medium was best, but noted that some microbes did not grow on gelatine–peptone. He regarded any suggestion that the medium was preferential for pathogens as not only untrue but 'mischievous.' He worried that contamination of plates frequently occurred, hence a plate culture did not necessarily reflect the bacteria in the sample. He, like Buchanan, was opposed to the use of the tests by water company personnel, on the grounds that such use 'might tend to [produce] erroneous and perhaps mischievous conclusions.' Bolton yielded, protesting still that water engineers badly needed a ready means to monitor their filters and that plate culturing was the obvious means.[41]

It is hard not to sympathize with Bolton on this point. He had a duty of ensuring effectual filtration which he had been carrying out without an adequate definition of what filtration was supposed to accomplish or any adequate means for measuring what the filters actually were accomplishing. Even though Frankland's chemical analyses were appended to his reports, they could scarcely be regarded as providing a useful guide for filter operation. Whatever their shortcomings, the new bacteriological methods seemed to provide a far sounder basis for managing filters than had hitherto been available.

Why then were Buchanan and Klein so hypersensitive? To understand their position we need to think briefly about the politics of London's water. Efforts to secure public control appeared to be progressing inexorably, if slowly, in the early '80s. Public takeover would take place, when Parliament settled on the compensation the companies would receive. As we have seen, in their handling of the Crookes, Odling, and Tidy reports and in their discussion of the feasibility of muzzling Frankland, the LGB did its best to maintain a balance of illusion in water propaganda. The public and Parliament had grown used to Frankland's stratagems and to those of his opponents. The strong claims of Frankland and Tidy neutralized one another to produce an equilibrium of uncertainty: the water was evidently not as safe as one would like but a good deal better than it might be. Bacteriology promised to disturb this equilibrium, and in a way that was difficult to predict. It presented an entirely new idiom for deception, one to which the public, goaded on by hysterical articles about invisible hordes, might be expected to be highly susceptible.

What was made clear to Percy Frankland in the autumn of 1885 was that his results could not stand alone as contributions to knowledge. Both sides would insist on imposing qualifications that would deprive his results of scientific significance and endow them with political significance. From the Buchanan–Klein perspective the apparent effectiveness of filters signified nothing because plate cultures might not detect pathogens and because filtration might favour some species over others. Percy Frankland might protest that chemical analysis entailed even more qualifications and assumptions, yet there was no room for the pragmatic empiricism he and Bolton offered. Proofs were one thing; additional data of unclear significance simply obscured matters. Water analysis was inescapably a partisan enterprise, and one's investigations either favoured one of the two sides or they meant nothing. Between autumn 1885 and winter 1886 Percy Frankland acknowledged this truth. He chose a side and left his father's camp to ally himself with Tidy, Crookes, and Odling.

We may never have a complete explanation for this turnabout. Family strains, a young scientist's ambitions, and Buchanan's censorship may all have been factors. But probably the greatest single factor was the need to find a significance for his filtration findings. His results were fully compatible with the perspective of either camp. One could, as would Tidy, take great pride that the London filters removed almost all water bacteria and see the plate cultures as confirming a long series of claims that London's water was safe, or one could continue to worry about bacteria that crossed the filters and argue that Londoners were as much at risk as ever. Yet while the results were compatible with either view, they were significant only in Tidy's. To the companies Percy Frankland was a vindicator, a scientist–hero who could finally end all the squabbling. From the Buchanan–E Frankland point of view the response was, in effect, 'So what?'

During the winter of 1885–86 Percy Frankland spoke up frequently in his own presentations and in discussions of others' about what he was finding and what it meant. In a December lecture to the Society of Chemical Industry he stressed the novelty of plate culturing as a test of filtration. He acknowledged and then bypassed those qualifications that were so important to Buchanan. Even though

TABLE I.

TOTAL NUMBER OF COLONIES OBTAINED BY CULTIVATION OF ONE CUBIC CENTIMETRE OF WATER.

Description of Water.	Jan.	Feb.	March.	April.	May.	June.	July.	August.	Sept.	October.	Nov.	Dec.	Average.
THAMES—													
Thames unfiltered....	45,400	15,800	11,415	12,250	4,800	8,300	3,000	6,100	8,400	8,600	56,000	63,000	20,255
Chelsea	159	305	299	91	59	60	59	303	87	31	65	222	146
West Middlesex......	180	80	175	47	19	145	45	25	27	22	47	2,000	231
Southwark............	2,270	284	1,562	77	29	94	380	60	49	61	321	1,100	524
Grand Junction	4,891	208	379	115	51	17	14	12	17	77	80	1,700	630
Lambeth..............	2,587	265	287	209	136	129	155	1,415	59	45	103	305	475
LEA—													
Lea unfiltered	39,300	20,600	9,025	7,300	2,950	4,700	5,400	4,300	3,700	6,400	12,700	121,000	19,781
New River............	363	74	95	60	22	53	46	55	17	10	32	400	102
East London..........	224	252	533	289	143	445	134	243	165	97	248	280	253
DEEP WELLS—													
Kent (well direct)	—	5	44	7	8	4	12	9	5	82	12	11	18
Kent (district)	43	149	38	47	101	39	48	13	25	344	196	66	92

TABLE II.

PERCENTAGE REDUCTION IN THE NUMBER OF DEVELOPABLE MICRO-ORGANISMS PRESENT IN THE RIVER WATERS BEFORE DELIVERY BY THE COMPANIES.

Description of Water.	Jan.	Feb.	March.	April.	May.	June.	July.	August.	Sept.	October.	Nov.	Dec.	Average.
THAMES—													
Chelsea	99·7	98·1	97·4	99·2	98·8	99·3	98·0	95·0	99·0	99·6	99·9	99·7	98·6
West Middlesex......	99·6	99·5	98·5	99·6	99·6	98·3	98·5	99·6	99·7	99·7	99·9	96·8	99·1
Southwark	95·0	98·2	86·3	99·4	98·9	98·9	87·3	99·0	99·0	99·3	99·4	98·3	96·7
Grand Junction	89·2	98·7	96·7	99·1	98·9	99·8	99·5	99·8	99·8	99·1	99·9	97·3	98·2
Lambeth..............	94·3	98·3	97·5	98·3	97·2	98·5	94·8	76·8	99·3	99·5	99·8	99·5	96·2
LEA—													
East London..........	99·1	98·8	94·1	96·3	95·2	90·5	97·5	94·3	95·5	98·5	98·0	99·8	96·5

Figure 9.1 Percy Frankland's reports on the numbers of microbes in raw and filtered waters were touted by the London water companies as proof of the effectiveness of their purification and the salubrity of the water supply. Most others felt they signified nothing (*J Soc Chem Ind* 6 [1887]: 317, 319).

the number of colonies that grew on the plate could not be regarded as 'absolutely' the same as the number of bacteria that existed in the water, plate cultures showed 'the value [of filtration] in removing micro-organisms.' Percy Frankland would not endorse London's river water, yet neither did he condemn it; instead the filtration results reflected a 'striking improvement.'[42] The overall message was that some remarkable piece of progress had been achieved even if no one could say quite what it was.

By this time Gustav Bischof had appeared on the scene to challenge this view of the significance of plate cultures. Bischof (1834–

1903) is much less well known than Percy Frankland though for a few years in the mid '80s his criticisms of inferences from plate culture counts were of great importance. Son of the pioneer geochemist G C G Bischof, he took a PhD at Berlin and held the post of professor of Technical Chemistry at the Andersonian University in Glasgow from 1871–75. For roughly the next fifteen years he was heavily involved in promoting his spongy iron filter. With the rise of confidence in sand filtration, Bischof turned to the manufacture of white lead for the remainder of his life. He does not appear to have been publicly involved in the water debates after 1888; the spongy iron system appears to have been a commercial failure by the mid '80s.[43]

In presentations to the Society of Chemical Industry in February and to the Society of Medical Officers of Health in May 1886 Bischof tried to flush Percy Frankland from his refuge of ambivalence. Bischof too had gone to see Koch. Since September he too had been investigating the bacteria in the London waters. Like Percy Frankland he was glad to find the water 'bacteriologically so much purer than ... anyone would have anticipated.' Yet microbe counts were absolutely without meaning since microbe populations would expand amazingly rapidly in suitable conditions. He told of a sample that had contained 53 microbes/cc on the day of collection and a horrifying count of 770,000 after six days' storage in a sterile flask. Initially it would have been judged acceptable according to Koch's standard of 100/cc as the limit of acceptability and it seemed absurd to think that it would have become any more dangerous during a week's storage.[44]

In chemical analysis each component had some significance, Bischof argued: measurements of organic nitrogen and carbon, of chlorine and phosphate, ammonia and nitrates told the skilled analyst what sorts of contamination a water had received, how much it had received, and how recently. All these 'point[ed] to something ... which is or may be injurious to health. *The numbers of colonies, excluding for the present specific germs of disease, can likewise claim significance only if they bear in some way an invariable, or at least practically invariable, ratio to wholesomeness.*' This they clearly did not do; no one, not even Percy Frankland, claimed that the number of colonies on a culture plate had any necessary relation to the harmfulness, or even to the danger of the water.[45] Bischof claimed that he too had held out great hopes for plate culturing, but that thus far these hopes had not been warranted. Only if Koch's practice of going beyond counting to determinative bacteriology were followed could bacteriology be useful.

A good many sanitarians were sympathetic with Bischof's views, but for a variety of reasons and with a variety of motives. Some shared the caution and skepticism of Klein. The chemist H E Armstrong asserted that at present 'they were absolutely unable to make any use whatever of this information,' and worried that publication of plate culture results 'could only have the effect of frightening ... [the public].' Charles E Cassal, public analyst for several London vestries (and Klein, who also spoke at one of Bischof's papers), agreed: bacteriology had fallen far short of what had been expected of it and what was still claimed for it. Too little progress had been made in the only important matter, detecting the presence of pathogens. John C Thresh, who would become an authority on water in the early twentieth century, summed up the views of many: 'we find a few more organisms in one than another, while yet we know nothing of the nature or properties of these organisms, is so illogical as to be absurd.'[46]

Bischof's challenge did force Percy Frankland to say more precisely what he thought his filtration results meant. He did so in comments on Bischof's papers and in a major address in the late spring of 1886 to the Institution of Civil Engineers. Under Bischof's criticisms Percy Frankland became bolder. As he would maintain repeatedly in the next decade, he had been the first to show the true effectiveness of sand filters. In the past the only rationale for filtration had been 'an innate feeling' or 'intuitive wisdom' that there was 'something invisible in the water which might be detrimental to health.' Chemical analysis had never been able to show that filters had any significant effect, 'but the improvement from a biological point of view was most striking ... sand filtration, hitherto ... regarded ... as only of little value, was really an exceedingly important process in rendering river water more fit for domestic use.' Thus the water companies were made to appear as the true guardians of public health, holding fast to a technology which others thought futile. Now with science to replace rules of thumb, their already excellent results might be recognized and improved upon.'[47]

By late 1886 it was quite clear that for Percy Frankland the practical conclusion of his bacteriological investigations was that London's water was safe. In public statements he remained cautious. At the Society for Chemical Industry in early May 1887, he spoke of his tests as indicating 'the efficiency of filtration' and 'variations in the amount of organic life present,' precisely the sorts of phrases that discomfited Buchanan and Klein. He then qualified his conclusion— 'I wish it to be most clearly understood that no conclusions what- ever, as to the relative excellence of the various waters, are to be drawn from their greater or less freedom from micro-organisms, any more than it is possible, on the strength of chemical composition, to say that one water is more wholesome than another'—and quickly unqualified it: 'On the other hand, these determinations undoubt- edly do indicate what would be the probable fate of any harmful organisms gaining access to the sources of supply, and what is the relative chance of their reaching the consumers; *for that method of treatment which abolishes the largest proportion of organisms of all kinds is also the most likely to abolish any pathogenic forms should they be present.*'[48]

The statement is a masterpiece of equivocation and it is char- acteristic of Percy Frankland's evaluations of his filtration studies. Yet within the ambiguity are two strong and partisan assertions: first, that what worked best on 'all kinds' of bacteria (not 'each and every kind') would be most effective on pathogens, and second, that the tests showed something significant about the likelihood of pathogens reaching consumers. By early 1887 a substantial body of research had accumulated on the fates of pathogens in filters, in filtered water, and elsewhere, but the results were ambiguous.

Throughout 1887 relations between Bischof and Percy Frankland remained poor. The efficacy of spongy iron as a filter medium even became a minor issue of strife, with Percy Frankland now claim- ing it was much less effective than Bischof claimed.[49] In January 1888 Bischof finally presented hard evidence of problems in Percy Frankland's techniques. A point on which he had long expressed skepticism was whether gelatine–peptone as commonly used was a suitable medium for the growth of the bacteria a water contained. Percy Frankland assumed that it was, at least for practical pur- poses. Bischof showed, however, that some species grew so rapidly in the medium that more slowly growing forms would never be ob- served. That anomaly might not be important if it were true that gelatine–peptone always brought forth roughly the same proportion of the microbes present and if it were true that the filters affected all species (including pathogens) to the same degree. But it was not clear that these assumptions were warranted. Indeed Bischof's sus- picion that there was a far greater range of microbe life than grew in gelatine–peptone turned out to be prescient. By the late '90s most bacteriological analysts would insist that use of several media was necessary if one was to speak confidently on the bacterial contents of a water.

The dialectic of challenge and response between Bischof and Percy Frankland might seem to provide a superb basis for scientific advance. Yet it did not. The dialogue between the two never became an experimental dialogue, but remained instead an exchange of contradictory assumptions that were ultimately beyond the reach of experiment. Bischof doubted filters could really be so effective and saw no reason why they should be awarded the benefit of the doubt. Frankland saw no reason why they should not be effective. Bischof saw no reason to think pathogens would be eliminated at the same rate as harmless bacteria; Frankland saw no reason to think that filters would not remove pathogens as effectively as any other forms.[50]

Partisanship Triumphant

Meanwhile, behind the scenes, Percy Frankland was losing his antipathy toward the water companies. In October 1886, he became one of the water company mercenaries he had so scorned two years earlier. Early in the month, eels, living and dead, had begun to issue from taps in houses served by the East London water company. There was fear in east London that the creatures were somehow connected with an outbreak of typhoid and the company commissioned Percy Frankland and Tidy to investigate. On the basis of their separate reports the company admitted the existence of the eels yet maintained that the water was good. In his report Percy Frankland focused on the typhoid, which he attributed to local unsanitary conditions.

It was not uncommon for a water company in such a situation to seek out reputable scientists to report to it nor was it uncommon for reputable scientists to accept such commissions. What is striking is not that Percy Frankland took up the commission (though this did represent a turnabout) but the uncritical character of his report. He had done no chemical or bacteriological analyses of the water in question. Although he was one of the few British scientists conversant with continental bacteriology, he had considered the outbreak without regard to the newer knowledge of typhoid transmission. Instead, he had founded his argument in old style sanitarianism: unsanitary conditions existed, their existence was to be sufficient explanation of the outbreak.[51] Though Percy Frankland never became the kind of kept scientist that Tidy was, and did not consistently preach an

Date.	Place where found.	Condition in which found.	No. on Map and on List.	Remarks.
1884.				
September -	16, Paul Street - -	Putrid -	20	House supply tainted.
„ -	Knight's Court, West Ham Lane.	Dead -	8	„ „
„ -	Hamfrith Road, Romford Road.	Putrid -	7	„ „
? 1885 †	"Duke of Cambridge" Public-house, Victoria Dock Road.	„ -	31	„ „
1886.				
April 21 -	Urinal, Plaistow - -	Not stated -	24	——
Spring - -	20, Romford Road - -	Putrid -	10	House supply tainted.
„ - -	Harrow Wharf, Stratford -	Dead -	15	——
July - -	54 and 56, Layton Road -	„ -	3	House supplies tainted.
„ 12 - -	Beaumont Road - -	Alive -	26	Pipe cut.
„ - -	Urinal, Barking Road - -	Putrid -	29	On three occasions.
„ 14 - -	"Green Man" Public-house, High Street, Stratford,	„ -	16	House supply tainted in ball valve.
„ 16 - -	"Green Man" Public-house, High Street, Stratford.	Dead -	15	House supply tainted, in supply pipe gate-way.
„ 20 - -	30, Union Street, Marsh -	Putrid -	18	House supply tainted.
July - -	Green Gate Street, Plaistow -	Dead -	25	„ „
„ - -	Solway Road Board School, Stratford.	Putrid -	6	„ „
„ 29 - -	Tucker Street, Canning Town	1 putrid, 4 alive.	28	Supply in main tainted.

* Not here reproduced.
† The time of occurrence of this case is very uncertain. It is referred to 1885 on the authority of Edward Bowden who, in the spring of 1887, stated that "about two years ago" he took a putrid eel from the supply pipe of the public-house in question.

Figure 9.2 The invasion of the east London water supply by a plague of eels in the mid '80s occasioned an outbreak of the partisan science which predominated in nineteenth century water matters (*17th Annual Report of the LGB*. Supplement containing the Report of the MO for 1887, p 124).

explicitly pro-companies doctrine, more often than not his interpretation took a direction favourable to the companies.

In the winter of 1887 Bolton died. His successor, Sir A deCourcy Scott, who had been a Local Government Board engineering inspector, had none of Bolton's naiveté. He advocated public takeover and was interested in bacteriological analysis so long as it was not subjected to simplistic and partisan interpretation.[52] Scott continued to publish Percy Frankland's filtration investigations until the end of 1888, by which time Frankland had left for Dundee.

The effect of these investigations and of Percy Frankland's continual parading of them before meetings of sanitarians was to give bacteria-counting far more prominence than it warranted. Even while it was admitted by all that numbers alone meant nothing, and even while techniques of determinative bacteriology were becoming available on the continent, British sanitarians remained preoccupied, even tantalized, by the simplicity of counting microbes. To a considerable extent Percy Frankland's equivocation about what his results meant was responsible for this. With each admission that counting bacteria had no sanitary significance came a reminder of the 'striking' efficiency of filtration that bacteria counts revealed. Even if one did not share Percy Frankland's optimism it was hard not to grant his results some significance, but how much significance? He had shown that filters were very good at removing a particular entity, but that entity was not the same as (though it was probably very similar to) the germs that all were concerned about. But did this partial significance signify anything?

Two illustrations may give some sense of the degree to which British water analysts had become preoccupied with quantification. The first is an 1886 paper on 'The Purification of Water' published by two chemists, A Gordon Salamon and W DeVere Mathew, in the *Journal of the Society for Chemical Industry.* The authors began with a common and paradoxical admission, that the proper approach to water analysis, determinative bacteriology, was 'in its infancy' and unsuitable for use, and they agreed that water ought not to be judged on the number of bacteria it contained. Like Bischof, they held that species that liquefied gelatine were probably the more harmful, but they recognized that this too was a matter of great uncertainty. But having made these admissions they went on to present a study of chemical and bacterial effects of methods of water purification, accepting at face value Percy Frankland's work and even using the same plate culture counts they had earlier dismissed. Bischof was outraged. How, he wondered, could analysts admit that their techniques were inconclusive and go on to draw conclusions? In relying solely on counting British analysts had betrayed Koch's intention. The gelatine–peptone method '*per se* was no test of the wholesomeness of a water,' he insisted.[53]

The Salamon–Mathew paper indicates the degree to which inadequate methods might become tolerable because they became customary. In acknowledging the limitations of the technique at the outset of an experimentally based argument, one was in effect dismissing them as inherent in the state of the art, as the price to be paid if the techniques were to be used at all. But it was not clear, argued Bischof, that bacteriology still warranted use with so many qualifications; no bacteriology might be better than bad bacteriology.

The cavalier tone of Salamon and Mathew stands in sharp contrast to the harsh criticism of a review editorial on 'Micro-organisms in Water' published in *Engineering* in early 1887. Taking a course that would be frequently followed in the next decade, the author began by reflecting on the sublime aspects of bacteriology, the revelation of the great good and great evil that bacteria did. All this was the big picture; yet knowing exactly which bacteria were the threatening ones, and under what circumstances they threatened raised much more difficult questions: 'that the authorities do not rarely disagree will not surprise us when we inquire into the methods and means of bacteriological study.' The sheer difficulty of technique, with its demand for sterility and for skill in arcane microscopic techniques, the author found staggering.

With this introduction he went on to consider the studies of bacteria in the London waters, drawing attention to the claim that pathogens were somehow reliably eliminated before the water reached consumers' cisterns. He sympathized with Bischof. Simply on grounds of prudence it seemed unwise to rely on probabilities or trust nature to provide safeguards: 'as long as we are not positive about these suicidal tendencies of bacteria, we [should] strive to purify our water from them as much as possible.' Last came a note of caution: 'there is a good deal of conflicting evidence; and the various figures [of microbes in the water] have as yet ... more scientific interest than practical value.'[54]

This concept of 'practical' as distinct from 'scientific' value is at the crux of the matter. Here science referred to empirical contributions, like the vast numbers of studies of the life spans of various bacteria species in various waters. Yet knowledge that provided a basis for action was what was demanded of water analysts, and such work provided no firm basis for action. In 1894 Percy Frankland would write that the 'obvious practical conclusion to be drawn from the numerous investigations on the behaviour of pathogenic bacteria in water, is that water which has been exposed to the possibility of contamination with noxious matters should be stored as long as possible before use, in order that the maximum opportunity may be afforded for any pathogenic forms that have been introduced to lose their vitality.' But the really practical issue was how long was long enough, and on this point Percy Frankland had nothing to say; he had only generalized from knowledge of mechanisms of bacterial removal.[55]

The Political Implications of the Possibility of Knowing

It should be clear that discussion of the uses of bacteriological water analysis was taking place in terms of two interconnected motifs. The first of these had to do with what qualifications it was necessary to make when accepting the results of such analyses. This was the issue between Bischof and Percy Frankland. Percy Frankland held that bacteriological results, like the results of any other not yet perfected techniques, warranted tentative acceptance and that it was appropriate to use them along with other kinds of knowledge in making public decisions. Bischof, on the other hand, saw the techniques as subject to such enormous errors that their results were no longer positive contributions to knowledge at all but pieces of misinformation capable of doing great harm if they were mistakenly taken as knowledge. Qualifying one's conclusions was no solution; the necessary qualifications were so enormous as to undermine any claim that a contribution to knowledge was being made. Percy Frankland argued that chemical processes had long been relied upon in making decisions about water supplies, even though they required one to make far greater qualifications. Bischof, on the other hand, worried that the false security that came from bacteriology might seduce people into forsaking common sense in choosing waters and even saw the gross and well-known inadequacies of chemical water analysis as one of its advantages, for they kept people from taking it too seriously.[56]

The second motif had to do with the politics of the London water supply. Here there was a tension (at least as far as water quality was concerned) between principle and pragmatism. Traditionally the political positions taken on the London supply by public figures, including scientists, had been extreme. Either the water was safe or it was dangerous, debilitating, and sometimes lethal. In bacteriology Bolton had seen a way of changing the character of London water politics, a way of replacing the clash of contradictory principles with pragmatic management of the existing supply. Let the politicians set what standards they wished; with bacteriology one could tell when the water was inferior and guarantee the effectiveness of remedial efforts. Yet Bolton was unable to bring about such a change. Those who opposed him included some, like Klein, who were honestly skeptical of bacteriological analysis. Others, however, for whom the principal end was securing public control of the supply found Bolton's pragmatic approach unacceptable. For them it was bacteriological analysis *per se* that posed the threat, not just the primitive versions currently in use. Like Edward Frankland, they denied that any analyses could show that the water was safe; analysis was for disclosing danger. Hence for them bacteriology provided no grounds for giving up the position Edward Frankland had maintained since 1868.

It might seem that the maturation of determinative bacteriology would end this war of rival assumptions. It did not. Determinative bacteriology brought with it a new and broader range of uncertainty. During the late '80s and early '90s, the unresolvable issue of what meaningless counts meant gave way to such questions as whether the pathogens of typhoid and cholera had in fact been discovered, whether they could be detected with confidence, whether they were rare species or mutations (perhaps random, perhaps environmentally caused) of common species, under what conditions they lived, multiplied, and died, and under what conditions they caused disease in a human being.

1   29 Jan 1886, PRO MH 29 10 12665K2/87.
2   PRO MH 29 8 Bolton to Owen, 26 Oct 1885, unnumbered; 'London Water Supply,' *JRSA* 30 (1881-2): 971–2.
3   PRO MH 29 8 Bolton to Owen, 15 Dec 1885.
4   cf Alfred Carpenter, 'First Principles of Sanitary Work,' *TSIGB* 1 (1880): 59; 'Review of C T Kingzett, *Nature's Hygiene*,' *SR* ns 2 (1880-81): 111–2; and reply from Kingzett, *Ibid*, pp 156–7; D Galton, 'The Public Health,' *SR* ns 4 (1882-83): 137–51.
5   J W Tripe in *SR* 8 (1878): 55–6; FSB François de Chaumont, 'On Certain Points with Reference to Drinking Water,' *SR* ns 1 (1879- 80): 163; Willis Tucker, 'The Sanitary Value of the Chemical Analysis of Potable Water,' *CN* 54 (1886): 135.

6   P Hinckes Bird, 'On Sewer Air, House Drain Ventilation, and Sewage Disposal,' *SR* 8 (1878): 3–5; Alfred Carpenter, 'Suggestions for Preventing the Spread of Infectious and Contagious Diseases,' *SR* 9 (1878): 226–8. See also the debate between Tripe and Wanklyn on the presumed nature of typhoid germs ('Typhoid Germs and their alleged destruction,' *SR* 4 [1876]: 387- 8; J A Wanklyn, 'On the Purification of Drinking Water by the Process of Filtration,' *SR* 4 [1876]: 391–2; Tripe, 'Water Analysis and Typhoid Germs,' *SR* 4 [1876]: 419–20).

7   But see J Lane Notter, 'The Filtration of Potable Water,' *SR* ns 2 (1880–1): 161; C Cassal and B H Whitelegge, 'Remarks on the Examination of Water for Sanitary Purposes,' *SR* ns 5 (1883–4): 427.

8   HLRO, Minutes of Evidence, House of Commons, 1878, v. 5 (Cheltenham Corporation Water Bill), pp 246–303. See also J H Balfour Browne, *Forty Years at the Bar*, pp 39–40.

9   Cheltenham Corp Water Bill, 12 March 1878, pp 6–40, 292–4; 13 March 1878, pp 25–43, 175–220. Thursfield was medical officer of health for Shrewsbury, Wilson was a physician at the Cheltenham Hospital Dispensary, Wright was the Cheltenham medical officer of health, while Hill was the Birmingham public analyst and a former student and follower of Frankland.

10   Rept of the Select Committee on Rivers Pollution (River Lea), QQ 3793–97.

11   C M Tidy, 'River Water,' (1880): 320–1.

12   *Ibid*, pp 321–6.

13   William Anderson, 'The Antwerp Water Works,' pp 24–83, esp pp 45–7, 61. See also Frank Hatton, 'On the Oxidation of Organic Matter in Water,' pp 258–76. On Frankland's connection see Imperial College, Lyon Playfair papers, General Correspondence, C 231, J Dewar to L Playfair, 15 Jan 1887; G Bischof, 'The Purification of Water,' *JRSA* 26 (1877–8): 486–96.

14   The argument was continued in the pages of the *Journal of the Royal Society of Arts*: J Hogg, 'Chemical Analysis of Drinking Water,' *JRSA* 31 (1882–3): 414–5; G Bischof, 'The Anti-Septic Action of Spongy Iron,' *Ibid*, pp 444–5; Hogg, 'The Anti-Septic Action of Spongy Iron,' *Ibid*, pp 467–8; J Lane Notter, 'Water Filtration,' *BMJ*, ii, 1878, pp 285, 556–7, 653.

15   Folkard, 'The Analysis of Potable Waters,' pp 65–6.

16   *Ibid*, pp 78–82.

17   *Ibid*, p 103.

18   A Wagner, 'Notes on Water Analysis,' *CN* 44 (1881): 176–7.

19   Smith in GBH *MWS*, App III, pp 94–5; R A Smith, 'On the Examination of Water for Organic Matter,' (Manchester), pp 39, 41, 45–6; same title, different text *CN* 19 (1869): 279; Eyler, 'The Conversion of Angus Smith,' p 225.

20 R A Smith, 'Rivers Pollution Prevention Act, 1876. Second Annual Report' (1884), pp 28–38; *idem* 'On the Development of Living Germs in Water,' *CN* 46 (1882): 288–9 (abstract); *SR* ns 4 (1882- 3): 308–9; *idem* 'Notes on the Development of Living Germs in Water by Dr. Koch's Gelatine Process,' *SR* ns 4 (1882–3): 344–7; Eyler, 'Smith,' pp 230–1.

21 'The New Method of Testing Water,' *SR* ns 4 (1882–3): 360.

22 Eyler, 'Smith,' pp 232–3; Gibson and Farrar, 'Robert Angus Smith and Sanitary Science,' pp 247–8; A Gibson, 'Robert Angus Smith and Sanitary Science,' p 1.25. See also DNB 18, pp 520–2; and E Schunk, 'Robert Angus Smith,' *CN* 51 (1885): 293–6.

23 William Bulloch, 'Biographical Notices,' in his *History of Bacteriology*, pp 348–406. See also W D Foster, *A History of Medical Bacteriology and Immunology*, pp 19–21, 57–8, 66–8; *idem, A Short History of Clinical Pathology*, pp 53–61. Both Burdon Sanderson and Klein were involved with these public health issues as consultants for the Local Government Board (see Burdon Sanderson, 'Introductory Report on the Intimate Pathology of Contagion,' in *Twelfth Report of the MOPC for 1869*, appendix 11, pp 229–56; and *idem*, 'Further report of Researches concerning the Intimate Pathology of Contagion,' in *Thirteenth Annual Report of the MOPC for 1870*, pp 48–69).

24 Burdon Sanderson, 'Further report of Researches,' pp 60-65. The test was included in Parkes' *Manual of Hygiene* 6th edn, I, p 71, as easily carried out but affording absolutely no useful information. See also GBH Medical Council, *Report of the Committee for Scientific Inquiries on the 1854 Cholera*, pp 36, 46–7; comments of James Ritchie and T Lauder Brunton in G Burdon Sanderson, *Sir John Burdon Sanderson: A Memoir* (Oxford: Clarendon Press, 1911), pp 75–8, 84–91; Maj Charles Smart, 'On the Present and Future of Sanitary Water Analysis,' p 85; Fox, *Sanitary Examinations of Water, Air and Food*, 2nd edn, p 25; 'Dr Sanderson's Experiments on the Growth of Microzymes in Water,' *Food Water Air* 1 (1871): 7–8. On Burdon Sanderson see *Proc. Royal Society* D 79 (1907): iii xviii.

25 Frankland, 'On the Development of Fungi,' p 74; Heisch, 'On Organic Matter in Water,' pp 371–5.

26 Bulloch, *History*, pp 227–30; Foster, *History of Medical Bacteriology*, p 59.

27 'The Biological Laboratory at the Health Exhibition,' *Lancet*, ii, 1884, pp 251–2, 332–3, 380–1, 557–8, 609–10, 705–6, 751–2, esp 705–6. See also 'Scientific Aspects of the Health Exhibition,' *Lancet*, ii, 1884, p 24.

28 C J H Warden, 'The Biological Examination of Water,' *CN* 52 (1885): 52.

29 *Ibid*, p 74.

30 *Ibid*, p 104.

31 W H Garner, 'Percy Faraday Frankland,' *J Chem Soc*, 1948, pt iii,
    pp 1996–8; P FFrankland in discussion of G Bischof, 'Notes on Dr
    Koch's Water Test,' *J Soc for Chem Industry* 5 (1886): 120; Colin
    Russell, 'Percy Frankland: The Iron Gate of Examination,' *Chemistry
    in Britain* 13 (1977): 425.

32 P Frankland, 'The Cholera and Our Water Supply,' pp 349–51.

33 *Ibid*, p 354.

34 P Frankland, 'The Upper Thames,' pp 432, 435, 565–6.

35 P Frankland, 'The Removal of Micro-organisms from Water,' *CN* 52
    (1885): 27–9, 40–2.

36 Percy Frankland, 'The Selection of Domestic Water Supplies,' *SR* ns 6
    (1884–5): 549–51.

37 'London Water Supply,' *CN* 52 (1885): 296; PRO MH 29 8 no number,
    P Frankland to Bolton, 7 Nov 1885; 110299/85 'October Report'; no
    number, Bolton to LGB 15 Dec 1885.

38 'London Water Supply,' *CN* 53 (1886): 91.

39 E Frankland, 'On Chemical Changes in their Relation to Microorgan-
    isms,' *CN* 50 (1884): 78–80. See the response of E Klein, 'Bacteriologi-
    cal Research from a Biologist's Point of View,' *J Chem Soc* 49 (1886):
    197–205.

40 On Klein, 'the father of bacteriology' in Britain, see 'Edward Emanuel
    Klein, 1844–1925,' *Proc Royal Society* B 98 (1925): xxix. But see Foster,
    History of Clinical Pathology, p 67.

41 PRO MH 29 8, P Frankland to LGB, 7 Nov 1885; 110299/85, 'October
    Report; Bolton to LGB, 15 December 1885; PRO MH 29 9 'Novem-
    ber report'; 8074/86 'December report'; PRO MH 29 10 12665K2/87
    'Memos of Buchanan, Klein, Owen, Bolton,' February 1886.

42 P Frankland, 'New Aspects of Filtration,' pp 701, 705, 709. See also dis-
    cussion of E Klein, 'Bacteriological Research from a Biologist's Point of
    View,' *CN* 53 (1886): 83. Compare with F Bolton and Percy Frankland,
    Lectures on Water.

43 'Obituary of Gustav Bischof,' *J Soc for Chemical Industry* 22 (1903):
    84; comments in P Frankland, 'Water Purification: Its Biological and
    Chemical Basis,' pp 230, 233; see also on Baker, *The Quest for Pure
    Water*, pp 313–5.

44 Gustav Bischof, 'Notes on Dr Koch's Water Test,* pp 116–9; *idem*,
    'Dr Koch's Gelatine-Peptone Water Test,' *CN* 53 (1886): 205–6; *idem*,
    'Dr Koch's Bacteriological Water Test,' *Lancet*, i, 1885, pp 382–3. For
    Bischof's investigations of the London waters see 'The Metropolitan
    Water Supply,' *Engineering* 41 (1886): 18, 116, 233, 384.

45 Bischof, 'Dr Koch's Gelatine-Peptone Water Test,'* pp 205–6. See also
    Bischof in discussion of Percy Frankland, 'Water Purification: its Bio-
    logical and Chemical Basis,' pp 224–7; 'The Bacterioscopic Examina-
    tion of Water,' *CN* 53 (1886): 232.

46 P Frankland, 'Water Purification: its Biological and Chemical Basis,' p 215; Bischof, 'Notes on Dr Koch's Water Test,'* p 120; *SR* ns 7 (1885–6): 550; Louis Parkes, 'Water Analysis,' *TSIGB* 9 (1887–8): 385–7; A Dupre, 'Presidential Address,' p 364; J Thresh, 'Water Analysis and the Defective Information afforded by it,' *BMJ*, ii, 1892, p 860.

47 In discussion of Bischof, 'Notes on Dr Koch's Water Test,'* pp 120–1.

48 P Frankland, 'Recent Bacteriological Research in connection with Water Supply,' p 319. See also his 'The Filtration of Water for Town Supply,' pp 276–84. Compare with his statement to the Civil Engineers a year earlier: 'investigations could be carried out without any reference to the influence of those micro-organisms upon health, the problem being simply to ascertain whether and to what extent the various processes of purification had the power of removing micro-organisms in general' ('Water Purification: its Biological and Chemical Basis,' p 244).

49 Frankland, 'Recent Bacteriological Research,' pp 323–6.

50 G Bischof, 'Extension of Time of Culture in Dr R Koch's Bacteriological Water Test by Partial Sterilisation, with special reference to the Metropolitan Water Supply,' *SR* ns 9 (1887–8): 325–32.

51 PRO MH 29 9, clipping from *Morning Post* 9 Oct 1886; 92059/86, East London Water Company to LGB; 109 560/86 W Ham Local Bd Works Ctte to LGB, 19 Nov 1886; MH 29 10, 212/87 W Ham Borough Clerk to LGB; 67728/87 Report of W H Power and A DeC Scott on the eel problem; *SR* ns 9 (1887–8): 111–2. According to the *Sanitary Record* at least six of the eels came through alive and were 'generally eaten by the finders.' Cf A DeC Scott and W H Power, 'Eels in Water Mains,' in *17th Annual Report of the LGB, Report of the MO for 1887*, 121–38.

52 *SR* ns 8 (1886–7): 355; 'The Control of the London Water Supply,' *The Engineer* 63 (1887): 422. On Scott see *CN* 80 (1899): 207. There was expectation that Edward Frankland would be appointed to the water examiner's post on Bolton's death.

53 A Gordon Salamon and W DeVere Mathew, 'The Purification of Water,' pp 201–7, 271–3.

54 'Micro-organisms and water,' *Engineering* 43 (1887): 184–6.

55 P Frankland, 'The Bacteriological Examination of Water and the Information it has Furnished,' p 7; E Duclaux, 'Les Microbes des eaux,' pp 568–9; Fox, *Sanitary Examinations of Water, Air, and Food*, 2nd edn, p 200; Armstrong in discussion of Bischof, 'Notes on Dr Koch's Water Test,'* p 120.

56 A few years later the American water analyst A R Leeds explored these issues sensitively in a paper entitled 'A Question of Water, Ethics, and Bacteria,' pp 259–68. See also E Duclaux, 'Les Microbes des Eaux,' p 569; Louis Parkes, 'Water Analysis,'* p 385.

10 What's Bacteriology For? Disenchantment and a New Realism, 1890–98

> *Whether the organisms did any harm, whether if some were harmful others were not innocent, whether there were not some organisms which destroyed others, and whether it would not be well to leave the destroyers in the water so that they might destroy. Nobody knew which were the bad and which were the good, or whether the bad would eat up the good, or the good eat up the bad.*[1]
>
> Frederick Bramwell

Bacteriology did not quickly change water analysis; indeed considering how widely the importance of bacteria was recognized, the science of bacteriology was remarkably uninfluential. The new science induced debate; it led to pronouncements about the meaning or meaninglessness of bacteria counts, and it led to some interesting studies about the properties of various microbe species, but it did not lead very many analysts to change the processes they used to analyse water or the criteria they used to assess results. At scientific meetings there was frequent talk of dashed hopes of bacteriological analysis. Bacteriological techniques were included in manuals for public health officers, yet chemical processes remained primary. By and large, those who used chemical processes of analysis did not give them up for bacteriology, and some of these people virtually ignored bacteriology. After the furor of the early '80s things were quiet at the Society of Public Analysts until the mid '90s when a fresh onslaught of undetectable typhoid epidemics brought the issue back into prominence. The indomitable Wanklyn continued to issue editions of his water analysis manual as if there were no science of bacteriology.[2]

Almost from the outset of plate culturing in Britain some bacteriologists were attending to a different issue. Their hope was to obviate much of the need for regular analysis by identifying the conditions under which pathogens could (and more significantly, could not) survive. Such findings might be conceived as meaningful either in a strong sense or a weak sense. In its strong sense this research might be presumed to lead to a time when the laws of the existence of pathogens had been determined and analysis, chemical or bacteriological, was therefore no longer necessary. With knowledge of such laws public health officials would know when and where to worry; they would know what climate, season, soil, and other environmental surroundings permitted the existence of particular germs. In epidemiology such a perspective had long been central; what was new was the focus on the disease agent rather than the disease itself. As Woodhead put it in 1893, since the *comma vibrio* of cholera had been discovered 'we now consider the conditions under which the *bacillus* can multiply and be carried from point to point ... instead of dealing with the cholera itself .... The epidemiologist has now assumed the role of biologist in the widest sense of the term.'[3]

This approach of deduction from laws of pathogen survival may seem unnecessarily exclusive in its rejection of complementary inductive approaches; why not complement and continually confirm one's deductions by testing for the microbe itself, one might ask. Yet we do find bacteriologists making relatively unqualified assertions of the safety of certain waters on the basis of what they regarded as laws of pathogen survival. Here, as with chemical analysis, the adversary context of British water science encouraged such assertions.

More commonly such researches were applied within a framework of inductive inference. Knowledge of the natural history of pathogenic bacteria along with the rest of the facts an analyst could gather helped one make an informed judgement on the safety of a water. In such an assessment no single piece of information need be considered primary, unless it be the actual discovery of the pathogen in the water; instead all made up a profile that more or less closely resembled a standard template laid out in a manual of analysis.

Most importantly such a perspective could be used to complement a search for the pathogen (almost always the typhoid organism) itself. Searches for the typhoid *bacillus* almost always failed and on the few occasions when success was claimed, the claims were likely to be rejected by more experienced bacteriologists as reflecting an inability to distinguish similar species. Hence the significance of the failures was unclear. Did they mean the bug was not, and never present? Or that it was occasionally or even commonly present but difficult to distinguish? It was hoped that research into the natural history of the organism might narrow the range of answers to these questions by giving analysts some sense of how much confidence to attach to negative results.

Because the degree of confidence analysts could have in their statements about water safety was a political issue, this research programme was also political. Among British bacteriologists, it was taken up most enthusiastically by those like Percy Frankland who were trying to expand the level of confidence with which one could speak about water safety. It was resisted by those like E E Klein, and by George Buchanan, long-standing skeptics of water analysis.

Issues of the degree of confidence that analyses warranted became increasingly important in the late '80s and early '90s, their importance climaxing during yet another public inquiry into greater London's water supply, that of the Royal Commission on Metropolitan Water Supply of 1892–93. This was an undisguisedly partisan affair, with the opposing sides (the companies and the London County Council) going to unprecedented lengths to mount a scientifically sophisticated case.

By the mid '90s it was evident that the 'laws-of-pathogen-survival' approach would not work. Bacteria were too 'fickle' as German Sims Woodhead put it; they refused to be pinned down to discrete habitats or conditions.[4] This failure left analysts with the direct approach: to analyse waters for pathogens themselves. By 1895 this finally could be done reasonably successfully, yet this technique too turned out to be of little practical use, for by the time one had mobilized the ponderous and sophisticated techniques of determinative bacteriology the typhoid *bacillus* had usually done its work and disappeared.

Crookes, Odling, and Tidy: Bacteriology for the Water Companies

The temptation to draw unambiguous conclusions about the conditions of bacterial existence was clearly underway by July 1886, when the companies' analysts, William Crookes, William Odling, and Charles Meymott Tidy, announced that they would eschew bacteria counts in favour of the more important issue of the survival of pathogens in water. Their first experimental subject was the anthrax *bacillus* which they found could last no more than a few hours. It was, as Percy Frankland pointed out, an irresponsible conclusion: Crookes, Odling, and Tidy had not taken into account that *B an-thracis* formed spores.[5] In the following months the three chemists reported their ongoing research. Even in sterilized London water (where an organism would not suffer from the competition of other species), the anthrax microbe was gone within two days. Such fragile microbes were reassuring. It was not that there was much likelihood of water-borne anthrax coming to London, but the public was regularly horrified by tales of germs and the chemists hoped to restore a balance of ignorance.[6]

The conclusions drawn from these experiments were clearly self-serving and as none of the authors was a reputable bacteriologist there was reason to be skeptical. The *Sanitary Record* reminded readers that the authors were committed to the principle that water automatically became pure. By early 1887 the three chemists had discontinued the experiments and their anthrax work quickly lapsed into obscurity.[7]

Percy Frankland and the Limits of Bacterial Activity, 1888–94

The bulk of the research on the conditions under which particular species of bacteria survived, multiplied, and died was carried out on the continent by German, French, and Italian bacteriologists. Percy Frankland was the British bacteriologist most heavily involved, and he served as one of the main conduits through which this research was brought to bear on questions of water purity in Britain. While Frankland had been experimenting on bacterial longevity as early as 1886, particularly on the fate of the *comma vibrio* (the cholera germ according to Koch), his first major address on the subject was in May 1887 at the Society of Chemical Industry.[8]

Previous research on the bacterial population of natural waters was of only 'indirect significance' he noted. It was all very well to catalogue the numbers of various species in different sorts of water, but what one needed to know was what became of pathogens. Yet questions of relevance plagued his own work too. 'To bring the investigation within the limits of experimental possibility' it was necessary to charge the species to be studied into sterilized natural waters. Pathogens would otherwise be too difficult to count since species that grew more quickly would mask their presence. But it was not clear what the results of such experiments would mean, for sterile 'natural' waters were not natural at all. In truly natural waters the fates of pathogens would be determined by their interaction with the totality of the aquatic environment, including their ability to compete with other bacteria for nutriment and to escape predation. Hence it was not clear that the results had any significance whatever for public health.

Notwithstanding the issue of relevance, the results were interesting. Percy Frankland reported on six organisms: the *comma vibrio*, the Finkler–Prior *comma bacillus* (morphologically similar to the *comma vibrio*), *B pyocyaneus* (associated with abscesses), the micrococcus of erysipelas, the anthrax organism, and the putative typhoid organism. The survival periods of these microbes varied greatly, from a single day to more than eleven months in the case of the cholera *vibrio*. These results dictated caution.[9] The 'very prevalent impression' that pathogens would succumb in competition with hardier native varieties had been 'much exaggerated,' he wrote. Continental work showed that while the cholera *vibrio* might quickly be swamped by the rapid growth of ordinary bacteria, it was not eliminated, and after a time would undergo a resurgence.

The main products of this line of research were two massive reports. The first, commissioned by the Water Research Committee of the Royal Society and co-authored with H Marshall Ward, Professor of Botany at the Royal Indian Engineering College, appeared in the Society's *Proceedings* between 1892 and 1894. The second was an 1894 treatise on 'Micro-organisms in Water: Their Significance, Identification, and Removal' co-authored with his wife Grace Toynbee Frankland. Each mixed Frankland's (or the Franklands'; Grace was a bacteriologist too) own research with a thorough, and indeed unprecedented, review of continental work.[10]

Indeed, the most striking characteristic of these works is their comprehensiveness. Percy Frankland had brought together the results of hundreds of experiments, literally from around the world. There was an enormous amount of information on the survival periods of various species in different waters: distilled water, spring water, sterile river or tap water, unsterilized river or tap water, and so on. The literature was full of conflicting accounts of the association (or lack of association) between outbreaks of diseases and the presence of species presumed responsible for them. Complicating this issue were taxonomic uncertainties: bacteriologists disagreed about whether failure to discover pathogens reflected the real morphological and physiological variability of bacteria or the inadequate technique of their colleagues. The latter possibility was further complicated by disagreement as to what exactly constituted adequate technique.[11]

			Behaviour of *Spirillum*		*cholera asiatica* in Water.			
Observers.	Temperature.	Distilled water.	Sterilised ordinary water.	Non-sterilised ordinary water.	Foul water.	Sea water or Concentrated Salt solution.	Mineral waters.	Remarks.
Straus and Dubarry	20°	14 days	{ 26 days* / 29 days†	..				* Water of Ourcq / † Water of Vanne
„	35°	..	30 days*	..				
Hochstetter	13—20°	24 hours to 7 days	267—382 to 392 days‡	..	..	..	3 hours§	‡ Berlin tap water / § Seltzer water / Berlin tap waters
Babes	..	1 day	7 days	..	..	..	..	
Wolffhügel and Riedel	16—22°	33 days	2 days to 7 months and even a year	1 to 20 days	7 months	..	..	Berlin tap water
Frankland (P. F.)	30°	..	9 days‖	..	Over 11 months¶	..	..	‖ Deep-well (Kent Co.'s water and Thames water) / ¶ Sterile sewage (London)
Nicati and Rietsch	Ordinary temp. of lab. in winter	20 days	**22—28 days††	..	32 days	64 days to 81 days‡‡	..	** Cale / †† Canal water / ‡‡ Old harbour } All sterile
Ringeling	..	..	..	..	37 days			
Hueppe	10°	Over 10 days	} Up to 20 days	..	..	..	..	Wiesbaden waters
„	16—20°	Over 5—10 days		..	..	..	..	Three waters employed as above
Kraus	10°5	1 day	..	1—2 days	..	..	..	Drinking waters of Innsbruck
Karlinski	8°	3 days	..	2—3 days	..	..	..	
De Giaxa	16—18°	..	..	..	..	2—4 days		
Freytag	..	..	..	..	..	6—8 hours		
Braem	..	24 hours	..	..				
Pfeiffer	..	..	..	7 months				
Maschek	..	40 days	60 days	..	..	..	..	A little common salt was added to the distilled water
Gärtner	11°	..	..	1—2 days				

Figure 10.1 By the early '90s bacteriological expertise meant familiarity with a vast and rapidly growing body of conflicting results on the fates of pathogenic bacteria in various kinds of waters. Note the range of results on the survival of the cholera *vibrio* in sterilized ordinary water; experimenters obtained results ranging from seven to 392 days (*Proc Royal Soc* 51 [1892]: 272–3).

In the main Percy Frankland's treatises focused on ecological aspects of bacteriology. The key implication for analysis came from recognition of the importance of inter-specific competition among bacteria. The first group of bacteria counters—Percy Frankland especially—had found it difficult to keep in mind the fact (which they knew to be true) that gelatine–peptone was not the ideal medium for all species. It was all too tempting (and all too easy) to equate what *grew* there with what was there *to grow*. Indeed, one of Bischof's concerns had been that any given medium would favour some species over others. Those that grew rapidly might completely obscure those that grew slowly, and some might not grow at all. It was recognized that something analogous must happen in the wild: natural waters were, after all, dilute media, and would suit some forms better than others. Since human pathogens were only accidental inhabitants of natural waters, it was argued that they would be less well suited to compete with native species, even though there was ambiguity as to exactly how one species of bacteria could 'overwhelm and suppress' another; whether the victims were starved out, poisoned with waste products, or consumed, but the concept was widely current nevertheless.[12]

Defenders of the use of river water for public supplies made much of such processes. Some went so far as to express a preference for dirty river water owing to the presence of these armies of microbe allies.[13] Yet not all the implications favoured the companies. While the competition concept made it seem likely that pathogens would be wiped out in naturally populated waters, it also made it plausible to think that they would flourish in pure or filtered waters, where the multiplication of the few remaining microbes would be extraordinarily rapid. If pathogens survived the river or if a large body of filtered water were accidentally infected (as in the east London cholera of 1866), a disaster would follow as the unopposed pathogens took over the water.

There was an equally serious implication for analysis itself. The growth of a few species suited to a particular medium could conceivably lead to the disappearance of pathogens in the sample. In such a case the process of analysis would have altered the constituents of the sample—precisely the problem John Murray had raised eighty years earlier in mineral water analysis. There was of course no one perfect medium that mirrored nature; searching for one factor meant ignoring others. This never emerged as an overwhelming practical problem for by the mid '90s, when bacteriologists had fully appreciated this conundrum, they had begun to rely on a number of different media. Yet as an inescapable condition of bacteriological research, it was a problem that in many forms periodically troubled bacteriologists. The French water bacteriologist Pierre Miquel, a pioneer in analytical technique, noted that refrigerating a sample prior to analysis, which was a vital step for an accurate bacterial count, would kill typhoid germs.[14]

What should be clear about these issues was that they lent themselves to a redefinition of what it meant to be an expert. When Crookes, Odling, and Tidy had taken up the study of the fate of the anthrax *bacillus* in London's water they were interested in finding an assurance that any organisms that ended up there were doomed.

They were working within a mode of expert advising that went all the way back to mineral water analysis. The expert was the one who asserted and got away with it. Experts didn't say maybe. In drawing together the wealth of bacteriological data, Percy Frankland was bringing forth a different kind of expertise. One was convincing not by being definite and defiant but by laying out the scenario that took into account the widest range of factors, that incorporated the widest and most current knowledge of the literature (including, especially, the foreign literature), and that reflected the greatest technical and methodological virtuosity. One who could successfully incarnate this wisdom could demolish any mere maker of assertions, simply by portraying the latter as simplistic and out-of-touch.

To some degree, British water analysts had already been engaged in this kind of scenario-generating expertise. It characterized Edward Frankland's presentation to the 1868 Royal Commission and some of Tidy's presentations. What Percy Frankland could offer, however, was an international context for such expertise. Because bacteriology was truly an international science, no longer did the expert have to find warrant for his claims in his own process of analysis; he could present his views as an assessment of the combined results of professors 'such-and-such' and 'so-and-so' in Berlin, Munich, and Paris. It need hardly be said that such bacteriologists held a supreme contempt for those who simply counted (a task suitable for 'an intelligent laboratory boy,' according to Miquel) and that the level of bacteriological mastery they aspired to was far beyond what the average local sanitarian, the sort of person Cornelius Fox was appealing to, could achieve.[15]

It should be noted also that the old mode of expertise did not disappear. Even in the early '90s, when he spoke of filtration Percy Frankland retained the old theme that filtration came very close to guaranteeing the safety of water. Ecological research had even led him to a new mechanism of bacterial removal, subsidence, a natural process of purification no less perfect than filtration.[16]

The Royal Commission on Metropolitan Water Supply, 1892–93

The forum which most clearly manifested this new mode of expertise was the investigation of the Royal Commission on Metropolitan Water Supply, appointed in early 1892. The commission was a product of two political changes that occurred in the late 1880s: the arrival in 1887 of an activist, Maj Gen A deCourcy Scott, as Bolton's replacement as Metropolis Water Examiner and the establishment in 1889 of the London County Council. Scott was bothered by the continued dumping of sewage into the upper Thames. The County Council shared his concern, but also viewed the water supply within a notion of comprehensive management of public services. By the early '90s there was evidence of the inadequacy of the short-term planning of private water companies: the East London Company was already running short of water and according to projections the Thames and Lea would be insufficient by 1931. The Council therefore revived a scheme considered in the late 60s of supplying London with water from the Welsh hills.[17] The Commission focused on quality and quantity, yet as in the past, the real issue was public takeover. Most of the witnesses represented either the LCC or the water companies and the proceedings took on a character of unrelenting partisanship as each side sought to bolster its case with the best science.

Despite being comprised of professionals in relevant fields, this was not an expert commission. Chaired by Lord Balfour of Burleigh, it consisted of a well-known geologist, Archibald Geikie; three civil engineers (only one of whom, James Mansergh, had much experience in water questions); a chemist, James Dewar of Cambridge; and a medical man, Dr William Ogle. Dewar and Ogle might be presumed most concerned with analysis and bacteriology yet neither had been much involved in the past decade's water controversies. Both sympathized with the companies.

The expert witnesses were as polarized as ever. According to most measures, London's water had improved in the previous quarter-century, but there was even greater concern about its safety. More upstream towns had built sewage purification plants, but there were more water closets and sewers too, and the treatment plants were often poorly run. In terms of chemical analysis, the river water the companies took was worse, Edward Frankland reported.[18] But thanks to Percy Frankland there was now greater confidence in the filters, even though some experiments had shown that pathogens could multiply prodigiously in reservoirs such as those used for storing filtered water. One might find relief in the fact that germs were no longer hideous conjectures but real entities with known properties, yet some bacteriologists remained skeptical that the germs of typhoid and cholera had indeed been discovered and warned that even if they had one could never be sure of their absence.

For bacteriological testimony, the water companies relied mainly on Edwin Ray Lankester, a prominent biologist and longtime editor of the *Quarterly Journal of Microscopical Science*, and Percy Frankland (though indirectly Frankland was actually working for both sides, since the Royal Society's Water Research Committee, which had commissioned the great reports he and Ward were producing, had been initiated and financed by the LCC).[19] Odling, Crookes (Tidy had died in late 1892), and William Robert Smith (a physician who taught forensic medicine at King's College) were minor bacteriological witnesses. Among the LCC experts were Edward Frankland, E E Klein, the Local Government Board's chief bacteriological consultant (who had taken issue with the claim that Koch's *comma vibrio* was the cause of cholera), and German Sims Woodhead, director of the joint bacteriological laboratories of the Royal Colleges of Physicians and Surgeons and editor of Britain's first bacteriological journal, *The Journal of Pathology and Bacteriology*.[20]

The experts agreed on the main facts: the supply was contaminated with sewage. The sewage probably included pathogens on occasion, yet the supposed typhoid and cholera germs had never been discovered in London's water. They disagreed about what the facts meant, and even then it was not so much a matter of conflicting interpretations as of conflicting scenarios of what might conceivably come to pass. The most expert experts—Percy Frankland and Lankester for the companies and Woodhead and Klein for the LCC—dealt with a number of questions on which continental bacteriologists had amassed an abundance of contradictory results. The first of these had to do with claims that the causal agents of cholera and typhoid had been identified. By the early '90s Koch's postulates were well established among British bacteriologists. To prove that an organism was responsible for a disease one had to obtain it in pure culture from fluids or tissues of a victim of the disease, inoculate it in an experimental animal, and recover it in pure culture from the inoculated animal which had to manifest the disease.[21] Neither the cholera *vibrio* nor the typhoid *bacillus* had been confirmed by this means; no one had been able to produce an unambiguous case of these diseases in an experimental animal. The 'discoveries' could therefore be accepted only tentatively. One need not go so far as to reject the claims; one could simply point out that there was a realistic possibility that the amassed evidence on their presence or absence signified nothing. Such a possibility, masterfully developed by Klein, left the issue just as Edward Frankland had left it 25 years earlier: in the face of ignorance it was wise to take no chances.[22]

More important was another issue of determinative bacteriology, the question of the relation of the presumed typhoid organism (*B typhosus-abdominalis* as it was usually called) to the more common colon bacterium *B coli*. It was hard to distinguish *B coli* from *B typhosus*, either morphologically or in culture, and some felt that the two organisms were not therefore separate species but forms of a single widely varying species, which was 'liable to the profoundest modifications through changes of environment and other causes.'[23] The possibility that the colon *bacillus* could change into the typhoid organism was worrisome simply because *B coli* was so common. If at any time (or even under some set of particular yet unknown conditions) it could shift into its pathogenic *typhosus* mode the threat to public health (or more properly the threat to the ability of science to protect public health) became much more serious. This possibility had been raised by several French bacteriologists and was taken seriously well into the first decade of the twentieth century. Such a theory would explain why the typhoid *bacillus* was difficult to find even in discharges of typhoid patients and how it survived in between outbreaks of the disease. It was plausible in a loosely Darwinian context that emphasized variability and adaptation and became even more so with the appreciation that there were a host of coliforms, differing only slightly from one another.[24] Though there was no clear evidence that such a transformation could happen, and even though experimenters, particularly Klein, tried and failed to make it happen, the possibility continued to be raised and worried about.[25] As with the earlier issue of the identification of pathogens it was not necessary that those who raised this issue actually believe that the organisms could change into one another; it was enough to maintain, as Klein did, that the possibility had to be taken seriously in deciding how London was to be supplied with water.[26]

The emphasis on plausible scenarios of what might happen in the water made actual analyses, chemical or bacteriological, less important in the 1892–93 hearings than in earlier investigations. They were drowned out by the flood of qualifications that could be made. This is ironic; having finally discovered the identity of the material they wished to find, scientists devoted more attention to hypothetical questions of the possibility of its presence than to demonstrations of its presence or absence. One might think that such speculations would have been more appropriate at an earlier period before pathogens had been discovered. Yet the contrary was true: that earlier period had been marked by massive analytical campaigns, as if making *a great many measurements* could make up for one's ignorance of what one was *measuring for*. Thus, what bacteriology had disclosed was how little bacteriological analyses could be relied upon. A negative result, even if accepted as an accurate indication of the state of a large body of water at a given time, told one nothing of what might be in the water in future; a positive result might well come too late for preventive measures to be taken.[27]

The Transformation of Edward Frankland

Among LCC witnesses, Edward Frankland's testimony was particularly striking. It was only to be expected that Frankland, the greatest advocate of uncorrupted water, would testify for the LCC. But Frankland's views were changing. In early 1891 he had begun a microbe-counting programme in conjunction with his analyses of the London waters. It is likely that the analyses were begun as a new stratagem in his long campaign against river water: at first he made only post-filtration counts on samples taken from standpipes, which probably contained (as he would later admit) more microbes than would have been present immediately after filtration. Frankland's first report (January) shows that he was still astute in choosing units to convey a vivid image. He supplemented the usual microbes per cc with microbes in 1/500 of a pint, and noted elsewhere that 'an ordinary tumbler contains about 250 cubic centimetres.' It was 'a plan eminently calculated to frighten people,' as Herbert Preston Thomas, the LGB official charged with explaining these returns to worried Members of Parliament, recognized. In the case of Percy Frankland's investigations, one could at least maintain the pretense that comparisons of the microbes before and after filtration revealed the effect of the filters, but these returns seemed intended only to shock. Advised of the concerns of LGB officers, Frankland omitted allusions to the number of microbes in water tumblers from subsequent reports.[28]

Throughout 1891 Frankland was kept busy by the LGB responding to complaints of vestry analysts and medical officers about water quality in their districts. Most persistent was Charles Cassal, prominent in the SPA and public analyst for Kensington, St George's Hanover Square, and Battersea, and said to have made his career 'by sheer force of character.' Frankland was coming to regard many of these complaints as trivial; Cassal and his friends were making much of vegetable pollutions, 'very undesirable, but only to the senses.'[29] In June 1892, about the time of his initial testimony to the Commission, Frankland began, at the request of the companies (and in some loose sense under their sponsorship) to list the microbe population of the raw waters from which the companies took their supplies. The reductions effected by filtration were even higher than Percy Frankland had recorded and he was struck as his son had been by the great effectiveness of the filters: 'if further observations confirm [that] the influence of filtration through a few feet of sand can be made to render river water bacterially as pure as deep well water, they must have a profound influence on the domestic water supply.'[30]

Hence when he testified to the Commission Frankland was working for both sides, though he officially represented only the LCC. During the enquiry he was continuing bacteriological studies for the companies, as well as consulting with them on filtering procedures. Over the months his views came increasingly closer to those of the companies' experts. In his first written statement to the Commission in the spring of 1892, before he began the pre- and post-filtration analyses, Frankland had taken his habitual line: the river was polluted by sewage; there could be no proof that 'noxious ingredients' had been removed. Hence the companies were wrongfully experimenting with peoples' health.[31] By June, when he testified on this statement he had begun comparative analyses and admitted to being 'somewhat astonished' by the results, yet still held that it was quite likely that pathogens might occasionally pass across the filters, multiply in the filtered water, and cause an outbreak of disease.[32]

When in February 1893 Frankland returned to testify on the filtration studies, his tone was quite different. Well-managed filters could, he asserted, give the public water as good as that from deep wells. He had also begun to reassess that network of possibilities and probabilities on which he had long condemned river water. He now thought pathogens probably were delicate and unlikely to multiply to any great extent in filtered water. The main objection to London's water was now only 'sentimental': it was unpleasant to drink purified sewage.[33] This 'moderation' would continue to the point where Frankland was virtually an apologist for Thames water. By the late '90s there was substantial concern among senior officials at the LGB that Frankland had deserted the side, and Richard Thorne Thorne, the Board's medical officer, demanded that Frankland omit bacteriological results from his official reports on the grounds that this work was financed by the companies.[34]

For the commissioners Frankland's moderating views were crucial. Without his righteous wrath, the LCC case was weak. Its representatives could do little more than rail against complacency. It did no good. In their report, the commissioners vindicated the companies, made much of nature's means of destroying pathogens, and pointed to the conversion of Edward Frankland, 'who, as well known, has been no sparing critic of the London water.'[35]

Stuck in the Mud, 1893–94

By no means did the Commission's verdict mark the end of agitation for better water and public control. The LCC kept the pressure on, securing another Royal Commission (to look into matters of finance) in 1897–98. Chemists and medical men kept complaining;[36] the companies' defenders (now Crookes and James Dewar, who had been one of the 1892 commissioners) kept reassuring.[37] In part, the campaign against the companies was based on a new analytical construct, the measurement of suspended solids, or 'mud.' The developer of the mud measure was the LCC chief chemist, William Dibdin, better known as a pioneer of biological sewage treatment. In the autumn of 1896 Dibdin began to study 'mud' in the water supplied to London consumers. From the amount of solid matter that could be filtered out he subtracted a constant figure representing ordinary inevitable mud, and multiplied the difference by 10 to indicate the wet weight of the mud. From these calculations he announced that metropolitan water consumers were subjected to 67.5 tons of preventable mud per year, precisely the sort of quasi-intelligible Franklandesque statistic that opponents would regard as unconscionably misleading. In fact, the measurement was not devoid of a bacteriological rationale. Dibdin argued that the enormous variability in the number of bacteria that appeared in a particular water was due in part to the fact that rather than floating freely bacteria frequently travelled in clumps, often clinging to some particulate raft. Pathogens might thus be smuggled into the water supply past analysts little concerned with suspended matters. In Dibdin's view, then, excess mud was inconsistent with a high quality water supply.[38] Dibdin's scheme was quickly rejected. His calculations of the metropolis' annual mud load were so full of unwarranted assumptions, and so patently partisan that he was 'laughed out of court.'[39]

Dibdin's initiative and its quick rejection are part of a more pervasive disenchantment with bacteriological water analysis during the early '90s. Sometime in those few years confidence that bacteriology would soon solve the problem of water safety peaked, and began to fall. Two factors were prominent in this disenchantment. The first was the co-option of bacteriology, its transformation into yet another medium for experts to disagree over the same set of issues they had been contesting with chemical theories and measurements. The second was the reappearance of major outbreaks of typhoid and cholera in circumstances that made it impossible to be so sanguine about the wonders of natural and artificial purification processes.

The first of these outbreaks was the typhoid that struck Darlington, Middlesborough, and Stockton in the Tees valley in September–October 1890 and again from December 1890 to February 1891. An LGB medical inspector, F D Barry, made an in-depth study of the epidemic and found the disease to be unmistakably water-borne, distributed in the filtered public water supplies of the three towns. In the first wave the attack rate of those using Tees water was ten times greater than those who did not; in the second wave it was twenty-eight times greater and there was ample evidence of sewage contamination upstream. Introducing the report, Thorne Thorne, the LGB medical officer, wrote 'seldom, if ever, has a case of the fouling of water intended for human consumption, so gross or so persistently maintained, come within the cognizance of the Medical Department, and seldom, if ever, has the proof of the relation of the use of water so befouled to wholesale occurrence of Enteric Fever been more obvious and patent.'[40]

The epidemic made untenable a number of facile generalizations— or at least it ought to have done so; its impact was lessened by a two-and-a-half year delay in getting the official report into print.[41] No longer could one legitimately claim that sand filters necessarily blocked out the microbes of typhoid. Nor could one put quite so much confidence in the subsidence of bacteria as a natural mechanism of purification; the first wave of disease had followed heavy rains which had probably re-suspended typhoid bacteria and sent them on downstream.

Among the most poignant inclusions in Barry's report were 'Reports of Analyses of Tees Water made in 1890–91.' Again and again during the epidemic analysts had found the water unobjectionable. On samples taken in late November, as the first wave was on the wane, W F K Stock of Darlington reported the presence of manifold traces of sewage in the raw water yet held the filtered water to be superb: 'As a matter of mechanical filtration the result *is simply perfection.*' At the end of January he again endorsed Darlington water, shortly after the second wave peaked. Edward Frankland analysed samples of Middlesborough water taken in August, shortly before the outbreak ('of an excellent quality for domestic use, and ... free from any trace of sewage or animal contamination'), in mid October, at the peak of the first wave ('free from every trace of previous sewage or animal contamination .... in all respects of excellent quality'), and late December ('of excellent quality'). A H Allen of Sheffield, a prominent member of the Society of Public Analysts, wrote that his results on a Middlesborough sample taken in mid October (likely a duplicate of the one sent to Frankland) 'negative any suspicion of contamination by sewage or cesspool drainage,' and that 'no suspicious results were obtained on bacteriological and other microscopical examination.'[42]

Still more sobering was a short report by Tidy of an 1887 investigation he had made on behalf of Darlington. As barrister, expert witness, water expert, chemist, and medical man, Tidy had been asked if legal action might successfully be taken under the 1876 Rivers Pollution Act against the town of Barnard Castle, the main source of pollution. He advised against action. In chemical terms the Tees was as pure or purer at Darlington as at Barnard Castle. These results, along with comparative health statistics and 'bacteriological examinations' convinced him 'that Darlington would not be prejudiced ... even if an outbreak of fever or cholera were to occur at Barnard Castle.'[43]

The contrast between this ongoing record of assurance and the typhoid in the three towns could scarcely be starker. To the medical chiefs at the Local Government Board, George Buchanan and later (in 1892) Richard Thorne Thorne, the epidemic confirmed longheld views of the utter futility of water analysis. During his tenure Buchanan had virtually ignored water analysis. In cases of waterborne disease, his inspectors confirmed their conclusions epidemiologically and through inspections of the circumstances of the water supply. Writing of the 1879 Caterham typhoid epidemic, which had spread over a large area from a minute contamination by a single well-digger with a mild case of typhoid, he had asserted that

> while we must ever be on the watch for the indications that chemistry affords of contaminating matters gaining access to our waters, we must (at any rate until other methods of recognition are discovered),

go beyond the laboratory for evidence of any drinking water being free from dangerous organic pollution. Unless the chemist is well acquainted with the origin and liabilities of the water he is examining, he is not justified in speaking of a water as "safe" or "wholesome," if it contain any trace whatever of organic matter; hardly indeed even if it contain absolutely none ... the chemist can, in brief, tell us of impurity and hazard, but not of purity and safety.[44]

Asked by the Royal Commission what was the best way to assess water quality, Buchanan told the commissioners that walking the river banks would be enough: chemists (or bacteriologists) could tell one little that was useful. Thorne Thorne too made little use of bacteriological or chemical analysis until 1894, when E E Klein perfected a technique for concentrating solid matters (including bacteria) in a sample, and culturing specifically for the typhoid *bacillus*.[45]

Even more important was the cholera that hit the German cities Hamburg and Altona in 1892–93. Like the 1854 cholera outbreak in south London, the Hamburg–Altona outbreak was a classic in the clarity of its epidemiological phenomena: two contiguous cities with very different water supplies—Hamburg drawing its water from upstream on the Elbe; Altona drawing its water from the tidal bore of the river, a few miles below the spot where both cities' sewers discharged, but carefully filtering that water. From August to November 1892 cholera flourished in Hamburg. There were more than 16,000 cases, over 8,000 deaths. In Altona there were a few hundred cases, most traceable to contact with Hamburg victims.

At first the epidemic seemed to vindicate Percy Frankland and the London water companies. Sand filters had protected Altona despite that city's reliance on an atrocious source of water. In February 1893, however, cholera returned to Altona, while Hamburg remained largely free of the disease. No longer did the filters seem so trustworthy. Thorne Thorne wrote in a vindictive tone of his own, of this vaunted sand filtration, a 'process which had formerly been deemed ... to remove ... the microorganisms.'

In fact, the implications of the Altona epidemic were unclear. It had been investigated by none other than Robert Koch, the premier bacteriologist of the day, who traced it to problems with the filters. This seemed to vindicate the companies, for Koch was confident that well-run filters would almost always prevent passage of the microbes of cholera and typhoid. But on the other hand, he admitted that failures of filters might be of brief duration and could not necessarily be avoided even by the best management (and the Altona engineers had been extremely scrupulous), a point Thorne Thorne saw fit to reiterate. Koch did, however, hold out the possibility that through more frequent bacteria counts (daily instead of weekly) filters could be monitored closely enough to prevent what happened at Altona (score another for the companies). But even this might not be sufficient, he finally declared: 'only subsoil water [such as the deep well water Edward Frankland had been insisting on] gives us absolute security ... and it should, therefore, ... be preferred under all circumstances to surface water.'[46]

Thus each side could draw support from Koch. In hopes that it would instill a healthy skepticism, Thorne Thorne had the report translated and published it in his annual report, yet Percy Frankland and others continued to cite the epidemic as proof of the benefits of filters and the importance of microbe counts.[47]

Perhaps the most important effect of Koch's dialogue between optimist and pessimist was to bring back to the debate a sense of the inescapability of living with imperfection and uncertainty. It had been hoped that bacteriology would bring certainty. But it could not, the experts admitted; the results of tests for the presence of the typhoid microbe had to be qualified nearly to the point of meaninglessness. For certainty they had substituted plausible fantasies, accounts of what might happen to bacteria, accounts that were so extreme, so dependent on fortuitous contingencies that they never could wholly be believed. Koch took neither extreme, or rather he took both. He showed that filtering was better than not filtering, that careful management of filters was better than *laissez faire*, that even careful management could not protect the public absolutely, and finally that when all was said and done, he, personally, would rather not drink this filtered water at all. Yet one had to live with uncertainty, to trust something less than rigorous demonstration, and be satisfied with estimates of risk. The central question among British water analysts in the mid '90s was what people, using what methods, and working within what sorts of institutions were to embody that trust.

The Society of Public Analysts and the Resurgence of Empiricism

In the view of many workaday public health officers and public analysts, Buchanan and Klein were right. Chemical and bacteriological techniques were of limited practical use. Analysts needed to give up boldness for wisdom, caution, and moderation. Evaluation needed to be more broadly based and less formulaic. Water science had to be decentralized, freed from its traditional partisan contexts, and made more flexible, better adapted for use in epidemiological hypothesis-testing. In early 1897 the editor of *The Engineer* caricatured 'the ardent bacteriological water examiner—commonly neither a chemist nor a bacteriologist—who is apt to be noisy and dogmatic, and from whom, most will agree, we have heard quite as much as is necessary for some time to come.'[48]

At the centre of this perspective was the Society of Public Analysts. In general its members took the view that too much had been expected of bacteriology and traditional methods had been too quickly rejected in its favour. But there were other components in their position. One was that bacteriological analysis usually represented analytical overkill: one could learn as much if not more from inspecting the source of the water. It was also felt that epidemiology had been wrongly neglected. The experts had been so full of theories of the conditions under which pathogens would be destroyed that they had failed to take into account the occasions when microbes had overcome the various barriers and gone on to cause epidemics.

The righteousness (and even belligerence) of the SPA is evident in Leo Taylor's 1892 paper on water analysis. A consultant analyst to the London suburb of Walthamstow, Taylor, like many metropolitan public analysts, had investigated company water during the years leading up to the RCMWS. He had followed SPA guidelines, which included neither bacteriology nor use of Frankland's combustion process. The Walthamstow authorities had sent his results to Frankland, who had objected that they were 'not complete' and hence useless in forming any 'trustworthy opinion as to the quality of the water.' As Taylor recognized, Frankland's appraisal carried legal implications; if Taylor were incompetent as Frankland implied, his analyses could no longer be accepted as grounds for closing polluted wells; he could no longer do his job effectively.[49] Outrage at Frankland's high-handedness was not new; what was new was bringing such a grievance to the SPA as the appropriate court of appeal. But to Taylor and other young water analysts like Charles Cassal, the SPA set the standards, and Frankland, however great his eminence and experience, had to be seen as shamefully and stubbornly clinging to the past. The SPA took no formal vote, but those who spoke did censure Frankland.[50]

The Taylor–Frankland clash reflects the seriousness of the split within the community of water analysts between those who served local health authorities and dealt with periodic epidemics or were involved in small-scale sanitary improvements such as securing the closing of polluted wells, and those who like Frankland and Tidy directed large and well-organized laboratories specializing in water work. As we have seen the former group needed easy, rapid, and adaptable processes; the latter stressed precision. The same split was also reflected in attitudes toward bacteriological analysis. To Klein and Woodhead the uncertainties of water bacteriology had simply indicated directions of research; imperfections in no way compromised the importance of bacteriology. To public analysts like Cassal the same uncertainties indicated that bacteriological water investigations were not worth the trouble.[51]

A pair of papers presented in the spring of 1895 illustrate the SPA perspective on bacteriological analyses. One was a sober and far-reaching assessment of the state of water analysis by John C Thresh, medical officer to the Essex County Council.[52] Thresh spoke as a health officer and was much less optimistic about any form of analysis than were many SPA members. Chemical analysis he regarded as 'useless' in arriving at a negative result (that the water was uncontaminated and hence safe), 'superfluous' for a positive result. Bacteriology was worse; there were too many species, the few deadly ones were too hard to find, and all species seemed too widely dispersed for the presence of any one of them to afford much information about pollution. The new science failed on the same grounds as the old one had: what mattered was the quality (species) of bacteria, but one could not test for that—waters known to cause disease repeatedly gave negative results. What one could measure was the number of bacteria, which meant nothing. The best one could do was take into account all evidence, giving precedence to a thorough inspection of the source of the water. But it was not just that such tests were uninformative, Thresh went on; substituting science for common caution could prove disastrous. He cited the case of Worthing, struck by typhoid in 1893. The source of the outbreak (water) had been detected and the populace advised to boil its water. When the epidemic had almost died out, a water sample was sent to a prominent London bacteriologist who pronounced it safe. Worthing began to drink the water again and typhoid promptly reappeared.[53]

The other paper was by August Dupre, for over a decade con-

TABLE OF ANALYSES OF POTABLE WATERS.

No.	Source of Locality	Appearance, etc.	Total Solids	Effect of Ignition	Nitric Nitrogen	Chlorine	Temporary Hardness	Total Hardness	Oxygen Consumed	Oxygen Consumed	Free Ammonia	Albuminoid Ammonia	Nitrites	Oxygen absorbed in four hours
			Results in Grains per Gallon.								In Parts per Million.			
1.	R. Ouse, Buckingham	Turbid and slight weedy odour	22·0	...	·014	1·2	...	...	...	...	·13	·30	Trace	1·30
2.	Ditto	Bluish tint, good in colour	37·5	...	·07	1·1	...	31·0	...	...	·09	·07	·0	·0
3.	Beverley Water Supply		26·0	...	...	1·55	19·9	25·6	...	...	·00	·01	...	...
4.	R. Trent, at Torksey	Turbid	26·4	...	·177	2·23	12·7	30·7	·12	·02	·07	...	...	1 34
5.	,, Korsith		26·6	...	·177	2·23	11·2	30·8	·12	·02	·09	...	...	1·29
6	Houghton-le-Spring supply	Colourless and nearly clear	34·4	...	...	2·1	...	...	...	...	·00	·03	...	·13
7.	R. Tees water, Middlesborough Supply, Aug., 1890	Pale brown, turbid, peaty taste	7·3	...	·00	·49	·0	3·9	·70	·047	·00	...	...	...
8.	R. Tees water, Middlesborough Supply, Oct. 22, 1890	Very slightly turbid, peaty taste	9·1	...	·00	·56	1·9	6·1	·30	·010	·00	...	...	...
9.	R. Tees water, Middlesborough Supply, Oct. 27, 1890	Cl- ar, dark yellow	10·8	...	·00	·50	...	...	...	...	·02	·12	·0	...
10.	R. Tees water, Middlesborough Supply, October 29, 1890	Light brownish yellow	12·0	...	...	·70	...	...	...	...	·01	·12	...	...
11.	R. Tees water, Darlington Supply, Dec. 2, 1890	Greenish yellow, clear and bright	10·5	...	·028	·7	...	...	...	...	·00	·04	...	3·0
12.	R. Tees water, Darlington Supply, Jan. 1, 1891	Clear, no peaty taste	13·4	...	·036	·81	4·8	8·2	·08	·020	·00	...	...	...
13.	R. Tees Water, Darlington Supply, Feb. 9, 1891	Brownish yellow, not quite clear	7·2	...	·911	·35	...	...	...	...	·03	·03	...	3·27
14.	R. Tees water, Stockton Supply, Aug., 1891	Dark brownish yellow, almost opaque	8·54	...	·00	·49	...	...	...	·49	·04	·11	...	10·4
15	R. Tees water, above Barnard Castle, 1887		9·3	...	...	·8	...	...	...	·354	...	...	...	4·16
16.	,, below ditto (Darlington), 1887		7·7	...	...	·65	...	...	...	...	...	...	...	4·01
17.	Mountain Ash Water Supply	Very pale brownish	6·2	Blackened	Trace	·67	1·5	5·0	...	...	·00	·014	·0	·31
18.	Massachusetts, Unpolluted Surface Water	Slightly turbid	3·6	...	·025	·16	...	...	...	...	·013	·033	...	...
19.	,, Polluted ditto		5·0	...	·020	·58	...	...	...	...	·002	·020	...	...
20.	,, Merrimack River		2·4	...	·010	·09	...	...	·8	...	·017	·12	...	...
21.	,, Chicopee River		2·8	...	·001	·09	...	...	·8	...	·002	·10	...	...
22.	,, Boston Supply. Purest reservoir		2·7	...	·004	·16	...	...	...	...	·014	·24	...	...
23.	,, Boston Supp'y. Mystic lake		7·4	...	·04	1·2	...	...	4·1	...	·21	·24	...	...
24	Essex. Deep well		95·0	Slightly charred	·13	25·6	4·0	4·0	...	...	·20	·05	·0	·75
25.	,, Same well, a few weeks later		113·0	Blackened	1·44	14·8	17·0	19·0	...	...	·01	·24·	·0	3·25
26.	,, Shallow well, near surface	Slightly turbid and yellow	94·0	Brownish	4·14	16·1	13·5	35·0	...	...	·02	·08	·0	1·75
27.	,, Sam? well, near bottom	Turbid and yellow	119·0	Charred	4·33	22·0	17·8	45·5	...	...	·02	·12	Trace	2·50
28.	Well at Melton Asylum, Oct. 11, 1893		42·	...	·0	12·7	...	...	...	...	·20	·09	...	·43
29	,, ,, Oct. 23, 1893		45·5	...	S.trace	6·4	...	...	...	...	·29	·24	...	·33
30.	,, ,, Feb. 9, 1894		23·	...	·0	7·2	...	...	...	...	·24	·20	...	·02

Figure 10.2 Lists like this one, of water-borne typhoid epidemics undetected by chemical analysis, brought home to some analysts at least, the enormous inadequacies of their science (*The Analyst*, 20 [1895]:103).

sultant on water analysis to the LGB. Dupre was militant, flatly maintaining the superiority of chemical water analysis. With chemical analysis one could detect sewage, which might sooner or later be contaminated with typhoid matter; bacteriology might allow isolation of the pathogenic organisms, but it was unlikely to do so in time to prevent an outbreak of disease. It could only confirm that water had been responsible, 'a fact which ... would in nine cases out of ten have been demonstrated by altogether independent investigation.' More troublesome was the fact that regardless of what the results of individual analyses might be, chemical and bacteriological approaches would tend to be utilized in public health administration in quite different ways. Bacteriological analysts emphasized the results of particular analyses as an indication of the general condition of a water, an unwarranted inference. This was true in regard both to filtration studies such as Percy Frankland had done and to attempts to detect pathogens. A high rate of microbial removal by a filter, or a failure to discover pathogens was taken to imply safety: because one was looking for the causal entities themselves (or close approximations), negative results were viewed as significant. On the other hand, by focusing solely on sewage contamination, almost always an antecedent to the presence of pathogens, chemists could legitimately make inferences from occasional analyses: the presence of sewage meant the water must be regarded as unsafe. 'Save us from the bacteriologist!' Dupre exclaimed.[54]

In the next three years members of the Society heard papers that confirmed them in their distrust of bacteriology. In 1896 bacteriologists T H Pearmain and G C Moor reviewed recent research on the relations of the typhoid *bacillus* to similar microbes. However much bacteriologists might be gaining confidence in their science, their view of nature was becoming more, not less complicated; new problems were being recognized faster than old ones were being solved. More types of colon bacteria had been discovered. *B typhosus* had still not yielded to Koch's postulates. Some research seemed to confirm that *B coli* and *B typhosus* were, if not altered varieties of a single species, very closely related in the immune response they produced in animals. The discovery of healthy carriers of typhoid complicated matters. Finally, microbes had also been discovered that produced symptoms similar to typhoid but had different characteristics in culture than the typhoid organism. 'It does not seem impossible that these pseudo-typhoid organisms may be the colon *bacillus* in transition stages,' they pointed out. Pearmain and Moor believed that it was possible and practical to determine the presence of *B typhosus* in water. Following the approach of Klein (and others), they suggested this could be done by filtering one to three litres of sample water through a bacterial filter and culturing the residue in a series of media uniquely suited to the typhoid microbe. But they did not think the procedure was necessary and denied that 'biological examination can in the smallest degree supplant ... chemical analysis.' The problem remained one of discovering the pathogen too late for the knowledge to be of any use.[55] Since typhoid had a two-week incubation period, analysts might only begin to search for the typhoid microbe after it had disappeared.[56]

Yet there remained a feeling among public analysts that some sort of bacteriological analysis ought to be done. Several spoke of using *B coli* as an indicator of sewage contamination but they were frustrated by the enormous variability and range of varieties of the species (if indeed there was only one). And already there was growing recognition that the microbe was not exclusively derived from the 'intestinal sources' that were the object of concern.[57]

In his 1898 annual address, Bernard Dyer, the SPA president, took up the problem—the feeling that some sort of bacteriology ought to be done, yet the recognition that bacteriological information fit into schemes of water evaluation in no recognizable way. Having reiterated the concerns raised in previous years Dyer noted that he nevertheless had 'the greatest possible faith in the value of bacteriology as an adjunct to the chemical analysis of water.' For several years he had been doing some simple bacteriological investigations of the waters he analysed—making the traditional microbe counts, counting colonies which grew in agar at blood heat (organisms that did well at body temperature would likely include those pathogenic to humans), and taking an interest in those that grew in an acid agar medium to which a small amount of phenol had been added (a rudimentary means for distinguishing the typhoid microbe). 'Any analyst who will systematically do this side by side with his chemical analysis, and steadily compare the results, will after a time feel himself very greatly strengthened in pronouncing his opinion on the great majority of cases that come before him,' Dyer predicted; any analyst who became accustomed to such a procedure 'would feel his judgement sadly lamed if he had for any reason to give up the practice, and to fall back upon his chemical results alone.'[58]

In some unexplained way, then, bacteriological analysis was useful. In part Dyer can be understood to have been defending the territory of the profession over which he presided: he noted that the simple tests he recommended were within the competence of public analysts, and properly their business. Yet Dyer's views belong also to a tradition of asystematic empiricism: additional information, no matter what kind, or how acquired, or what it signified, was to be regarded as intrinsically valuable and needing no further justification for its collection. Another SPA member, A H Allen, made this explicit: 'In water analysis it was impossible to have too many data upon which to base an opinion. ... Obtain as many chemical factors as possible, and, if bacteriology was likely to be of service, make use of it also.'[59]

Earlier I invoked a faith in empiricism to account for the mineral water chemists' willingness to accept apparently contradictory analytical results. Here it warranted inclusion of uninterpretable and hence unusable information. In both cases pragmatism underlay the perspective. However boldly they might boast of their chemical prowess, the mineral water analysts of the first half of the century recognized how rapidly their science was changing and were truly in doubt as to how much the composition of mineral waters varied. Dealing with matters of life and death in the communities they served, water analysts in the '90s faced a similar dilemma. They were torn between the temptation of finally fixing on an analytical system in which all possible complexions of danger were anticipated and every parameter had a definite meaning and level of significance, and the fear that for lack of looking for everything they might miss some unforeseen fingerprint of water-borne disease. It is this fear that is reflected in the comments of Dyer and Allen. They had greater faith that the true phenomena of water-borne typhoid would somehow surface out of the tangle of data than in their ability to sort out from that tangle the unmistakable signs of deadly germs.

In one sense, bacteriological water analysis, so heralded in the mid '80s as *the* solution to *the* problem, was a failure, relegated to a minor role in general water analysis and to the restricted context of filtration monitoring. Clearly it had not lived up to expectations: the recognition of the entities that caused the diseases feared for so long and even the development of means for discovering their presence had not significantly changed water analysis. Yet major and positive changes in the climate of water analysis in Britain did occur in the last few years of the century, and some of these were due to bacteriology and ironically to its failure. Water analysis was being taken more seriously and being more fully utilized in programmes for monitoring local water quality. Those programmes depended on (and augmented) the expertise of rank-and-file water analysts and medical men. No longer were analysts relying on (or seeking) simple means or arbitrary standards for distinguishing safe from unsafe water. The dispelling of the belief that there could be simple tests and universal standards led to far more careful and sober consideration of each case. It is in the '90s that one finds deep concern among analysts with the ethics of water analysis. As they became experienced in water analysis and began to understand the practical complexity and theoretical subtlety of judging water, analysts began more and more to acknowledge and worry about their responsibility to the public. As they recognized that that judgement lay in their own assessment of a range of evidence and could not be reduced to a bacteriological or chemical algorithm they took their judgements more seriously and made them more carefully.

There was also a decline in the intermixture of scientific questions about the accuracy or utility of processes of water analysis and political conflicts over control of water supplies. The experts still disagreed, but no longer was debate polarized between rivals who claimed that their unique access to truth implied the particular political and administrative arrangements they favoured. The moderation of Edward Frankland and the ability of the LGB and the water companies to arrive at a mutually acceptable means of filtration monitoring reflect this new civility. Also the great political conflict over control of the London water supply that had prevailed almost throughout the century would soon end with public purchase of the companies and establishment of the Metropolitan Water Board in 1904.[60]

1   Comments in P Frankland, 'Water Purification: Its Biological and Chemical Basis,' p 246.
2   Henry R Kenwood, *Public Health Laboratory Work*, pp 425–44; J A Wanklyn, 'Water Analysis,' *CN* 59 (1889): 46–7; Wanklyn's comments in P Frankland, 'Water Purification: Its Biological and Chemical Basis,' p 241.
3   G S Woodhead, *Bacteria and their Products* (New York, Scribners, 1891), p 190. On types of water bacteriology see Percy Frankland, 'The Application of Bacteriology to Questions relating to Water Supply,' p 370.
4   Woodhead in RC Metropolitan Water Supply, [*RCMWS*], Q 10676.
5   Crookes, Odling, and Tidy in *CN* 54 (1886): 44–5; P Frankland, 'Pathogenic Micro-organisms in Water,' *CN* 54 (1886): 72.
6   Crookes, Odling, and Tidy, in *CN* 54 (1886): 185. The experiments were also presented to the British Association (William Odling, 'Micro-organisms in Drinking Water,' p 544). On the microbe scare see 'Review of C E Parker-Rhodes, *Our Daily Water-Supply*, *CN* 55 (1887): 185–6; W B Carpenter, 'The Germ Theory of Zymotic Diseases,' pp 317–36; Thomas Watson, 'The Abolition of Zymotic Disease,' pp 78–96.
7   'Biological Examinations of Water Supplies,' *SR* ns 8 (1886–7): 120. But see Fox, *Sanitary Examinations of Water, Air, and Food*, 2nd edn, p 74.
8   P Frankland, 'Pathogenic Micro-organisms in Water,' *CN* 54 (1886): 72.
9   P Frankland, 'Recent Bacteriological Research in connection with Water Supply,' pp 319–22. See also his 'On the Application of Bacteriology to Questions relating to Water Supply,' pp 369–77.
10 Frankland and Frankland, *Micro-organisms in Water*.
11 For an excellent review see E Duclaux, 'Action sur l'eau sur les bactéries Pathogènes,' pp 109–24.
12 Frankland, 'The Bacteriological Examination of Water and the Information it has Furnished,' p 6.

13 P Miquel, *A Practical Manual of the Bacteriological Analysis of Water*, p 502; PRO MH 29 17, 79769K2/92, A deC Scott to H Owen, 31 August 1892. But see Frankland and Ward, 'First Report to the Water Research Committee of the Royal Society,' p 187.

14 Miquel, *Practical Manual*, p 419.

15 *Ibid*, 460–1; Fox, *Sanitary Examinations of Water, Air, and Food*, 2nd edn, p 74.

16 P Frankland, 'The Bacterial Purification of Water,' pp 90–6.

17 Luckin, *Pollution and Control*, pp 148–9; Owen, *The Government of Victorian London*, pp 135–40; A Mukhopadhyay, *Politics of Water Supply*, ch 3; PRO MH 29 14, Scott to LGB 14 February 1890, Scott to Owen 5 July 1890; vestry complaints 29 July 1890; PRO MH 29 15, 11265/91, 11448/91, 23787/91, 45003/91; PRO MH 29 16, 84364/91, 97333/91, 98128/91; 'Impure London Water,' *BMJ*, ii, 1892, pp 1069–70; 'Dangers of Rivers as Sources of Public Water Supply,' *BMJ*, ii, 1893,, pp 1112–3; 'The London Water Supply,' *The Times* 19 Sept 1893, 6 a–c; 20 Nov 1893, 4a; 'Metropolitan Water Supply,' *Engineering* 53 (1892): 265–6, 383–4; H W Dickinson, 'Water Supply of Greater London,' *The Engineer* 186 (1948): 432–3.

18 *RCMWS*, Appendices, C-10, C-11, C-13.

19 Royal Society of London, Water Research Cttee, Minutes and Letters and Papers, 1891–6.

20 On Woodhead (1855–1921) see *Proc Royal Soc Edinburgh* 42 (1921–2): 394–5. On Lankester (1847–1929) see Gavin de Beer, 'Lankester, Edwin Ray,' in *DSB*, vol 8, pp 27-7. On Klein (1844–1925) see *Proc Royal Soc London* B98 (1925): xxv–xxix.

21 G S Woodhead and A Hare, *Pathological Mycology*, I, pp 22–3; W Coleman, 'Koch's Comma Bacillus: the First Year,' *Bull Hist Med* 61 (1987): 315–42.

22 *RCMWS*, QQ 11009–11. Cf George Newman, *Bacteriology and Public Health*, p 282; Carl Fraenkel, *Text-Book of Bacteriology*, pp 267–8, 289–93; Frankland and Appleyard, 'Third Report to the Water Research Committee,' pp 396–7; Frankland and Frankland, *Microorganisms in Water*, pp 264–70.

23 Frankland and Appleyard, 'Third Report,' p 398. Cf Newman, *Bacteriology and Public Health*, p 50; Miquel, *Practical Manual*, pp 471–2; Fraenkel, *Text-Book of Bacteriology*, p 119.

24 Newman, *Bacteriology and Public Health*, pp 28–9, 50, 302–3; E E Klein, 'Report on the Etiology of Typhoid Fever,' in *22nd Annual Report of the LGB, Report of the MO for 1892–3*, pp 345–9; C Flügge, *Microorganisms*, p 334.

25 Turneaure and Russell, *Public Water Supplies*, p 182; F W Richardson, 'The Bacteriological Analysis of Water,' *J Soc Chem Ind* 13 (1894): 1159; Miquel, *Practical Manual*, pp 471–2; E Klein,'Report on the Eti-

ology of Typhoid Fever,' pp 345–65.

26 *RCMWS*, QQ 13235–7, 11009–11.

27 Cf Percy Frankland, 'The Application of Bacteriology to Questions relating to Water Supply,' p 370.

28 PRO MH 29 15 23787/91, 32189/91; *RCMWS*, Appendix C 14, p 205.

29 PRO MH 29 16 84364/91, Frankland to Owen, 31 Oct 1891; C H Cribb, 'Charles E Cassal (1858–1922),' *The Analyst* 47 (1922): 102–4.

30 PRO MH 29 17 61974/92, 73598/72.

31 *RCMWS* Appendix C 13, pp 201–2.

32 *RCMWS* Evidence, QQ 4244, 4259–65, 4283, 4403–6, 4523, 4554, 4561–5, 4588.

33 *RCMWS* Appendix C 14, p 208; RCMWS Evidence, QQ 12600–2, 12614–6, 12775, 12778.

34 PRO MH 29 34 129365/98 Scott to Owen 23 Oct 1898; 132373/98, 131311m/98; *Lancet*, ii, 1897, p 1055.

35 *RCMWS* Evidence, QQ 12873, 12876–9, 12885, 12927–9; *RCMWS* Report, pp 58–69.

36 *Times* 20 Nov 1893 4a; 'The London Water Supply,' *Engineering* 59 (1895): 281–5; 'London Water Supply,' *Lancet*, ii, 1896, pp 1165–6; P Frankland, 'The London Water Supply and its Bacterial Contents,' *Lancet*, ii, 1896, pp 1414–5; Crookes and Dewar on the London Water Supply, *Lancet* ii, 1896, p 1478; 'Water Supply,' *The Engineer* 83 (1897): 18–9; *The Engineer* 83 (1897): 496; 84 (1897): 81; *BMJ*, i, 1897, p 679.

37 'London Water,' *The Engineer* 83 (1897): 115.

38 W J Dibdin, 'The Character of the London Water Supply,' *J Soc Chem Ind* 16 (1897): 9–15; 'London Water,' *The Engineer* 83 (1897): 116. Cf C Hamlin, 'William Dibdin and the Idea of Biological Sewage Treatment,' *Technology and Culture* 29 (1988): 189–218.

39 'London Water,' *The Engineer* 83 (1897): 196.

40 R Thorne Thorne, comments on F D Barry, 'Enteric Fever in the Tees Valley,' in *Supplement to 21st Annual Report of the LGB, Report of the Medical Officer for 1891*, pp vi–viii.

41 Barry delayed until he could get fresh census data for accurate epidemiological comparison (Report of the MO for 1891, p viii).

42 Barry, 'Enteric Fever,' pp 119–23.

43 *Ibid*, p 136, see also pp 127–33 for the bacteriological investigations to which Tidy is probably referring.

44 Buchanan in *11th Annual Report of the LGB, Report of the MO for 1879*, p xxi.

45 See Theodore Thomson, 'Report on an Epidemic of Enteric Fever in the Borough of Worthing,' in *23rd Annual Report of the LGB, Report of the MO for 1893–4*, pp 47–80; E E Klein, 'Report on the Etiology of Typhoid Fever,' pp 345–65; *idem*, 'On the Behaviour of the Typhoid Bacillus and Koch's Vibrio in Sewage,' in *24th Annual Report of the*

LGB, Report of the MO for 1894, pp 407–10. For views on water analysis see *15th Annual Report of the LGB for 1885–86*, p cxxvi; and *Supplement containing the Report of the MO for 1885*, p iv; *17th Annual Report of the LGB*, pp cliv–clv, and *Supplement containing the Report of the MO for 1887*, p xvii; *18th Annual Report of the LGB for 1888–89*, pp cxlvii, 242–3; *RCMWS*, App E, 528–32.

46 R Koch, 'Water Filtration and Cholera,' trans A J A Ball, in *22nd Annual Report of the LGB, Report of the MO for 1892–3*, App C; Thorne Thorne in *ibid*, pp xix–xx, and Thorne Thorne in *23rd Annual Report of the LGB, Report of the MO for 1893–4*, p xxiii. Cf Richard J Evans, *Death in Hamburg: Society and Politics in the Cholera Years, 1830–1910* (Oxford: Clarendon, 1987).

47 For reactions see 'Water Filtration and Cholera,' *Lancet*, ii, 1893, p 96; G Frankland, 'How cholera may be spread,' *SR* ns 14 (1892–3): 421–2; P Frankland, 'The Bacterial Purification of Water and Sand Filtration,' *SR* ns 19 (1897): 8–9, 29–30; Wolf Defries, 'The Purification of Water Supplies,' *SR* ns 19 (1897): 111; Frankland and Frankland, *Micro-organisms in Water*, pp 152–4.

48 'London Water,' *The Engineer* 83 (1897): 116, 196–7.

49 On Taylor see D T Lucke, 'Leo Taylor,' (1861–1940), *The Analyst* 66 (1941): 184.

50 L Taylor, 'Reports on Water Analysis,' *The Analyst* 17 (1892): 89–95. Cassal had had a similar run-in with Frankland in the summer of 1891 (PRO MH 29 16 84364/91).

51 C E Cassal, 'Chemical Analysis and the Purity of Water,' *CN* 64 (1891): 249.

52 See also Thresh, 'The Interpretation of Results in Water Analysis,' *TSIGB* 13 (1892): 316–7. On Thresh see 'John Clough Thresh,' *The Analyst* 57 (1932): 549–50.

53 J C Thresh, 'The Interpretation of the Results obtained upon the Chemical and Bacteriological Examination of Potable Waters,' *The Analyst* 20 (1895): 80–91, 97–111.

54 A Dupre, 'Note on the Chemical and Bacteriological Examination of Water, with remarks on the Fever Epidemic at Worthing in 1893,' *The Analyst* 20 (1895): 73–9.

55 T H Pearmain and C G Moor, 'The Bacteriological Examination of Water for the Typhoid Bacillus,' *The Analyst* 21 (1896): 117–22, 141–48. For other views on the complexity and futility of searching for pathogens see M A Adams, 'Water Supply in Relation to the Maidstone Epidemic,' *The Analyst* 23 (1898): 153; 'Review of J H Fuertes, *Water and Public Health*,' *The Engineer* 84 (1897): 112; N F Bunshaw, 'The Examination of Water suspected of being contaminated by the Cholera virus,' *Lancet*, ii, 1896, 329–30; William Dibdin, *The Purification of Sewage and Water*, 3rd edn (London: Sanitary Publishing Co,

1903), pp 257–8; Newman, *Bacteriology and Public Health*, pp 28–9; E Klein, 'The Etiology of Typhoid Fever,' *J Sanitary Inst* 15 (1894): 343–52; Flügge, *Microorganisms*, p 334; 'The Scope and Methods of Water Examination,' *BMJ*, ii, 1893, 1118; Foster, *History of Clinical Pathology*, pp 70–1.

56 Pearmain and Moor, 'Bacteriological Examination,'* p 147. Those taking elementary bacteriology courses at the London hospitals would have learned procedures for distinguishing *B typhosus* from *B coli*, but not the great range of techniques Pearmain and Moor considered. See A A Kanthack and J H Drysdale, *Elementary Practical Bacteriology* (London: Macmillan, 1895), pp 53–5.

57 Adams, 'Maidstone Epidemic,'* pp 153, 156, 159.

58 B Dyer, 'Annual Address of the President,' *The Analyst* 23 (1898): 63–4.

59 A H Allen in Adams, 'Maidstone Epidemic,'* p 159.

60 But see Tebb, *Metropolitan Water Supply*, p 19.

Conclusion: What are Experts For?

Let me suggest four generalizations about the role of water analysis, particularly in the second half of the century during which a scientific breakthrough—bacteriology—is usually thought to have occurred. It would appear that:

(1) Good water analysts at the end of the century were probably about as good at detecting dangerous contamination as were good water analysts fifty years earlier. In a great many cases contamination was obvious without analysis, in a few it was virtually undetectable (except epidemiologically), and in the rest it was clearly indicated in a number of different analytical variables—high levels of chlorine, nitrates, and especially nitrites and ammonia, high albuminoid ammonia, a high bacterial count, and a characteristic microscopic appearance. All but two of these methods—bacterial counts and albuminoid ammonia—were available in 1850. To a substantial degree analysts in 1850 would have made the same decisions on the same bases as analysts in 1900.

(2) At the most important level of evaluating sources of water, views on which waters were good and which were bad (and in a broad sense what made water bad) changed relatively little during the period. Faecally contaminated water was held to be bad in 1828 (and doubtless earlier); it was still the problem in 1900, although admittedly it was seen as a more serious problem. The definition of bad water was more refined and the recognition of bad water less accessible to lay persons.

The more important changes were in the circumstances and institutions of water analysis. These are as follows.

(3) There were by the end of the century many more medically qualified people taking a regular interest in water quality and holding one another to a more severe standard of accountability. There had been a transformation from the making of blanket statements about

the safety of a particular source of water to accounting for individual outbreaks of water-borne disease. This made for a much more intimate level of assessment and for an agenda more concerned with small-scale deviations in disease incidence than was characteristic of the grand legal, financial, and political conflicts that had provided the occasion for analysis earlier in the century. Water analysis was also more closely tied to practical technical actions—the closing of a well, or the repair of a main, in contrast with the earlier period, when analysts had attempted to magnify issues by claiming either that things were supremely satisfactory or verging on disaster. It is not clear that more water analyses were being done, but more water supplies were probably being regularly monitored.

(4) Analysts in the 1890s were more united than they had been two decades earlier, and more modest. As the contexts of analysis changed the attitudes of analysts changed too. It is hard to take seriously the claims of Wanklyn and Brande that they had achieved perfection in their analytical processes; their statements appear to reflect professional machismo and self-promotion, and even an arrogant cynicism toward their roles as advisors on matters of health. They may have known that their advice was not being taken seriously, did not deserve to be taken seriously, and felt free of moral burden. Surely, this is too simple, but it is clear that the forces that had led analysts of the '60s and '70s to make extreme statements guaranteeing safety or prophesying doom were no longer dominant, perhaps because the issues occasioning analyses were no longer so large.

One might make a number of observations about this story but for me the most important relate to the nature of expertise and of the application of science. In that regard the stance of the Society of Public Analysts toward water analysis as a matter of the experienced application of a number of available techniques to problems of epidemiology is a proper place to end this work. John C Thresh, the main British authority on water analysis of the early twentieth century, emerged from this tradition and it is the hallmark of the first edition of his great treatise on the *Examination of Waters and Water Supplies*. What this means is that the great breakthrough never did occur; one can find in the water literature of the early twentieth century a number of significant developments, yet none of them significant enough to warrant the sense of absolute security from water-borne dread that most of us in the developed world now hold, or have held until recently. Analysts never did find the simple and readily discoverable signature of deadly water, but they did become confident in the use of coliform counts as an indicator of danger. For a time, during the 1920s and '30s, there was hope that a particular coliform associated exclusively with faecal contamination could be found, but all the many types of coliforms turned out to be too widely distributed for this to be possible. New purification technologies also relieved much of the anxiety about water quality. Cheap and effective biological sewage treatment processes were available after 1895, and chlorination after 1912. Some of the more explosive political issues, such as ownership of London's water supply, were also finally resolved (that supply became public in 1904).[1]

But even if there was no breakthrough there was a great increase in knowledge of the effects of water on health during the nineteenth century. Regarded as of little importance in 1800, faecally contaminated water had been recognized by 1900 as one of the principal sources of disease in humans. The causal agents of diseases barely defined at the beginning of the century were by its end recognized and detectable. And most importantly a much larger population had access to purer waters; the water-borne diseases that had ravaged the population during the nineteenth century have been insignificant in the developed world for most of the twentieth.[2] But what was the relationship between discovery and these political and technical actions?

It has been easy to think of this achievement in terms of a sequence of discovery, recognition of implications, and action, in the way the application of science is usually conceived to take place. For example, Colin Russell writes 'at first chemical and then biological analysis of water supplies, together with new methods of purification and an understanding of the aetiology of many infectious diseases led to insights into what was necessary for a healthy population to exist under urban conditions.'[3] This view presumes that knowledge (of water contaminants or of the effects of contaminated water) guides action (to secure better water). Yet for several reasons it is a view difficult to apply to the history of water supply in Britain and particularly to the history of water analysis. First of all is the problem of what 'knowledge' was at any given point in the story. The old view attributed to water science a false unity, when the period was in fact one of great conflict over who would speak for science. There is also the problem, both historically and in the present, of whether scientific discoveries implied certain technical actions. Discoveries that appear to us clear and compelling in their implications—Snow's discovery of the transmission of cholera by water, for example—did not quickly lead to the safeguarding of public water supplies. It may be argued that this lag is quite understandable: the inertia of invested capital had to be overcome. Unquestionably this is true, but it begs the question, for one of the means by which the resisters resisted was to appeal, successfully, to science for arguments that would undermine the arguments of those urging radical change. Science did sanction action (and inaction), but all parties in conflicts over water appealed to it for sanction, and almost always scientists were able to oblige. Finally it is usually thought that the growth of knowledge will strengthen the mandate for positive social changes, in this case the securing of better water. Yet this was not always the case. Percy Frankland's empirical studies of the effects of filtration on bacterial populations were truly significant contributions to knowledge, yet they discomfited proponents of purer water supplies because they seemed to invite reliance on a fragile technology. A much stronger mandate for better water came from Edward Frankland's speculations on germs, even though these were formulated on the basis of very little knowledge of the nature of pathogenic microbes.

Thus the claim that better water came from science is true, but not very informative. In the institutions in which water policy was made in nineteenth century Britain not only were there not the means for determining which knowledge claims were reliable, but those institutions—Parliament, the Local Government Board, the courts, and official arbitrators—actively encouraged the proliferation of conflicting knowledge claims by allowing the adversary process to become central in the making of water policy.

But it would be just as wrong to say that science was unimportant in these changes because it supplied no firm and uncontested answers. The very fact that each party found it necessary to ground its case in science, that enormous numbers of water analyses were done in the name of making sound policy, is of great significance. What science contributed was confidence. Having options battled over by means of scientific arguments assured that a satisfactorily rational process of decision-making had been gone through. In the context of policy-making science therefore served as a symbolic technology, a tool useful in securing a social end of taking a technical action with respect to the provision of water.

That confidence was in the ideal of science. Scientists themselves were struggling continually—as individual rivals, or as representatives of different methodologies or disciplines—to embody the authoritative ideal of science. Would being scientific mean upholding the chemical reductionism of Bergman, the holism of the spa doctors, or the devastating criticisms of Murray? Could Pearson, Gardner, Brande, and Taylor maintain the prestige of mineral water chemistry in the new context of evaluating public water supplies, or would they be forced to yield to the microscopy of Smith and Hassall, or to Chadwick's attempt to undercut the chemists by proposing hardness as the significant standard of quality? Would organic nitrogen prevail over that widely popular artefact of analysis, albuminoid ammonia? And finally, was it scientific (or more scientific) to make pronouncements on water policy on the basis of the number of culturable bacteria in a cubic centimetre of water or the failure to grow typhoid microbes in an acid agar culture at blood heat? In each case proponents of these processes were pressing for symbolic authority. They were arguing that the recommendations they gave represented, hence could be taken as a symbol for, some process of investigating water that validated their advice.

It is hard, especially in retrospect, not to think of symbolic as 'only symbolic,' and to see the role that science had in nineteenth century water policy as a poor substitute for the role it ought to have had, if only someone had taken the responsibility for making water analysis the subject of an independent, systematic, well-funded and long-term inquiry even larger and longer than that which Mallet and his associates undertook in the United States. Yet if it is true that the problems facing water analysts are ultimately problems of policy, and consequently trans-scientific, then we must accept that their resolutions will continue to be symbolic. No one, now or then, will be able to guarantee an unqualified assurance that our environment is safe, always and in all respects. We will always be stuck with having to decide whether the science we have is a strong enough symbol of certainty to undertake the action proposed. If we do re-acquire a complacency in our water, it will have come from a decision that the mode of decision-making we have comprises due process and symbolizes rationality.

What is remarkable then about the state of water analysis in Britain at the end of the nineteenth century was the appeal to embody symbolic authority in persons rather than in processes. The sort of water analyst envisioned by Thresh and the leaders of the SPA was one who recognized the advantages, disadvantages, and particular uses of all processes, and who understood too, that analysis was at best an imperfect science limited to a single sample, and that pronouncing on the safety (or danger) of any water had grave moral consequences. This perspective toward authority is significant enough to warrant a brief comment. It represented the acknowledgement by experts that making decisions about water involved much the same weighing of uncertainties and values as decision-making in any other important area of ordinary life. Bruno Latour has described the adoption of the Pasteurian outlook in France as an extension of the laboratory into the world, and the concomitant acceptance of its definitions, possibilities, and explanations.[4] The reverse took place in British water analysis. During the '60s, '70s, and '80s the categories of the laboratory had been primary. Ordinary people were expected to define what could happen to them by drinking a particular water in terms of some infinitesimal amount of ammonia or organic nitrogen that some chemist claimed to find in that water (or later, to believe that what grew on a gelatine–peptone culture plate was what determined whether a water was harmful). By the mid '90s, water analysts had become less interested in determining what could happen, and more concerned with what was happening, or had happened, and with how to make wise judgements. They became increasingly explicit about acknowledging that giving advice involved a moral responsibility. So long as it had been possible to maintain the illusion of certainty, rendering a verdict had been morally uncomplicated: the analyst told the truth, nature was responsible for the state of things. Writing in 1893, the American analyst A R Leeds expressed the new view: 'with our daily growing knowledge of the etiology of disease, and the discovery of the fact that the water which we daily use in our homes ... is the most widely diffused and general bacterial culture fluid, the feeling of responsibility connected with water diagnosis steadily increases.' Though Leeds had twenty years' experience in water analysis, he wrote that he felt 'a more painful sense of this responsibility, and am at times more embarrassed in arriving at a decision than when I made my first analysis.'[5] We must act, Leeds was recognizing; doing nothing was itself an action and we cannot wait for confirmation or certainty.

With respect to issues of environmental health policy, the historical present in which we now exist resembles the historical present of the mid nineteenth century. Although we claim now to know a great deal about the relations of science and society, the insights and perspectives of contemporary historians and philosophers of science are of little help; even in this age of relativism and constructivism we continue to look to science for certainty—very much the kind of certainty bacteriology has been seen to exemplify—when questions arise of what we are to do to protect ourselves and maintain our world. We too find in science only a welter of conflicting personalities, methodologies, and institutions. Daily we encounter sophisticated attempts to manipulate our thinking couched in the idiom of science, frequently in some form of analytical science. It is appealing to think that there might be a more satisfactory form of symbolic authority, but in modern western nations it is not clear in what institutions such expertise would be embodied.

1   J C Thresh, *The Examination of Waters and Water Supplies*, (Philadelphia: Blakiston, 1904); S C Prescott, C-E A Winslow and M H McCrady, *Water Bacteriology with Special Reference to Sanitary Analysis*, 6th edn (New York: John Wiley, 1946), pp 142–206; Baker, *The Quest for Pure Waters*; C Hamlin, 'William Dibdin and the Idea of Biological Sewage Treatment,' *Technology and Culture* 29 (1988): 189–218.

2   It should be noted that recent scholarship (J A Hassan, 'The Growth and Impact of the British Water Industry in the Nineteenth Century,' *Economic History Review*, 2nd series, 38 [1985]: 531–47) has confirmed the view of the most important nineteenth century commentator (Arthur Silverthorne, *London and Provincial Water Supplies* [London: Crosby, Lockwood, and Co, 1884]) that the quest for better water was fuelled by economic rationality rather than a concern for health. The argument remains an ambiguous one, however.

3   Colin Russell, *Science and Social Change, 1700–1900* (London: MacMillan, 1983), p 257.

4   B Latour, 'Give me a Laboratory and I will Raise the World,' in K Knorr-Cetina and M Mulkay, eds, *Science Observed: Perspectives on the Social Study of Science* (London: Sage, 1983), pp 141–70.

5   A R Leeds, 'A Question of Water, Ethics, and Bacteria,' *American J of the Medical Sciences* 105 (1893): 259, 266.

Appendix   Edward Frankland's Justification of Analytical Interpretations, 1868–76

Examination of Edward Frankland's explanations of his interpretations of water analyses during these years makes it clear that in principle any given analytical profile could be taken either as evidence that a water was good or that it was bad. As Frankland admitted, a water's history was what mattered; knowledge of that history determined the routes of inference he would make in interpreting a water analysis.

Frankland regarded the sewage contamination of the Thames and Lea as 'self-evident from the circumstance that sewage flows into the Thames and Lea.' Table 1 shows how he interpreted his results to confirm that conclusion. First he considered the concentration of organic nitrogen, presumed to represent 'actual sewage contamination.' Its presence he regarded as analytical justification for condemning the water.[1]

But such 'unoxydised sewage' as he called it was only rarely present in the London water supply, which was known to be polluted with sewage. Therefore Frankland looked for evidence of previous sewage contamination, a test 'second only in importance to [the search for] actual sewage contamination.' If these nitrates, nitrites, and ammonia were present in quantities greater than could be explained by rainfall input there were grounds for condemnation. While these substances were not regarded as harmful in themselves, they were regarded as indicators of sewage pollution. Frankland argued that undetectable germs that had entered a river with sewage might remain intact long after dead organic matter in the sewage had been converted to inorganic forms. He also argued that conver-

Table 1 Frankland's explanation of analytical results for waters presumed bad (composite diagram).

Test	Result	Verdict	Explanation
1. Organic nitrogen	Present	Bad	Represents sewage contamination
	Absent	Go to test no 2, PSC	
2. Previous Sewage Contamination (nitrates, nitrites, ammonia)	Present	Bad	Organic nitrogen may not always be entirely oxidized, or resistant and undetectable germs may still be present
	Absent	Bad	PSC is a minimum; vegetation may have removed all nitrates, etc, but germs may still be present

sion might not always be complete, so that, even if one believed the dangerous matter in water were some form of putrefying matter, the finding of nitrates was still ground for condemnation.[2]

In some cases, however, Frankland found little evidence of previous sewage contamination in waters known to be contaminated. The water could still be condemned in such cases since PSC was to be understood as indicating the minimum amount of sewage and animal wastes with which a water was contaminated. Aquatic vegetation would remove nitrates from water; it was also possible that in a very heavily polluted environment with low oxygenation levels nitrates might be reduced in oxidizing more unstable matters. In such a case an increase in putrescible matter would be marked by a decrease in PSC, though such contamination would probably be evident in the finding of an unusually large quantity of ammonia. The polluted Lea water supplied by the East London Water Company—the very water that had caused the cholera of 1866—sometimes showed no previous sewage contamination during the summer months, yet Frankland continued to condemn it (much to the astonishment and outrage of the companies' chemists, it need hardly be said).[3]

There were also cases of waters with good histories and bad analyses; such was the case with most of the waters Frankland advocated as alternatives for the London supply. In the cases of the mid Wales and Cumberland watersheds, these waters were heavily contaminated with peat. In the case of deep chalk aquifers, they were heavily laden with fossil nitrates. Frankland's interpretations

Table 2 Frankland's explanation of analytical results for waters presumed good (composite diagram).

Test	Result	Verdict	Explanation
1. Organic nitrogen	Present	Go to test no 2; organic carbon	
	Absent	Go to test no 3, PSC	
2. Organic carbon	High	Good	C:N ratio is high for harmless vegetable organic matter
	Low	Good	During oxidation of vegetable organic matter carbon oxidizes faster than nitrogen. During oxidation of animal organic matter nitrogen oxidizes faster than carbon. Therefore, ratios of organic elements of animal and vegetable extracts will become increasingly similar. Old sewage looks analytically like old peaty water.
3. Previous Sewage Contamination (nitrates, nitrites, ammonia)	High	Good	As with Kent Company water sometimes PSC is still high after extensive filtration which can be assumed to have removed all disease germs.
	Low or absent	Good	

of such waters are outlined in table 2.

If the waters had substantial previous sewage contamination levels (but no organic nitrogen) he argued that in some cases nitrates, etc, were simply not indicators of danger. The water the Kent Company obtained from deep wells in the chalk Frankland regarded as safe on the grounds that it was highly unlikely that germs could percolate through great thicknesses of chalk. Nevertheless, these high PSC levels remained an embarassment, and it is likely that his switch in 1876 from PSC to Table E, in which the organic elements in the Kent's water were set at unity and the organic elements in the other companies' waters represented on that scale, was prompted by the need to find a less ambivalent measure with which to represent the superiority of the Kent's waters.

In the cases of the northern and western waters the problem was not so much high levels of PSC but high levels of organic nitrogen, derived from peat. Inasmuch as the nitrogen was of vegetable origin and the water came from regions sparsely inhabited by humans or animals there was every reason to regard it as good water. In such cases Frankland turned to the carbon:nitrogen ratio. Peaty waters usually contained far more carbon than nitrogen and Frankland used this characteristic to distinguish waters polluted by vegetation from waters polluted by sewage. Yet as he gained experience Frankland began to encounter cases where this rule did not hold. Sometimes wholly blameless waters produced carbon:nitrogen ratios giving them an analytical profile identical to sewage-contaminated water. The reason for this, Frankland believed, was to be found in the contrasting courses of decomposition of vegetable organic matter as opposed to animal organic matter. In peaty waters organic carbon disappeared more rapidly than organic nitrogen while in waters contaminated with animal matter such as sewage or manure organic nitrogen disappeared more rapidly than organic carbon. Hence as time went on vegetable and animal extracts gave increasingly similar analytical returns. On this basis Frankland could argue that an unpolluted peaty water with the analytical characteristics of a sewage-polluted water was simply old.[4]

1   R C Water Supply, Evidence, QQ 6223, 6405–8.
2   R C Water Supply, Appendix D, p 20.
3   R C Water Supply, Appendix AK, pp 78–9.
4   R C Water Supply, Evidence, Q 6291; R C Rivers Pollution, 1868, 6th Report, pp 6–8.

Bibliographic Essay

The character of the sources of this work makes a normal bibliographic format unworkable. Brief articles in trade papers, either untitled or entitled 'Water Analysis,' 'London Water,' or something equally unuseful, make up a large portion of the sources, and in many cases the most important part. This bibliographic essay leaves out biographical material, primary sources considered only briefly, and most secondary sources, which are fully cited in chapter endnotes. It also omits a great many brief articles in such periodicals as the *Sanitary Record*, *Chemical News*, the *British Medical Journal*, the *Lancet*, the *Engineer*, *Engineering*, the *Builder*, and the *Times* newspaper. As these periodicals are well indexed (in the case of *Chemical News* there is a cumulative index for the first hundred volumes) and as so much of the material in them reflects the day to day business of sanitary science, including all citations would have left a bibliography prohibitively long. I have hoped to give some sense of bibliographic possibilities (mainly with respect to printed materials) and to provide full citations for less well-known materials. All author's names are listed in the index; place of publication is London unless otherwise listed.

1. Parliamentary Papers

1828   Royal Commission on Metropolitan Water Supply, Report and Minutes of Evidence, 9, (267.) [*RCMWS*]

1834   S C Metropolis Water, Report and Minutes of Evidence, 15, (571.)

1840   S C (Lords) on the Supply of Water to the Metropolis, Report and Minutes of Evidence, 12, (354)

1850   Report of the General Board of Health on the Epidemic Cholera of 1848 and 1849, 21, [1273]

General Board of Health, Report on the Supply of Water to the Metropolis, 22, [1281]

1851   S C Metropolis Water Bill, Report and Minutes of Evidence, 15, (647)

1852   S C Metropolis Water Bills, Report and Minutes of Evidence, 12, (395 and 395-I)

1854–5   General Board of Health Medical Council, Report of the Committee for Scientific Inquiries in Relation to the Cholera Epidemic of 1854, 21 [1980] and Appendices [1996]

T E Blackwell, Report to the President of the GBH on the Drainage and Water Supply of Sandgate in Connexion with the Outbreak of Cholera in that Town, 45, (82)

1856   Reports to the Rt Hon William Cowper M P, president of the General Board of Health, under the Provisions of the Metropolis Water Act, 52, [2137]

1857   Report to the Rt Hon William Cowper M P, president of the General Board of Health, on the Microscopical Examination of the Metropolitan Water Supply, under the Provisions of the Metropolis Water Act, sess 1, 13, [2203]

1860   S C on the Serpentine, Report and Minutes of Evidence, 20, (192)

Second Annual Report of the Medical Officer of the Privy Council, 29, [2736]

1865   Royal Commission on the Cattle Plague, Third Report and Minutes of Evidence, 22, [3656]

1866   Royal Commission on Rivers Pollution (1865), First Report, with Minutes of Evidence, appendix, and plans, 33, [3634 and 3634 I]

S C on the Thames Navigation Bill, Report with Minutes of Evidence, 12, (391)

1867   Ninth Annual Report of the Medical Officer of the Privy Council, 37, [3949]

Royal Commission on Rivers Pollution (1865), Second Report (River Lee) with Minutes of Evidence, 33, [3835 and 3835-I]

S C on the East London Water Bills, Report, proceedings, minutes of Evidence and appendix, 9, (399)

Correspondence between the Board of Trade and the East London Water Works Company with reference to

Captain Tyler's Report on the water supplied by the Company, 58, (574)

Report by Capt Tyler to the Board of Trade on the Quantity and Quality of the Water supplied by the East London Water Works Company, 58, (339)

1867–8   William Farr, Report on the Cholera Epidemic of 1866 in England, supplement to the 29th Annual Report of the Registrar General of Births, Deaths, and Marriages in England, 37, [4072]

S C on the River Lea Conservancy Bill, Report and Minutes of Evidence, 11, (306)

1868–9   Royal Commission on Water Supply, Report with Minutes of Evidence and Appendices, 33, [4169] and [4169-I] and [4169-II]

1870   Royal Commission on Rivers Pollution (1868) First Report with Evidence and Plans, 40, [C-37] and [C-109]

Twelfth Report of the Medical Officer of the Privy Council for 1869, 38, [C-208]

1871   Thirteenth Annual Report of the Medical Officer of the Privy Council for 1870, 31, [C-349]

1872   Copy of any Reports to the Board of Trade made by the Water Examiner under the Metropolis Water Act, 1871, 49, (82)

Copies of a Letter from the Royal Commission on Rivers Pollution to the Board of Trade on the 10th Day of April [1872] and A Report Thereon by the Water Examiner on the 26th Day of April [1872], 49, (186)

1873   S C on the Pollution of Rivers, Lords Papers, 9, (132)

1874   Royal Commission on Rivers Pollution (1868) Sixth Report, 33, [C-1112]

1880   Ninth Annual Report of the Local Government Board. Supplement containing the Report of the Medical Officer for 1879, 27, [C-2861-I]

1881   R Angus Smith, Rivers Pollution Prevention Act, 1876. Report to the Local Government Board, 23, [C-3080]

1882   Eleventh Annual Report to the Local Government Board. Supplement containing the report of the Medical Officer for 1881, 30 pt ii, [C-3337-I]

1884   R Angus Smith, Rivers Pollution Prevention Act, 1876. Second Annual Report to the Local Government Board, 19, [C-4085]

1884–5   Fourteenth Annual Report of the Local Government
            Board for 1884–5. Supplement containing the report
            of the Medical Officer for 1884, 33, [C-4516]

1886   S C on Rivers Pollution (River Lee), Report and Minutes
            of Evidence, 11, (207-sess 1)

1888   Return of all Royal Commissions issued from the year
            1866 to the year 1874, 81, (426)

            Seventeenth Annual Report of the Local Government
            Board, Supplement containing the Report of the Medical
            Officer for 1887, 49, (C-5526-I]

1893–4   Royal Commission on Metropolitan Water Supply,
            [*RCMWS*], Report, 40 pt 1, [C-7172]

            Royal Commission on Metropolitan Water Supply, Min-
            utes of Evidence, 40 pt 1, [C-7171-I]

            Royal Commission on Metropolitan Water Supply, Ap-
            pendices, 40 pt 2, [C-7172-II]

            Twenty-first Annual Report of the Local Government
            Board, Supplement containing the Report of the Medical
            Officer for 1891 on Enteric Fever in the Tees Valley, 42,
            [C-7054]

1894   Twenty-second Annual Report of the Local Government
            Board, Supplement containing the Report of the Medical
            Officer for 1892–3, 39, [C-7412]

            Twenty-third Annual Report of the Local Government
            Board, Supplement containing the Report of the Medical
            Officer for 1893–4, 40, [C-7538]

1895   Twenty-fourth Annual Report of the Local Government
            Board, Supplement containing the Report of the Medical
            Officer for 1894, 51, [C-7906]

2.   General

A unified history of water matters in nineteenth century Britain is
much needed. Useful, but brief are F T K Pentelow, *River Purification, A Legal and Scientific Review of the Past 100 Years, being the
Buckland Lectures for 1952* (Edward Arnold, 1953) and Bill Luckin,
*Pollution and Control: A Social History of the Thames in the Nine-
teenth Century* (Bristol: Adam Hilger, 1986). A re-issue of my dissertation, *What Becomes of Pollution: Adversary Science and the
Controversy on the Self-Purification of Rivers in Britain, 1850–1900*

(New York: Garland, 1987), gives fuller accounts of the geography of pollution, the contemporary theory of purification, and sewage treatment. On the popular sensibility toward water purity Dorothy Hartley's *Water in England* (MacDonald, 1964) contains interesting material. M N Baker's *The Quest for Pure Water, The History of Water Purification from the Earliest Times to the Twentieth Century* (New York: Engineering News, 1948) is an extremely useful though mostly undocumented history of water purification technology. J A Hassan, 'The Growth and Impact of the British Water Industry in the Nineteenth Century,' *Econ History Review* 2nd ser 38 (1985): 531–47, is the modern authority on the growth of the water industry. Also useful are J H Balfour Browne's *Water Supply* (MacMillan, 1880) and R W Rennison, *Water to Tyneside: A History of the Newcastle and Gateshead Water Company* (Newcastle: Newcastle and Gateshead Water Company, 1984). Roy MacLeod's 'Government and Resource Conservation: The Salmon Acts Administration, 1860–1886,' *J British Studies* 7 (1968): 115–50, is an exemplary account of the development of a multi-use resource policy. Legal aspects of water are fully developed in Clement Higgins, *A Treatise on the Law relating to the Pollution and Obstruction of Watercourses: together with a brief summary of the Various Sources of Rivers Pollution* (Stevens and Haynes, 1877) and Frederick Clifford, *A History of Private Bill Legislation*, 2 vols (1885: reprinted, New York: Cass, 1968). See also Anthony S Wohl, *Endangered Lives: Public Health in Victorian Britain* (Cambridge: Harvard University Press, 1983) and two papers by P J Smith: 'The Foul Burns of Edinburgh: Public Health Attitudes and Environmental Change,' *Scottish Geographical Magazine* 91 (1975): 25–37, and 'The Legislated Control of River Pollution in Victorian Scotland,' *Scottish Geographical Magazine* 98 (1982): 66–76.

3. The Rise of the Chemists

A number of recent works have examined the rise of chemistry to a position of social prominence in which chemists became widely ranging technical experts. Most useful are J K Crellin, 'The Development of Chemistry in Britain through Pharmacy and Medicine, 1700–1850', PhD thesis, University of London, 1969, for the 18th century, and W A Campbell, 'The Analytical Chemist in Nineteenth Century English Social History,' MA Thesis, University of Durham, 1971 and M Berman, *Social Change and Scientific Organization:*

The Royal Institution, 1799–1844 (Ithaca: Cornell University Press, 1978), for the early 19th century. The best source on the technical foundations of this achievement remains Archibald Clow and Nan L Clow, *The Chemical Revolution: A Contribution to Social Technology* (Batchworth, 1952). C A Browne, 'The Life and Chemical Services of Frederick Accum,' *J Chem Ed* 2 (1925): 829–51, 1008–35, 1140–8, gives an in-depth portrait of a practical chemist in the early 19th century but lacks documentation. On the pretensions of chemists to academic status see R Bud and G K Roberts, *Science versus Practice: Chemistry in Victorian Britain* (Manchester: Manchester University Press, 1984). On the professionalization of chemistry see A Chaston Chapman, *The Growth of the Profession of Chemistry during the past Half Century (1877–1927)* (The Institute of Chemistry, 1927), Colin Russell, N G Coley, and G K Roberts, *Chemists by Profession—The Origins and Rise of the Royal Institute of Chemistry* (Milton Keynes: Open University Press, 1977), Robert Bud, 'The Discipline of Chemistry: The Origins and Early Years of the Chemical Society of London,' PhD thesis: U of Pennsylvania, 1980, and W H Brock, 'The Spectrum of Scientific Patronage,' in G L'e Turner, ed. *The Patronage of Science in the Nineteenth Century* (Leyden: Noordhoff, 1976).

Specifically on the Society of Public Analysts see R C Chirnside and J H Hamence, *The Practising Chemists: A History of the Society for Analytical Chemistry, 1874–1974* (Society for Analytical Chemistry, 1974), Ernst W Steib, in collaboration with Glenn Sonnedecker, *Drug Adulteration in Nineteenth Century Britain* (Madison: University of Wisconsin Press, 1966), Bernard Dyer, *The Society of Public Analysts and Other Analytical Chemists, Some Reminiscences of its First Fifty Years with a Review of its Activities by C Mitchell Ainsworth* (Cambridge: the Society/W Heffer, 1932).

On the activity of chemists as expert witnesses see June Fullmer, 'Technology, Chemistry and the Law in Early Nineteenth Century England,' *Technology and Culture* 21 (1980): 1–28, and C Hamlin, 'Scientific Method and Expert Witnessing: Victorian Perspectives on a Modern Problem,' *Social Studies in Science* 16 (1986): 485–513. J H Balfour Browne's *Forty Years at the Bar* (London: H Jenkins, 1916) offers a fascinating account of expert witnessing from the viewpoint of a barrister with extensive practice in water matters.

4.   Mineral Water Analysis

Historians have given little attention to the enormous literature on the chemical composition and medical properties of mineral waters. W H Dalton, 'A List of Works referring to British Mineral and Thermal Waters,' *Report of 52nd Meeting of the BAAS, 1888* (1889), reports, pp 858–97 is a fine bibliography, especially when supplemented by E H Guitard, *Le Prestigieux Passé des Eaux Minerales* (Paris: Société d'Histoire de la Pharmacie, 1951), and H Bolton's *A Select Bibliography of Chemistry 1492–1892*, Smithsonian Miscellaneous Collections #850 (Washington D C: Smithsonian Institution, 1893) and First Supplement (1899). F R Peddie's *Subject Index to Books to 1880* contains well over 2000 titles on mineral waters, including guide books, classified geographically. All these should be consulted: there is remarkably little overlap among them. As for histories of mineral water science and medicine Guitard's *Le Prestigieux Passé* is an immensely useful and neglected starting point. Popular histories of British spas are helpful, especially P J Neville Havins, *The Spas of England* (Hale, 1976) and A B and M D Anderson, *Vanishing Spas* (Dorchester: Friary Press, 1974), as are local histories, e.g. T B Dudley, *From Chaos to the Charter: A Complete History of Royal Leamington Spa, from the Earliest Times to the Charter of Incorporation* (Leamington: Tomes, 1901) and Gwen Hart, *A History of Cheltenham* (Leicester: Leicester University Press, 1965). Helpful for making sense of the humoral medicine of the spas is Lester King, *The Philosophy of Medicine: The Early Eighteenth Century* (Cambridge: Harvard University Press, 1978).

Aspects of the social history of mineral water chemistry have been well developed by Noel Coley in 'Physicians and the Chemical Analysis of Mineral Waters in 18th Century England,' *Med Hist* 16 (1982): 123–44, and in 'The Presentation and Uses of Artificial Mineral Waters, ca 1680–1825,' *Ambix* 31 (1984): 32–48. See also W Kirkby, *The Evolution of Artificial Mineral Waters* (Manchester: Jevons and Brown, 1902). Aspects of the analytical techniques used by mineral water chemists have been developed by Allen Debus, 'Solution Analyses Prior to Robert Boyle,' *Chymia* 8 (1962): 41–61, and Debus, 'Sir Thomas Browne and the Study of Colour Indicators,' *Ambix* 10 (1962): 29–36, Frederic L Holmes, 'Analysis by Fire and Solvent Extractions: The Metamorphosis of a Tradition,' *Isis* 62 (1971): 129–48 and Holmes, 'From Elective Affinities to Chemical Equilibria: Berthollet's Law of Mass Action,' *Chymia* 8 (1962):

107–47, Uno Boklund, 'Torbern Bergman as Pioneer in the Domain of Mineral Waters,' in Torbern Bergman, *On the Acid of Air* (Stockholm: Almqvist and Wiksell, 1956) pp 105–28, and Jon B Eklund, *Chemical Analysis and the Phlogiston Theory, 1738–1772: Prelude to Revolution*, PhD thesis, Yale University, 1971. General histories of chemistry, especially Ferenc Szabadvary, *History of Analytical Chemistry* trans by Gyula Svehla, (Oxford: Oxford University Press, 1966), J R Partington, *A History of Chemistry* (MacMillan, 1961–7) and H Kopp, *Geschichte der Chemie*, 4 vols (Braunschwig: Vieweg, 1843–7) have much to offer.

Important primary sources on mineral water chemistry include Torbern Bergman's 'Of the Analysis of Waters,' in his *Physical and Chemical Essays*, trans by Edmund Cullen, 3 vols (J Murray, 1784), I, 91-192, 'Of the Artificial Preparation of Cold Medicated Waters,' in his *Physical and Chemical Essays*, I, 232–79, and his 'Treatise on Bitter, Seltzer, Spa and Pyrmont Waters and their Synthetical Preparation,' trans Sven M Jonsson, in T Bergman, *On the Acid of Air* (Stockholm: Almqvist and Wiksell, 1956), pp 29–104. John Murray's major papers on water analysis are 'An Analysis of the Mineral Waters of Dunblane and Pitcaithly; with General Observations on the Analysis of Mineral Waters, and the Composition of Bath Water,' *Annals of Philosophy* 6 (1815): 256–69, 347–63, 'Analysis of Sea Water, with Observations on the Analysis of Salt Brines,' *Phil Mag* 2nd ser 51 (1818): 10–25, 91–103 and 'A General Formula for the Analysis of Mineral Waters,' *Annals of Philosophy* 10 (1817): 93–98, 169–77. Other significant contributions, particularly on the question of the state of combination of salts in solution are R Kirwan, *An Essay on the Analysis of Mineral Waters* (Myers/D Bremner, 1799), A B Northcote, 'On the Water of the River Severn at Worcester,' *Phil Mag* 4th ser 34 (1867): 249–70, J Berzelius, 'Examen chimique des eaux de Carlsbad, de Toplitz, et de Konigswart,' *Annales de Chimie et de Physique* 2nd ser 28 (1825): 225–63, 366–406, and J H Gladstone, 'On the Salts Actually present in the Cheltenham and other Mineral Waters,' *26th Report of the BAAS* (Cheltenham, 1856) (1857), sections, 51–2, and George Merck and Robert Galloway, 'Analysis of the Water of the Thermal Spring at Bath,' *Phil Mag* 3rd series 31 (1847): 56–67.

Much of the mineral water literature was promotional, either making claims for a place or taking issue with such claims. Classic examples are F Accum, 'Analysis of the lately discovered mineral waters at Cheltenham: and also of the medicinal springs in its Neighbourhood,' *Phil Mag* 31 (1808): 14–28, and F Accum, 'Analysis of the Chalybeate Spring at Thetford,' *Phil Mag* 53 (1819): 359-65. Sometimes the names of well-known chemists were included even when their involvement was slight. See G W Pigott, *On the Harrogate Spas and Change of Air: Exhibiting a medical Commentary on the Waters founded on Professor Hofmann's analysis*, new and enlarged edn (Churchill, 1856) and *Synopsis of the Analyses of the Mineral Springs of Harrogate extracted from Dr Hofmann's report, with Practical Remarks by the Medical Section of the Water Committee* (n p, 1854). See also Frederick Slare, *An Account of Pyrmont Waters, dedicated to Sr. Issac Newton* (n p, 1717), J Barker, *A Treatise on Cheltenham Water and its Great Use in the Present Pestilential Constitution* (Birmingham: Pearson, 1786), Dr Evans, 'An Account of Sutton Spa, near Shrewsbury,' *Phil Mag* 22 (1805): 61–8, R Phillips 'An Analysis of the Salts prepared by Mr Henry Thompson from the Cheltenham Waters,' *Annals of Philosophy* 11 (1818): 28–31, W T Brande and S Parkes, 'A descriptive Account of Mr Thompson's Laboratory at Cheltenham, for the Preparation of the Cheltenham Salts; with a Chemical Analysis of the Waters whence they are produced,' *Q J Science, Literature and the Arts* 1 (1817): 54–71, Edwin Godden Jones, 'Chemical Analysis of the Mineral Waters of Spa,' *Trans Medico-Chirurgical Association* 7 (1816) pt 1, 1–69, A Walcker, 'Analysis of the Mineral Water of Bath,' *Q J Science, Literature and the Arts* 4 (1829): 78–89, R Phillips, 'Analysis of the Hot Springs at Bath,' *Phil Mag* 24 (1806): 342–61, Henry Freeman, *The Thermal Baths of Bath: their History, Literature, Medical and Surgical Uses and Effects* (Hamilton, Adams, 1888), Charles Perry, *An Account of an Analysis of the Stratford Mineral Water* (Northampton: Duay, 1744), F A Abel and Thos Rowney, 'Analysis of the Water of the Artesian Wells, Trafalgar Square,' *Q J Chem Soc* 1 (1848): 97–103, F A Abel and Thos Rowney, 'On the Mineral Waters of Cheltenham,' *Q J Chem Soc* 1 (1848): 193–212.

A number of works were also put forth as impartial guides or compendia to the mineral water literature to counter the excesses of much of the partisan literature. Among the most important are Charles Lucas, *An Essay on Waters*, 3 vols (A Millar, 1756), William Saunders, *A Treatise on the Chemical History and Medical Powers of Some of the Most Celebrated Mineral Waters: with Practical Remarks on the Aqueous Regimen* (Phillips, 1800), Meredith Gairdner, *Essay on the Natural History, Origin, Composition, and Medical Effects of Mineral and Thermal Springs* (Edinburgh:

Blackwood, 1832), C G B Daubeny, 'Report on the Present State of our Knowledge with Respect to Mineral and Thermal Waters,', *6th Report of the BAAS* (Bristol 1836) (1837), reports, 1–95, and A B Granville, *The Spas of England and Principal Sea Bathing Places with a new introduction by Geoffrey Martin*, 2 vols (1841, rpt; Adams and Dart, 1971). Also see Diederick Wessel Linden, *A Treatise on the Origin, Nature, and Virtues of Chalybeat Waters and Natural Hot Baths, with a Description of the Mineral Waters in England and Germany*, 2nd edn (D Browne, 1755), John Elliot, *An Account of the Nature and Medicinal Virtues of the Principal Mineral Waters of Great Britain and Ireland*, 2nd edn, corrected and enlarged (A Johnson, 1789), and John Rutty, *An Essay towards a Natural, Experimental, and Medical History of the Mineral Waters of Ireland* (Dublin: privately printed, 1757).

5. The London Water Companies

Several histories of London government in the nineteenth century take up the question of public takeover of the water supply. Most useful for political aspects is Asok Mukhopadyay, *Politics of Water Supply: The Case of Victorian London* (Calcutta: World, 1981). See also David Owen, *The Government of Victorian London, 1855–1889: The Metropolitan Board of Works, the Vestries, and the City Corporation* edited by Roy MacLeod with contributions from David Reeder, Donald Olsen, and Francis Sheppard (Cambridge: Harvard University Press, 1982) and Ken Young and Patricia L Garside, *Metropolitan London: Politics and Urban Change, 1837–1981* (Edward Arnold, 1982). A description of the variety of official reports on London's waters is 'Reports on the Examination of Thames Water,' *JRSA* 31 (1882–3): 74–6, 87–90. There were numerous quasi-historical accounts of the London water works during the period, many of them published in connection with attempts at public takeover. See G Phillips Bevan, *The London Water Supply: Its Past, Present, and Future* (Edward Stanford, 1884), Francis Bolton, *London Water Supply, including a History and Description of the London Water Works, Statistical Tables, and Maps*, new ed. entirely revised and enlarged with a short exposition of the law relating to water companies generally ... by Philip A Scratchely (Clowes and Sons, 1888), 'A Civil Engineer', *The London Water Supply, being an examination of the alleged Advantages of the Schemes of the Metropolitan Board of Works and of the inevitable Increase of Rates which would be required thereby* (Spon 1878), and W Scott Tebb, *Metropolitan Water Supply* (n p, [1907]).

On early expressions of concern for the quality of public water supplies, see Lucas, *Essay on Waters*, Thomas Percival, *Experiments and Observations on Water: particularly on the Hard Pump Water of Manchester* (J Johnson, 1769), and William Lambe, *An Investigation of the Properties of Thames Water* (Butcher, 1828). Good secondary sources on the 1828 controversy are D Lipschutz, 'The Water Question in London, 1827–1831,' *Bull Hist Med* 42 (1968): 510–26, and A Hardy, 'Water and the Search for Public Health in London in the Eighteenth and Nineteenth Centuries,' *Medical History* 28 (1984): 250–82. See also 'The Thames Water Question,' *Westminster Review* 12 (1830): 31–42, W T Brande, 'The Supply of Water to the Metropolis,' *Q J of Science, Literature and the Arts* 5 (1830): 350–6, Michael Ryan, *Remarks on the Supply of Water to the Metropolis, with an Account of the Natural History of Water in its simple and combined states: and of the chemical composition and medical uses of all known mineral waters* (Longmans, 1828), and [Charles Wall], 'Metropolis Water Supply,' *Fraser's Magazine* 10 (1834): 561–72. Relevant parliamentary papers are the reports and evidence of the Royal Commission on Metropolitan Water Supply (1828), the Select Committee on Metropolis Water (1834), and the Select Committee (Lords) on the Supply of Water to the Metropolis (1840).

The controversies in the early '50s are well described in the two main biographies of Edwin Chadwick: S E Finer, *The Life and Times of Sir Edwin Chadwick* (Methuen, 1952) and R A Lewis, *Edwin Chadwick and the Public Health Movement, 1832–1854* (Longmans, Green, 1952). Also see Chadwick's famous *Report on the Sanitary Condition of the Labouring Population of Great Britain* ed with an introduction by M W Flinn (Edinburgh: Edinburgh University Press, 1965) for the marginality of his concern with water quality. Important primary sources are [W H Wills] 'The Troubled Water Question,' *Household Words* 1 (1850): 49–52, Thos Graham, W A Miller, and A W Hofmann, 'Chemical Report on the Supply of Water to the Metropolis,' *J Chem Soc* 4 (1851): 375–413, [F O Ward], 'Metropolitan Water Supply,' *Quarterly Review* 87 (1850): 468–502, Samuel Homersham, 'Review of the Report by the General Board of Health on the Supply of Water to the Metropolis, contained in a report to the directors of the London (Watford) Spring Water Company' (Weale, 1850), Edwin Lankester, 'Drinking Waters of the

Metropolis,' *Proc R I* 2 (1854–8): 466–70, [Charles Kingsley], 'The Water Supply of London,' *North British Review* 15 (1851): 228–53, [W O'Brien], 'The Supply of Water to the Metropolis,' *Edinburgh Review* 91 (1849-50): 377–408, W T Brande, 'Analysis of the Well-Water at the Royal Mint with Some Remarks on the Waters of the London Wells,' *J Chem Soc* 2 (1850): 342–52, [N Beardmore], 'Water Supply,' *Westminster Review* 54 (1851): 185–96, G R Burnell, 'On the Present Condition of the Water Supply of London,' *JRSA* 9 (1860–1): 169–77, and William Ranger, Henry Austin, and Alfred Dickens, 'Report on the Examination of the Thames,' in *Reports to the General Board of Health* (PP 1856). Many of the citations in the 'Microscopic Approaches' section below also deal with the controversies of the '50s. Relevant parliamentary papers include the reports and evidence of the General Board of Health's *Report on the Epidemic Cholera of 1848 and 1849* (1850), and its *Report on the Supply of Water to the Metropolis* (1850), the Select Committee on the Metropolis Water Bill (1851), the Select Committee on the Metropolis Water Bills (1852), the report of the GBH Medical Council on *Scientific Inquiries in Relation to the Cholera Epidemic of 1854* (1854–5), and the *Reports to the General Board of Health, under the Provisions of the Metropolis Water Act* (1856).

The alternatives considered in the late '60s are considered in J F Bateman, 'On the Present State of our Knowledge of the Supply of Water to Towns,' *25th Report of the BAAS* (Glasgow 1855) (1856), reports 62–77. The most important parliamentary paper is the *Report of the Royal Commission on Water Supply* (1868-9). Other relevant parliamentary papers include the reports and evidence of the first and second reports of the first (1865) Royal Commission on Rivers Pollution (1866 and 1867), of select committees on the Thames Navigation Bill (1866), on the River Lea Conservancy Bill (1867–8), and on the East London Water Bills (1867). Citations in sections 7 and 9 below also deal with the controversies of the '60s.

Later reports of interest include the investigation of the eel epidemic: A deC Scott and W H Power, 'Eels in Water Mains being a Report on an Inquiry into the Quality of the Water supplied by the East London Waterworks,' in *17th Annual Report of the Local Government Board, Report of the Medical Officer for 1887*, pp 121–38. Archives of the Local Government Board's regulation of the water companies are in the Public Record Office as PRO MH 29. Finally the Royal Society holds documents relating to the formation, at the request of the London County Council, of its Water Research

Committee (Royal Society of London, Water Research Committee, Minutes, and Letters and Papers, 1891–6).

Parliamentary papers relevant to the controversy in the '80s and early '90s include the reports and evidence of the Select Committee on Rivers Pollution (River Lee) (1886), and the Royal Commission on Metropolitan Water Supply (1893–4).

6.   Microscopic Approaches to Water Analysis

The development of this approach is reflected in the works of Arthur Hill Hassall: *A Microscopic Examination of the Water Supplied to the Inhabitants of London and Surrounding Districts* (S Highley, 1850), 'Memoir on the organic analysis or microscopical examination of the water supplied to London and suburban districts,' *Lancet*, i, 1850, pp 230–5, and (as the 'Lancet Analytical Sanitary Commission'), 'Records of the results of microscopical and chemical analyses of the solids and fluids consumed by all classes of the public,' *Lancet*, i, 1851, pp 187–93, 216–25, 253–6, 279–84. Also important are two parliamentary papers, his 'Report on the Microscopical Examination of different waters (principally those supplied to the Metropolis) during the cholera epidemic of 1854,' to the GBH Medical Council's *Report of the Committee for Scientific Inquiries on the Cholera Epidemic of 1854* (1854–5), pp 217–81, and 'Report to the GBH on the microscopical examination of the Metropolitan Water supply, under the provisions of the Metropolis Water Act' (1857). See also Hassall's journal *Food, Water and Air* (1871–4) and his autobiography: *The Narrative of a Busy Life* (Longmans, Green, 1893). For other pioneering approaches in microscopical analysis see J Brittan and R Etheridge in Blackwell's *Report on the Cholera . . . in Sandgate* (P P 1854–5), R Angus Smith, 'On the Air and Water of Towns,' *Report of the 18th Meeting of the BAAS*, (Swansea, 1848) (1849), reports, pp 16–31, and E Lankester and P Redfern, *Reports made to the Directors of the London (Watford) Spring Water Company on the Results of Microscopical Examination of the Organic Matters and Solid Contents of Waters supplied from the Thames and Other Sources* (n p 1852). The orthodoxy of such perspectives in France is reflected in Alphonse Gerardin, 'Alteration, corruption, et assainissement des rivieres,' *Annales d' Hygiène Publique et de Medecine Légale* 2nd ser 43 (1875): 5–41, 261–29. The perspective toward the relation between microscopic life and purity that underlay such analysis is developed in C Hamlin, 'Robert Warington and the Balanced Aquarium,' *J History of Biology* 19 (1986): 131–53. For later attempts to make microscopical analysis serve political purposes see J Hogg, 'River pollution with special reference to impure water supply,' *JRSA* 23 (1874–5): 579–92, and Hogg, *A Microscopical Examination of Certain Waters submitted to Jabez Hogg and a Chemical Analysis by Dugald Campbell, with introductory notes by S C Homersham* (Trounce, 1874). For the development of microscopic water analysis after 1880 see section 10 below. The main source in the later period was J D MacDonald, *Guide to the Microscopical Examination of Drinking Water*, 2nd edn (Churchill, 1883).

7.   The Germ Theory and the Zymotic Theory

The seminal work on relations between various filth theories of disease and the germ theory is Margaret Pelling, *Cholera, Fever, and English Medicine, 1825–1865* (Oxford: Oxford University Press, 1978). Other secondary sources are W M Frazer, *A History of English Public Health, 1834–1939* (Balliere, Tindall, and Cox, 1950), C-E A Winslow, *The Conquest of Epidemic Disease: A Chapter in the History of Ideas* (1943, rpt; Madison: University of Wisconsin Press, 1980), C Hamlin, 'Providence and Putrefaction: Victorian Sanitarians and the Natural Theology of Health and Disease,' *Victorian Studies* 28 (1985): 381–411, John M Eyler, 'The Conversion of Angus Smith: The Changing Role of Chemistry and Biology in Sanitary Science, 1850–1880' *Bull Hist Med* 54 (1980): 216–24, Eyler, *Victorian Social Medicine: The Ideas and Methods of William Farr* (Baltimore: Johns Hopkins University Press, 1979), A Gibson and W V Farrar, 'Robert Angus Smith, F R S and Sanitary Science,' *Notes and Records of the Royal Society of London* 28 (1974): 241–62, J K Crellin, 'Airborne particles and the germ theory: 1860–1880,' *Annals of Science* 22 (1965–6): 49–66, and Crellin, 'The Dawn of the Germ Theory: Particles, Infection, and Biology,' in F N L Poynter ed, *Medicine and Science in the 1860s, Proc Sixth British Congress on the History of Medicine, University of Sussex, 6–9 Sept 1967* (Wellcome Institute, 1968) pp 57–76.

John Snow's investigations of cholera are also relevant. See Snow, *On the Mode of Communication of Cholera*, 2nd edn (1855) and *On Continuous Molecular Changes, more particularly in their Relation to Epidemic Diseases*, (1853) in *Snow on Cholera—A Reprint of Two Papers by John Snow, M D, together with a biographical memoir by B W Richardson M D, and an introduction by Wade Hamp-*

ton Frost (New York: Commonwealth Fund, 1936). See also E A
Parkes, 'Mode of Communication of Cholera,' *British and Foreign
Medico-Chirurgical Review* 15 (1855): 449–63, H Whitehead, 'The
Broad Street Pump: An Episode in the Cholera Epidemic of 1854,'
MacMillan's Magazine 13 (1865–6): 113–22, Whitehead, 'The influ-
ence of impure water on the Spread of Cholera,' *MacMillan's Mag-
azine* 14 (1866): 182–90, and two important papers by P E Brown,
'Another Look at John Snow,' *Anesthesia and Analgesia* 43 (1964):
646–54, and 'John Snow—the Autumn Loiterer,' *Bull Hist Med* 35
(1961): 519–28.

Liebig's zymotic theory is considered at length in my *What Becomes of Pollution* (section 2) and by Pelling. Classic statements
of the zymotic theory are Henry Letheby, 'Report to the City of
London Commissioners of Sewers, Sept 9 '58: Sewage and Sewer
Gases,' *JPH&SR* 4 (1858): 275–96 and A W Hofmann and Lyndsay
Blyth, 'Report on the Chemical Quality of the Water supplied to
the Metropolis,' in *Reports to the General Board of Health under
the Provisions of the Metropolis Water Act* (1856). Variations on
the zymotic theory are developed in B W Richardson, *The Field
of Disease, A Book of Preventive Medicine* (MacMillan, 1883), T
H Barker, 'The Influence of Sewer Emanations,' *JPH&SR* 4 (1858):
70–82, and T H Barker, *On Malaria and Miasmata and their Influence in the Production of Typhus and Typhoid Fevers, Cholera
and the Exanthemata: founded on the Fothergillian Prize Essay for
1859* (John Davies, 1863), and Charles Murchison, *A Treatise on the
Continued Fevers of Great Britain* (Parker, son, and Brown, 1862).
The incorporation of these ideas into manuals of public health is
represented in A H Church, *Plain Words about Water* (Chapman
and Hill, 1877), George Wilson, *A Handbook of Hygiene and Sanitary Science*, 4th edn (Churchill, 1879), C A Cameron, *A Manual
of Hygiene, Public and Private* (Dublin: Hodges and Foster, 1874)
W T Gairdner, *Public Health in Relation to Air and Water* (Edinburgh: Edmonston and Douglas, 1862). Discussions of the diffuse
border between living and non-living ferments are J Burdon Sanderson, 'Introductory Report on the Intimate Pathology of Contagion,'
in *Twelfth Report of the M O P C*, pp 229–56, Burdon Sanderson,
'Further report of Researches concerning the Intimate Pathology of
Contagion. The Origin and Distribution of Microzymes (Bacteria)
in Water and the Circumstances which determine their existence in
the tissues and fluids of the Living Body,' in *Thirteenth Annual Report of the M O P C*, pp 48–69, William Farr, *Report on the Cholera*

Epidemic of 1866 in England (P P 1867–8), and the *Third Report of the Royal Commission on the Cattle Plague,* (1865). The capacity of the germ theory to instill fear of one's surroundings is evident in W B Carpenter, 'The Germ Theory of Zymotic Diseases considered from a Natural History Point of View,' *Nineteenth Century* 15 (1884): 317–36, and Thomas Watson, 'The Abolition of Zymotic Disease,' *Nineteenth Century* 1 (1877): 380–96.

8. The Wanklyn–Frankland Controversy

The long war between proponents of the ammonia and combustion processes began with announcement of the two processes in 1867–8: J A Wanklyn, E T Chapman, and Miles H Smith, 'Water Analysis: Determination of the nitrogenous Organic Matter,' *J Chem Soc* 20 (1867): 445–54, and E Frankland and H E Armstrong, 'On the Analysis of Potable Waters,' *J Chem Soc* 20 (1868): 77–108. For improvements and validation of the ammonia process see E T Chapman, 'The Relation between the results of water analysis and the Sanitary Value of Water,' *CN* 16 (1867): 275, J A Wanklyn, 'Verification of Wanklyn, Chapman, and Smith's Water Analysis on a Series of Artificial Waters,' *J Chem Soc* 20 (1867): 591–95, and J A Wanklyn and E T Chapman, 'On the Action of Oxidizing agents on Organic Compounds in the Presence of Excess Alkali,' *J Chem Soc* 21 (1868): 161–72. Frankland presented improvements in the combustion process in E Frankland, 'On Some Points in the Analysis of Potable Waters,' *J Chem Soc*, 3rd ser 1 (1876): 825–51. Summaries of the discussions of these papers in *Chemical News*, along with letters, reviews, and suggestions in that journal and the *British Medical Journal* reflect the development and issues of the controversy. Most of Wanklyn's attacks on Frankland were published in *Chemical News*.

Important reviews of the two processes and of water analysis in general were [B H Paul], 'Water Analysis for Sanitary Purposes,' *BMJ*, i, 1869, pp 427–8, 495–7, 543–4, 1869, ii, pp 32–3, and C Meymott Tidy, 'Processes for Determining the Organic Purity of Potable Waters,' *J Chem Soc* 35 (1879): 46–106. Important in making clear the inadequacies of both processes were R D Cory, 'On the results of the examination of certain samples of water purposely polluted with excrements from fever patients, and with other matters,' in *Eleventh Annual Report of the Local Government Board, Report of the Medical Officer for 1881*, pp 127–65, and J W Mallet, 'Reports on Water

Analysis,' *Annual Report of the National Board of Health [US]*, 1882, Appendix D, pp 189–306 and shorter version, 'Determination of Organic Matter in Potable Water,' *CN* 46 (1882): 63–6, 72–5, 90–2, 101–2, 108–12. For other contemporary perspectives on water analysis see W A Miller, 'Observations on Some Points in the Analysis of Potable Waters,' *J Chem Soc* 18 (1865): 117–32, R A Smith, 'On the Examination of Water for Organic Matter,' *Proc Manchester Literary and Philosophical Society* 3rd ser 4 (1871): 37–81, same title, different text *CN* 19 (1869): 278–82, 304–6; v 20 (1869): 26–30, 112–15, and Smith, 'Rivers Pollution Prevention Act, 1876, Report to the Local Government Board,' (P P 1881). Contemporary views of water quality are well reflected in the discussions following papers by E Byrne, 'Experiments on the Removal of Organic and Inorganic Substances in Water,' *MPICE* 27 (1867–8): 1–54 and by C N Bazalgette, 'The Sewage Question,' *MPICE* 48 (1856–7): 105–250.

9.   Edward Frankland and the Tradition of Analytical Activism

The development of Frankland's activism is detailed in C Hamlin, 'Edward Frankland's Early Career as London's Official Water Analyst, 1865–1876: The Context of "Previous Sewage Contamination",' *Bull Hist Med* 56 (1982): 56–76. Important primary sources are E Frankland, 'Water supply of the Metropolis during the year 1865–1866' *J Chem Soc* 19 (1866): 236–48, Frankland 'On the water supply of the Metropolis,' *Proc R I* 5 (1866–9): 109–26, Frankland 'The Water Supply of London,' *Q J of Science* 4 (1867): 313–29, and his reports and testimony to the Royal Commission on Water Supply (1868–9). The struggle over responsibility for the East London cholera of 1866 is recorded in Bill Luckin, 'The Final Catastrophe: Cholera in London, 1866,' *Medical History* 21 (1977): 32–42, the Select Committee on East London Water Bills (1867), J Netten Radcliffe, 'Cholera in London especially in the eastern districts' in *9th Annual Report of the M O P C*, pp 264–331, Farr, *Report on the Cholera Epidemic of 1866 in England* (P P 1867–8), the *Report by Capt Tyler on the Water supplied by the East London Company* (P P 1867) and the associated correspondence with the company, and the Lancet Analytical Sanitary Commission, 'On the Epidemic of Cholera in the east end of London, *Lancet*, ii, 1866, pp 157–60, 217–9, 273–6, 293–4.

Early expressions of outrage at Frankland's tactics are H Letheby, 'Methods of Estimating Nitrogenous Matter in Potable Waters,' *Medical Times and Gazette*, i, 1869, pp 429–33, [William Pole], 'The Water Supply of London,' *Quarterly Review* [American edn] 127 (1869): 234–51, and *Our Water Supply. A Discussion for and Against the Fitness of Thames and River Water for Domestic Use, reprinted for the Surrey Comet* (Trounce, 1880). The later development of the controversy is reflected in C Meymott Tidy, *The London Water Supply, being a Report submitted to the Medical Officers of Health on the Quality and Quantity of the Water supplied to the Metropolis during the past ten years* (n p, 1878), Tidy, 'River Water,' *J Chem Soc* 37 (1880): 267-327, Tidy, 'River Water,' [second paper], *CN* 43 (1881): 113–4, E Frankland, 'On the Spontaneous Oxidation of Organic Matter in Water,' *J Chem Soc* 37 (1880): 517–46, Frank Hatton, 'On the Oxidation of Organic Matter in Water by Filtration through various media; and on the Reduction of Nitrates by Sewage, Spongy Iron, and other Agents,' *J Chem Soc* 39 (1881): 258–76, Charles W Folkard, 'The Analysis of Potable Waters with special Reference to Previous Sewage Contamination,' *MPICE* 68 (1881–2): 57–113, and W N Hartley, 'The Self-Purification of Peaty Rivers,' *JRSA* 31 (1882–3): 469–84. The records in PRO MH 29 contain much pertinent information on this issue. Many brief attacks on Frankland's position were also published in *Chemical News*, especially in the early '80s. See also W C Young, 'A Comparison of the Organic Carbon and Organic Nitrogen Results obtained by Dr Frankland and the Companies' Analysts from the waters supplied by the Metropolitan Water Companies,' *The Analyst* 20 (1895): 159–64.

That the same issues of debate arose outside of London is clear in House of Lords Record Office, Minutes of Evidence, House of Commons 1878, v. 5 (Cheltenham Corporation Water Bill). Percy Frankland's early activism as an opponent of river water is reflected in his papers on 'The Cholera and Our Water Supply,' *Nineteenth Century* 14 (1883): 346–55, 'The Upper Thames as a Source of Water Supply,' *JRSA* 32 (1883–4): 428–53, and 'The Selection of Domestic Water Supplies,' *SR* ns 6 (1884–5): 547–51. William Dibdin's unsuccessful attempt to initiate a new analytical activism on the basis of suspended contents in London's waters is reflected in his 'The Character of the London Water Supply,' *J Soc Chem Ind* 16 (1897): 9–15, and 'The Microscopical Examination of Water,' *The Analyst* 21 (1896): 2–12.

10.   The Programme of the Society of Public Analysts

Early criticism of the dominance of the London elite is reflected in the manuals for medical officers of C B Fox, *Sanitary Examinations of Water, Air, and Food: A Handbook for the Medical Officer of Health* (Philadelphia: Lea and Blakiston, 1878) and second edn (Churchill, 1885), and E A Parkes, *Manual of Hygiene*, 6th edn, 2 vols, ed by F S B Francois De Chaumont (New York: Wood, 1883), and papers by J Carter Bell, 'Water Analysis' *CN* 32 (1875): 246–7, A Ashby, 'Water Analysis,' *The Analyst* 6 (1881): 108–9, W Lauder Lindsay, 'The Estimation of the Quality of Potable Waters,' *BMJ*, ii, 1876, pp 783–5, and FSB Francois de Chaumont, 'On Certain Points with Reference to Drinking Water,' *SR* ns 1 (1879–80): 163–5. The activities of the SPA were a result of Wigner's paper 'On the Mode of Statement of the Results of Wwater Analysis and the Formation of a Numerical Scale for the Valuation of the Impurities in Drinking Water,' *The Analyst* 2 (1878): 208–20, and its revision 'On the Valuation of the Relative Impurities of Potable Waters,' *The Analyst* 6 (1881): 111–25. Much space in *The Analyst* between 1881 and 1884 is devoted to discussion of the program of analyses itself. For responses to Wigner's ideas see C Cassal, 'Hygienic Analysis,' *TSIGB* 7 (1885–6): 272–80, C Cassal and B H Whitelegge, 'Remarks on the Examination of Water for Sanitary Purposes,' *SR* ns 5 (1883–4): 427–9, 479–82, Louis Parkes, 'Water Analysis' *TSIGB* 9 (1887–8): 377–94, A Dupre and O Hehner, 'On District Standards in Water Analysis,' *The Analyst* 8 (1883): 53–8, and John Muter, 'On the most simple and generally useful mode of expressing the results of water analysis so as to be universally comprehensible: with examples drawn from London Water and also from a case of typhoid Epidemic,' *The Analyst* 8 (1883): 93–8.

On the resurgence of SPA concern with water analysis in the early '90s see C Cassal, 'Chemical Analysis and the Purity of Water,' *CN* 64 (1891): 249, J C Thresh, 'The Interpretation of the results obtained upon the Chemical and Bacteriological Examination of Potable Waters,' *The Analyst* 20 (1895): 80–91, 97–111. A more sophisticated epidemiology was central in this resurgence. See A Dupre, 'Note on the Chemical and Bacteriological Examination of Water, with remarks on the Fever Epidemic at Worthing in 1893,' *The Analyst* 20 (1895): 73–9, and M A Adams, 'Water Supply in Relation to the Maidstone Epidemic,' *The Analyst* 23 (1898): 142–61. The perspective was a response to epidemiological studies by the Local Government Board medical staff. See R Thorne Thorne, 'On an Extensive Epidemic of Enteric Fever at Redhill, Caterham and Adjoining Places,' in *9th Annual Report of the Local Government Board, Report of the Medical Officer for 1879*, pp 75–92, F D Barry, 'Enteric Fever in the Tees Valley' in *21st Annual Report of the Local Government Board, Report of the Medical Officer for 1891*, and Theodore Thomson, 'Report on an Epidemic of Enteric Fever in the Borough of Worthing,' in *23rd Annual Report of the Local Government Board, Report of the Medical Officer for 1893–4*, pp 47–80. On the Local Government Board medical staff see C Fraser Brockington, *Public Health in the Nineteenth Century* (Edinburgh: Livingstone, 1965) and Royston Lambert, *Sir John Simon and English Social Administration* (McGibbon and Kee, 1963). Important discussions of the future and ethical problems of water analysis that are consistent with the SPA outlook are Charles Smart, 'On the Present and Future of Sanitary Water Analysis,' *Reports and Papers of the American Public Health Association* 20 (1884): 79–85, and A R Leeds, 'A Question of Water, Ethics, and Bacteria,' *Am J of the Medical Sciences* 105 (1893): 259–68.

11.   The Filtration Question

On the issue of the use of analyses, particularly bacteriological analyses, to monitor filtration performances see G Bischof, 'The Purification of Water,' *JRSA* 26 (1877–8): 486–96, William Anderson, 'The Antwerp Water Works,' *MPICE* 72 (1882–3): 22–83. A great many of Percy Frankland's papers during the '80s and '90s dealt with this question. See P Frankland, 'The Removal of Micro-organisms from water,' *CN* 52 (1885): 27–9, 40–2, 'New Aspects of Filtration and other methods of water treatment; the Gelatine Process of Water Examination,' *J Soc Chem Ind* 4 (1885): 698–707, 'The Filtration of Water for Town Supply,' *TSIGB* 8 (1886–7): 276–84, 'Water Purification: its Biological and Chemical Basis,' *MPICE* 85 (1886–7): 197–263, 'The Application of Bacteriology to Questions relating to Water Supply,' *TSIGB* 9 (1887–8): 369–77, 'Recent Bacteriological Research in connection with Water Supply,' *J Soc Chem Ind* 6 (1887): 316–26, 'The Bacteriological Examination of Water and the Information it has furnished,' *J State Med* 2 (1894): 1–12, 'The Bacterial Purification of Water,' *MPICE* 127 (1896–7): 83–159, 'The London Water Supply and its Bacterial Contents,' *Lancet*, ii, 1896, pp 1414–5, 'The Bacterial Purification of Water and Sand

Filtration,' *SR* ns 19 (1897): 8–9, 29–30, and F Bolton and Percy Frankland, *Lectures on the Collection, Storage, Purification and Examination of water, delivered to the School of Military Engineering at the Royal Engineers Institute, Chatham, on the 24th*

and 25th of March, 1886 (Harrison and sons, 1886). Numerous documents in PRO MH 29 also bear on this issue. Most important were the views of Robert Koch 'Water Filtration and Cholera,' trans A J A Ball, in *22nd Annual Report of the Local Government Board, Report of the Medical Officer for 1892–3*, pp 439–59.

12.   Water Bacteriology and Bacteriological Water Analysis

For central issues in the history of bacteriology see William Bulloch, *The History of Bacteriology* (1938, rpt; New York: Dover, 1979), W D Foster, *A History of Medical Bacteriology and Immunology* (Heinemann, 1970), and Foster, *A Short History of Clinical Pathology* (Edinburgh: Livingstone, 1961). Early attempts at some form of bacteriological water analysis are Charles Heisch, 'On Organic Matter in Water,' *J Chem Soc* 23 (1870): 371–5, and E Frankland, 'On the Development of Fungi in Potable Water,' *J Chem Soc* 24 (1871): 66–76, the reports of Burdon Sanderson (section 7), and A Dupre, 'On Changes in Aeration of Water as indicating the Nature of the Impurities present therein,' in *Fourteenth Annual Report of the Local Government Board, Report of the Medical Officer for 1884*, Appendix B11, Dupre, 'On Changes effected by the Aeration of Certain waters by the Life Processes of particular Micro-organisms under different conditions of temperature, light, and nutrient material,' in *17th Annual Report of the Local Government Board, Report of the Medical Officer for 1887*, pp 272–9, and his 'Presidential Address to the section on chemistry, meteorology, and geology,' *TSIGB* 9 (1887–8): 352–67. R Angus Smith's attempts to culture water bacteria are recorded in his 'On the Development of Living Germs in Water,' *CN* 46 (1882): 288–9, and 'Notes on the Development of Living Germs in Water by Dr Koch's Gelatine Process,' *SR* ns 4 (1882–3): 344–7, as well as in his second report as a rivers pollution inspector: 'Rivers Pollution Prevention Act, 1876. Second Annual Report to the Local Government Board by Dr R Angus Smith, one of the inspectors appointed under the act on the Examination of Waters' (P P 1884).

Koch's plate culturing techniques are introduced in C J H Warden, 'The Biological Examination of Water,' *CN* 52 (1885): 52–4, 66–8, 73–6, 89, 101–4. Their application in the polemics on London water supply is already evident in William Odling, 'Micro-organisms in Drinking Water,' *J Soc Chem Ind* 5 (1886): 544, and is abundantly clear in the evidence and appendices of the Royal Commission on Water Supply (P P 1893–4). See also L C Parkes, 'The Possibilities of the Spread of Disease Through River Waters supplied to London: A Review of the Evidence given by the Bacteriological Witnesses before the Royal Commission on Water Supply, 1893,' *TSIGB* 15 (1894): 243–56.

Gustav Bischof's critiques of the inferences being made on the basis of plate cultures are 'Notes on Dr Koch's Water Test,' *J Soc Chem Ind* 5 (1886): 114–21, 'Dr Koch's Gelatine-Peptone Water Test,' *CN* 53 (1886): 205–6, 'Dr Koch's Bacteriological Water Test,' *Lancet*, i, 1885, pp 382–3, and 'Extension of Time of Culture in Dr R Koch's Bacteriological Water Test by Partial Sterilisation, with special reference to the Metropolitan Water Supply,' *SR* ns 9 (1887–8): 325–32. See also his remarks in A Gordon Salamon and W DeVere Mathew, 'The Purification of Water,' *J Soc Chem Ind* 5 (1886): 261–7, 271–3.

The fates of pathogens in natural waters and their detection are dealt with extensively in most contemporary treatises on bacteriology. For reviews see E Duclaux, 'Les Microbes des eaux,' *Annales De l'Institut Pasteur* 3 (1889): 560–9, Duclaux, 'Action sur l'eau sur les bacteries pathogènes,' *Ann Inst Pasteur* 4 (1890): 109–24, and P Frankland and H Marshall Ward, 'First Report to the Water Research Committee of the Royal Society, on the present state of our Knowledge concerning the Bacteriology of Water, with especial reference to the Vitality of Pathogenic Schizomycetes in Water,' *Proc Royal Society* 51 (1892): 183–279. The latter contains an extensive bibliography. See also G Sims Woodhead, *Bacteria and their Products* (Scott, 1891), G Sims Woodhead and Arthur W Hare, *Pathological Mycology*, section 1, Methods (Edinburgh: Pentland, 1885), Percy Frankland and Mrs Percy Frankland, *Micro-organisms in Water: their Significance, Identification, and Removal* (Longmans, 1894), P Miquel, *A Practical Manual of the Bacteriological Analysis of Water*, trans in *Wood's Medical Monographs*, 1891: 399–528, Carl Fraenkel, *Text-Book of Bacteriology*, 3rd edn, trans and ed by J H Lindsay (New York: Wood, 1891), C Flügge, *Micro-organisms with Special Reference to the Etiology of Infective Dis-*

eases trans from the 2nd edn of Fermente und Mikroparisiten by W Watson Cheyne (New Sydenham Society, 1890), George Newman, *Bacteriology and Public Health*, 3rd edn (Philadelphia: Blakiston, 1904), J C Thresh, *The Examination of Waters and Water Supplies* (Philadelphia: Blakiston, 1904), and F B Turneaure and H L Russell, *Public Water Supplies, Requirements, Resources, and the Construction of Works*, 2nd edn (New York: John Wiley, 1908).

A great deal was also written specifically on the stability and detectability of the typhoid bacillus. See P F Frankland (assisted by J R Appleyard), 'The Behaviour of the Typhoid Bacillus and of the Bacillus coli commune in Potable Waters,' in P F Frankland and H M Ward, 'Third Report to the Water Research Committee of the RSL,' *Proc RSL* 56 (1894), T H Pearmain and C G Moor, 'The Bacteriological Examination of Water for the Typhoid Bacillus,' *The Analyst* 21 (1896): 117–22, 141–48. Especially important were the studies of E Klein, 'Report on the Etiology of Typhoid Fever,' in *22nd Annual Report of the Local Government Board, Report of the Medical Officer for 1892–3*, pp 345–65, 'Further Report on the Etiology of Typhoid Fever,' in *23rd Annual Report of the Local Government Board, Report of the Medical Officer for 1893–4*, pp 345–65, and 'On the Behaviour of the Typhoid Bacillus and Koch's Vibrio in Sewage,' in *24th Annual Report of the Local Government Board, Report of the Medical Officer for 1894*, pp 407–10. For a summary see Klein, 'The Etiology of Typhoid Fever,' *J Sanitary Inst* 15 (1894): 343–52.

A sense of what was being taught in elementary bacteriology courses and courses for medical officers can be gained from A A Kanthack and J H Drysdale, *Elementary Practical Bacteriology* (MacMillan, 1895) and Henry R Kenwood, *Public Health Laboratory Work* (H K Lewis, 1896).

Finally, the considerable and ongoing problems of distinguishing among different types of coliforms is discussed in S C Prescott, C-E A Winslow and M H McCrady, *Water Bacteriology with Special Reference to Sanitary Water Analysis*, 6th edn (New York: John Wiley, 1946).

Index

Italicized entries indicate citations.

ABC sewage treatment
    process, 172, 183
Abel F, 39–40, 188, *318*
Accum F, 19, 52–5, 65, 110,
    315, 317, 318
Adams M A, *328*
Aikin A, 50, 117
albumenoid ammonia, 185,
    187, 190, 219, 226, 227,
    233, 299
Allen A H, 285, 292, 293
Altona, 286, 287
ammonia process (*see also*
    Wanklyn, albumenoid
    ammonia), 185–90, 215,
    232, 234
Anderson W, 244, 245, *329*
animalculae, 75, 99, 102–5,
    116, 119
anthrax, 242–5, 273, 274, 276
Aristotle, 23
Armstrong H E, 163, 258, *325*
Arrhenius S, 23
Association of Metropolitan
    Medical Officers; 192,
    215
Austin H, 139, 140, *321*
Ayres P B, 132

Baconianism in water
    analysis, 52, 55, 61–4
bacteria, pathogenic, 252,
    254, 255, 258–60, 263,
    265, 271–4, 276, 278–80,
    282, 283, 288, 290, 291
bacteriological culture
    techniques, 10, 13, 241,
    246–50, 270
bacteriological water analysis,
    3–5, 10, 13, 95, 153, 174,
    237
  qualitative, 271–3, 276,
    277, 279–93, *330–2*
  quantitative, 241, 244–60,
    262–5, 277, 304
bacteriology as a science, 207,
    241–3, 248–51, 258, 260,
    263, 270–7, 279–81, 305,
    330–2
Baker M N, *314*
Bamber J K, 196
Barker J, 54, 57, 60, 64, *318*
Barker T H, 131, *324*
Barnard Castle, 285
Barry F D, 284, *329*
Bateman J F, 154, *321*
Bath, 16, 18, 20, 31, 66, *317,*

333